Silicon Wet Bulk Micromachining for MEMS

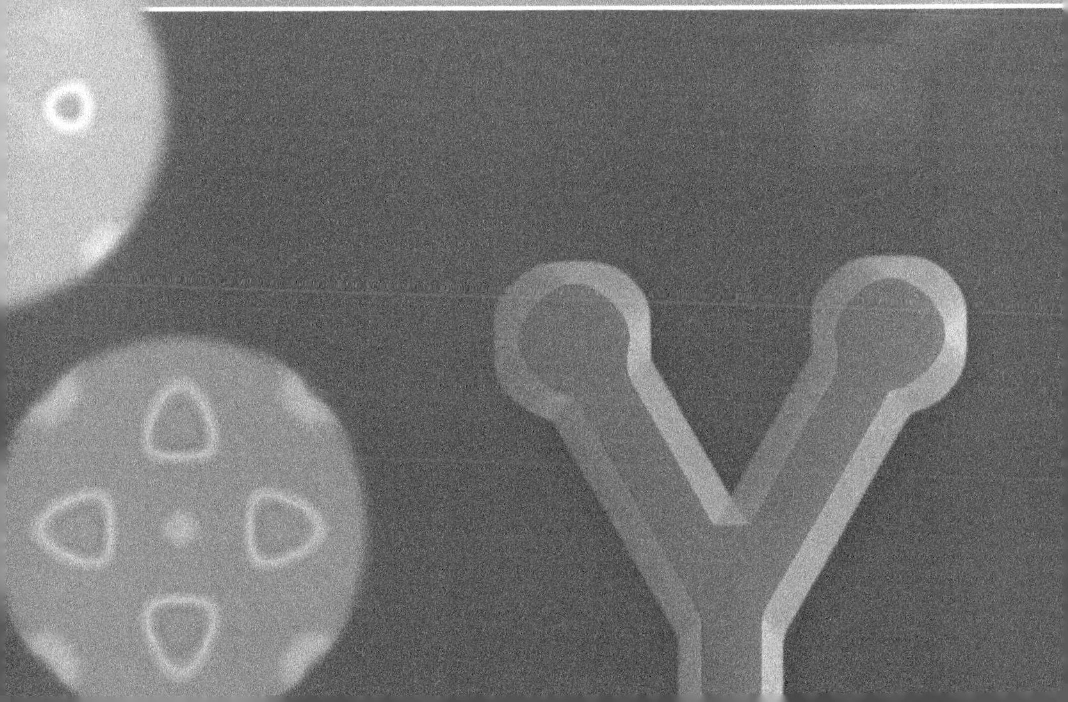

Silicon Wet Bulk Micromachining for MEMS

Prem Pal
Kazuo Sato

Pan Stanford Publishing

Published by

Pan Stanford Publishing Pte. Ltd.
Penthouse Level, Suntec Tower 3
8 Temasek Boulevard
Singapore 038988

Email: editorial@panstanford.com
Web: www.panstanford.com

British Library Cataloguing-in-Publication Data
A catalogue record for this book is available from the British Library.

Silicon Wet Bulk Micromachining for MEMS

ISBN 978-981-4613-72-9 (Hardcover)
ISBN 978-1-315-36492-6 (eBook)

Printed in the UK by Ashford Colour Press Ltd.

Contents

Preface xi

1. A Brief Introduction of the Crystal Structure **1**
1.1 Introduction 1
1.2 Crystal Structure 2
1.3 Unit Cell: Primitive and Nonprimitive 4
1.4 Symmetry Operations 5
1.5 Types of Lattices 7
1.6 Index System for Crystal Planes and Directions 9
1.7 Cubic Structures 12
1.7.1 Simple Cubic or Primitive Cubic Structure 12
1.7.2 Body-Centered Cubic Structure 13
1.7.3 Face-Centered Cubic Structure 13
1.8 Stereographic Projection 14

2. Brief Overview of Silicon Wafer Manufacturing and Microfabrication Techniques **23**
2.1 What Is Silicon? 23
2.2 Why Is Silicon Used as a Material for ICs and MEMS? 24
2.3 What Are Microelectromechanical Systems? 25
2.4 Wafer Manufacturing 26
2.4.1 Crystal Growth and Doping 26
2.4.2 Ingot Trimming, Grinding, and Slicing 29
2.4.3 Polishing, Cleaning, and Packaging 32
2.5 Silicon Microfabrication Processes 33
2.5.1 Wafer Cleaning 34
2.5.2 Thin-Film Formation 36
2.5.2.1 Thermal oxidation (or reactive growth) 36
2.5.2.2 Chemical vapor deposition 38
2.5.2.3 Physical vapor deposition 40
2.5.3 Photolithography 41
2.5.4 Etching 45

2.5.4.1 Silicon wet isotropic etching 48
2.5.4.2 Silicon wet anisotropic etching 49
2.5.4.3 Silicon dry etching 50
2.5.5 Local Oxidation of Silicon 54
2.5.6 Micromachining 55
2.5.6.1 Bulk micromachining 55
2.5.6.2 Surface micromachining 56
2.5.7 Doping 57
2.5.7.1 Diffusion 57
2.5.7.2 Ion implantation 60
2.5.8 Silicon Direct Wafer Bonding 62
2.5.8.1 Hydrophilic bonding 63
2.5.8.2 Hydrophobic bonding 64

3. Isotropic Etching of Silicon and Related Materials 67
3.1 Isotropic Etching Properties 67
3.2 Isotropic Etching Solutions 68
3.3 Conditions for Isotropic Etching 71
3.4 Applications 72

4. KOH-Based Anisotropic Etching 79
4.1 Etching in Pure KOH Solutions 79
4.1.1 The Etching Mechanism 79
4.1.2 Advantages and Disadvantages of KOH 82
4.1.3 Characterization of Etch Rate Anisotropy 84
4.1.4 Effects of KOH Concentration 88
4.1.5 Effects of Etching Temperature 90
4.1.6 Etch-Stop Technologies 92
4.1.6.1 Etch stop by heavy boron doping 93
4.1.6.2 Electrochemical etch stop 94
4.1.6.3 Use of a SOI wafer 96
4.1.7 Etched Surface Roughness 97
4.1.8 Comparison of KOH and TMAH Etching Characteristics 101
4.2 Effect of Alcohols and Surfactant Additives 106
4.2.1 Etching in KOH Solutions Saturated with Alcohols 106

4.2.1.1 Properties of KOH solutions saturated with alcohols 106
4.2.1.2 Results of etching in KOH solutions containing alcohols with one hydroxyl group 113
4.2.1.3 Results of etching in solutions containing alcohols with two hydroxyl groups (diols) 121
4.2.2 Etching in KOH Solutions with the Alcohol Concentration below Saturation 125
4.2.2.1 Effect of alcohol concentration on the course of etching 125
4.2.2.2 Using unsaturated solutions for texturing Si(100) substrates 138
4.2.2.3 Analysis of the etching mechanism in KOH solutions containing alcohols 142
4.2.3 Etching in KOH Solutions Containing Surfactants 148

5. TMAH-Based Anisotropic Etching 163
5.1 Introduction 163
5.2 The Etching Mechanism 165
5.3 Etch Rates 165
5.3.1 Etch Rates of Various Crystallographic Planes Using the Wagon-Wheel Structure 169
5.3.2 Etch Rates of the Whole Range of Crystallographic Planes Using a Convex Hemispherical Specimen 172
5.4 Surface Morphology 174
5.5 Undercutting and Etched Profiles 178
5.6 Why Are Etching Characteristics Affected When a Surfactant Is Added to TMAH? 182
5.6.1 FTIR Measurement 184
5.6.2 Ellipsometric Measurement 187
5.6.3 The Physical Model 188
5.7 Etched Profile Control 191
5.7.1 P^+ Silicon Etch Stop 191
5.7.2 Electrochemical Etch Stop 192

5.7.3 Protection of Convex Corners 194
5.7.4 Effect of Mechanical Agitation 194
5.7.5 Role of the Open-Circuit Potential 195
5.8 Summary 198

6. Convex and Concave Corners in Silicon Wet Bulk Micromachining 211
6.1 Introduction 211
6.2 Etch Rate, Underetching, and Undercutting 213
6.3 Role of Corner Undercutting in MEMS Fabrication 219
6.4 Why Does Undercutting Start at Convex Corners? 219
6.5 Etched Profiles at Sidewalls and Corners 226
6.6 Fabrication Techniques of Convex Corners 236
6.6.1 The Corner Compensation Method 236
6.6.1.1 Corner compensation geometries for a Si{100} wafer 237
6.6.1.2 Corner compensation geometries for a Si{110} wafer 262
6.6.2 Perfect Convex Corners Using Two-Step Etching Techniques 269
6.6.2.1 Sharp-edge convex corners by lithography on anisotropically etched patterns 269
6.6.2.2 Perfect convex corners using the LOCOS process 271
6.7 Summary 275

7. Alignment of Mask Patterns to Crystallographic Directions 287
7.1 Introduction 287
7.2 Role of Precise Alignment in Wet Bulk Micromachining 288
7.2.1 Microchannels 288
7.2.2 Cantilever Beams 290
7.2.3 Diaphragms 292
7.3 Alignment Techniques 294
7.3.1 Cleaved-Edge Alignment 295
7.3.2 Pre-etched Patterns 296
7.3.2.1 Identifying crystallographic directions on a Si{100} wafer 297

7.3.2.2 Identifying crystallographic directions on a Si{110} wafer 313

7.4 Summary 320

8. Simple to Complex Structures Using Wet Bulk Micromachining 323

8.1 Introduction 323

8.2 Cavities, Channels, and Mesa Structures 326

8.3 Microstructures with 45° Slanted Sidewalls 328

8.4 Silicon Gratings 330

8.5 Suspended Microstructures 331

8.5.1 Time-Controlled Back-Side Etching 335

8.5.2 P^+ Silicon 337

8.5.3 Electrochemical Etch Stop 338

8.5.4 Silicon-on-Insulator Wafers 338

8.5.5 Micromachined Cavities Sealed by Wafer Bonding 339

8.5.6 Silicon Nitride–Based SOI Wafers 339

8.6 Microstructures with Perfectly Sharp Edges and Corners 342

8.7 Suspended Microfluidic Channels 348

8.8 AFM Cantilevers 355

8.9 Microvalves and Micropumps 357

8.9.1 Microvalves 358

8.9.2 Micropumps 359

8.10 Microstructures Inside a Silicon Wafer 361

8.11 Dry-Assisted Wet Etching 368

8.12 Silicon Molds for the Fabrication of PDMS Structures 372

8.13 Complex Patterns for Very-High-Aspect-Ratio Microstructures Using Photoelectrochemical Etching 372

Index 383

Preface

Microelectromechanical systems (MEMS)-based sensors and actuators have become remarkably popular in the past few decades. Rapid advances have taken place in terms of both technologies and techniques of fabrication of MEMS structures. Wet chemical–based silicon bulk micromachining continues to be a widely used technique for the fabrication of microstructures used in MEMS devices. Researchers all over the world have contributed significantly to the advancement of wet chemical–based micromachining, from understanding the etching mechanism to exploring its application to the fabrication of simple to complex MEMS structures. In addition to its various benefits, one of the unique features of wet chemical–based bulk micromachining is the ability to fabricate slanted sidewalls, such as 45° walls as micromirrors, as well as freestanding structures, such as cantilevers and diaphragms. This makes wet bulk micromachining necessary for the fabrication of structures for myriad applications. Considering the importance of wet bulk micromachining in the fabrication of MEMS, all that we wanted to do is to write a deep book that can cover topics from the basic to the advanced level and can be used as a reference and as a textbook.

This book provides a comprehensive understating of wet bulk micromachining for the fabrication of simple to advanced microstructures for various applications in MEMS. It includes introductory to advanced concepts and covers research on basic and advanced topics on wet chemical–based silicon bulk micromachining. The book thus serves as an introductory textbook for undergraduate- and graduate-level students of physics, chemistry, electrical and electronics engineering, materials science, and engineering, as well as a comprehensive reference for researchers working or aspiring to work in the area of MEMS and for engineers working in microfabrication technology.

To understand the wet anisotropic etching for silicon micromachining, an elementary understanding of crystallography is essential. Hence we have included a separate chapter to cover the basics of the crystal structure and stereographic projection.

In addition, a basic understanding of microfabrication techniques employed in semiconductor industries is required in order to explore wet bulk micromachining for the fabrication of MEMS components. Therefore, a separate chapter, "Brief Overview of Silicon Wafer Manufacturing and Microfabrication Techniques," is also included.

We gratefully acknowledge financial support from the Department of Science and Technology (Project No. SR/S3/ MERC/072/2011), India, the Japan Society for the Promotion of Science (JSPS, Fellowship ID No. L14538), and the Japan International Cooperation Agency (JICA). JICA supported Prof. Prem Pal in visiting Sato Lab (AIT Toyota) in October 2015 and Prof. Sato in visiting the Indian Institute of Technology (IIT) Hyderabad, India, in December 2015. These visits have been useful for discussing some topics covered in this book.

We sincerely thank two distinguished contributors Prof. Irena Zubel (Wrocław University, Poland) for writing Section 4.2, Chapter 4, and Mr. Sajal Sagar Singh (IIT Hyderabad, presently at the University of Michigan) for coauthoring Chapter 7. We are highly thankful to Ms. Michiko Shindo (secretary to Prof. Sato) for her assistance in obtaining permissions to reproduce a few figures from published papers. At Pan Stanford Publishing, we wish to thank Mr. Stanford Chong for inviting us to write a book on silicon bulk micromachining and his team that has helped in refining the material in this book and extended all possible support at every stage of the manuscript. We welcome comments or suggestions on this textbook by email at prem@iith.ac.in and sato@aitech.ac.jp.

Prem Pal

Kazuo Sato

Chapter 1

A Brief Introduction of the Crystal Structure

1.1 Introduction

All of the solids, liquids, and gases that we encounter in our daily life are classified as a common type of matter. In solids and liquids the distance between neighboring atoms is of the order of a few angstroms. In the case of the gases, an average distance between molecules is approximately 30 Å at room temperature under 1 atm. Solids are one of the major states of matter. On the basis of the atomic arrangement, solids are classified into three categories: crystalline, polycrystalline, and amorphous [1–5]. In the crystalline structure (or single crystal or monocrystalline), the periodicity of atoms (or molecules) extends throughout the material (e.g., diamond, quartz, etc.). The opposite of a single crystal is an amorphous structure where the atomic position is completely random. In between the two extremes exist polycrystalline structures, which are made up of a number of small crystals known as crystallites. The crystallites in polycrystalline structures are randomly oriented. The small crystallites are known as grains and the boundaries separating them as grain boundaries. The crystalline, polycrystalline, and amorphous structures are schematically illustrated in Fig. 1.1.

Silicon Wet Bulk Micromachining for MEMS
Prem Pal and Kazuo Sato

ISBN 978-981-4613-72-9 (Hardcover), 978-1-315-36492-6 (eBook)
www.panstanford.com

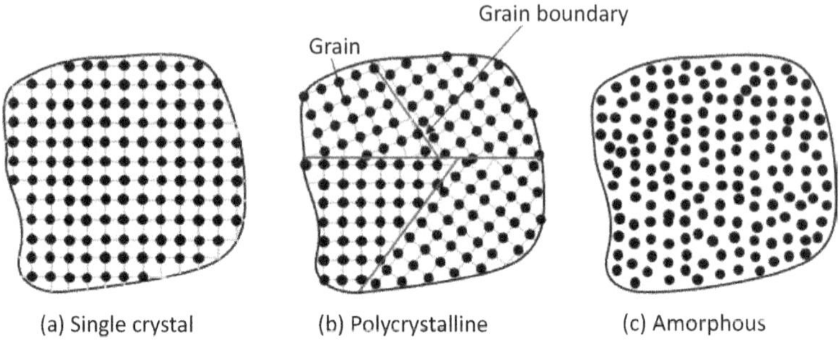

Figure 1.1 Schematic illustration of three types of solids: (a) crystalline, (b) polycrystalline, and (c) amorphous.

1.2 Crystal Structure

To understand the crystal structure, one must know the answer of following questions:

- What is a space lattice?
- What are the lattice translation vectors?
- What is the basis?

A **space lattice** is a regular periodic array of infinite number of imaginary points in three-dimensional space [1–5]. It can be defined in three dimensions by three fundamental **translational vectors** $\vec{a}$, $\vec{b}$, and $\vec{c}$ such that the atomic arrangement looks exactly the same in every respect when viewed from the point r as when viewed from the point r', where

$$\vec{r}' = \vec{r} + u_1\vec{a} + u_2\vec{b} + u_3\vec{c} \tag{1.1}$$

where u_1, u_2, and u_3 are arbitrary integers. The set of points r' defined by Eq. 1.1 for all u_1, u_2, and u_3 defines a lattice. We can say that a lattice is a mathematically concept. The aforementioned periodic arrays of the infinite number of imaginary points in one dimension and two dimensions are called **line lattice** and **plane lattice**, respectively.

The crystal structure is always described in terms of atoms rather than point. Hence, to achieve a crystal structure, an atom or a group of atoms (i.e., molecule) must be placed on each lattice point in a regular fashion. Such an atom or a group of atoms is called the

basis and acts as a building unit or a structural unit for the complete crystal structure. The crystal structure is formed when a basis of atoms is attached identical to every lattice points. Every basis should be identical in composition, arrangement, and orientation. Consequently, the crystal looks the same when viewed from any equivalent lattice point (Fig. 1.2). Mathematically, it is expressed as

Space lattice + Basis = Crystal structure.

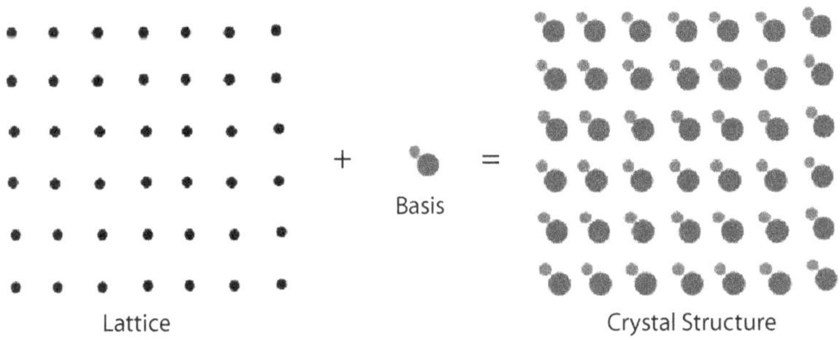

Figure 1.2 Schematic representation of the formation of a crystal structure by the addition of the basis to every lattice points of the lattice.

A lattice translation operation T is defined as the displacement of a crystal by a crystal translation vector

$$\vec{T} = u_1\vec{a} + u_2\vec{b} + u_3\vec{c} \tag{1.2}$$

From Eqs. 1.1 and 1.2

$$\vec{r}' = \vec{r} + \vec{T} \tag{1.3}$$

It indicates that $\vec{r}'$ can be obtained from $\vec{r}$ by the application of Eq. 1.2.

A lattice translation vector ($\vec{r}'$) with translation operation ($\vec{T}$) for a two-dimensional arrangement of the lattice points is shown in Fig. 1.3.

The lattice and the translation vectors $\vec{a}$, $\vec{b}$, and $\vec{c}$ are said to be **primitive** if any two points r, r' from where the atomic arrangement looks the same always satisfy Eq. 1.1 with a suitable choice of the integers u_1, u_2, u_3. With this definition of the **primitive translation vectors**, there is no cell of smaller volume than $\vec{a}\cdot(\vec{b}\times\vec{c})$ that can serve as a building block for the crystal structure.

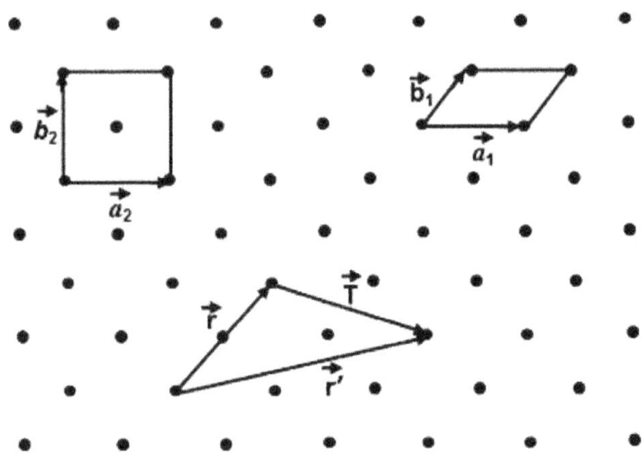

Figure 1.3 A two-dimensional lattice. The atomic arrangement at $\vec{r}$ and $\vec{r}'$ looks the same to observers at these points, since $\vec{T}$ is an integer number of the **primitive translation vectors** $\vec{a}_1$ and $\vec{b}_1$ ($\vec{T} = 2\,\vec{a}_1 - \vec{b}_1$). $\vec{a}_2$ and $\vec{b}_2$ are the **nonprimitive translation vectors**.

1.3 Unit Cell: Primitive and Nonprimitive

The **unit cell** is a parallelepiped formed by the translation vectors, $\vec{a}$, $\vec{b}$, and $\vec{c}$ of a crystallographic basis of the lattice and contain one or more lattice points [1–5]. It is regarded as building blocks for constructing the complete lattice by its continuous repetition. Both primitive and nonprimitive translation vectors, as shown in Fig. 1.3, may be used to construct a unit cell. Accordingly, a unit cell is named as a primitive and a nonprimitive unit cell. The **primitive cell** is a minimum-volume cell and there is always one lattice point per primitive cell. Generally lattice points are shared by several cells and the sum of the fractions of lattice points within one primitive cell is always equal to 1.

The volume of the primitive cell is given by scalar triple product of translation vectors ($\vec{a}$, $\vec{b}$, and $\vec{c}$):

$$V_c = \vec{a} \cdot (\vec{b} \times \vec{c}) \tag{1.4}$$

The basis of a primitive cell is called the **primitive basis** and contains the least number of atoms.

There are many ways of choosing the primitive axes and primitive cell for a given lattice. The most common way of choosing the primitive cell is known as the **Wigner–Seitz cell**. The procedure to construct a Wigner–Seitz cell, as shown in Fig. 1.4, is as follows:

(i) Draw lines connecting a given lattice point to all nearby (closest) lattice points.
(ii) At the midpoint of each line, draw another line (or plane) normal to each of the first set of lines.
(iii) The smallest area (or volume) enclosed in this way is called the Wigner–Seitz primitive cell

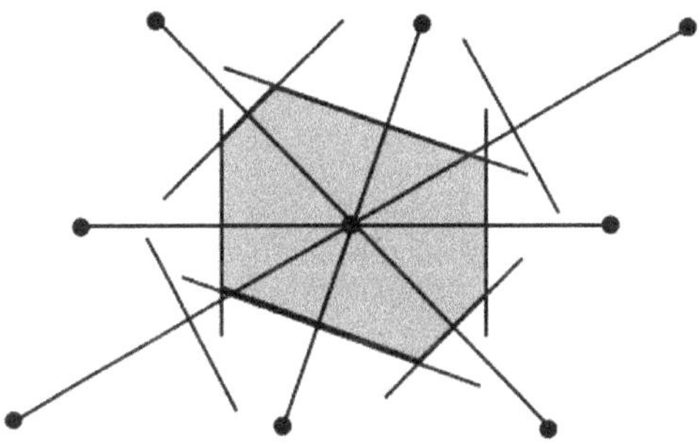

Figure 1.4 Wigner–Seitz primitive cell of a two-dimensional lattice.

1.4 Symmetry Operations

A symmetry operation is an operation that can be performed either physically or imaginatively that results in no change in the appearance of an object, that is, a crystal remains invariant under a symmetry operation [1–5]. The symmetry operations performed about a point (or a line) are called **point group**, for example, rotations, reflections, and inversions (no translations). If the symmetry operations performed by translation as well (i.e., point group operations + translations) are categorized as **space group**. Crystals exhibit both types of symmetries independently and in compatible combinations. These operations are briefly described below:

- **Translation:** A move of one cell in each of three axis directions restores the structure.

- **Rotations:** A lattice is said to possess the rotation symmetry if its rotation by θ about an axis (or a point in a two-dimensional lattice) transform the lattice into itself. If an object can be rotated about an axis and repeats itself every 90° of rotation then it is said to have an axis of **4-fold** rotational symmetry. The axis along which the rotation is performed is an element of symmetry referred to as a rotation axis. The possible rotational symmetry which are compatible with the requirement of translation symmetry to build the long-range order of a crystalline solid are 1-fold, 2-fold, 3-fold, 4-fold, and 6-fold. Long-range order is synonymous with periodicity, requiring some unit structure which repeats itself by translation in all directions infinitely. Although objects themselves may appear to have 5-fold, 7-fold, 8-fold or higher-fold rotation axes, these are not possible in crystals as it is not possible to fill the area of a plane with a connected array such as pentagons, octagons, etc., as shown in Fig. 1.5. It means a single molecule can have any degree of rotational symmetry, but an infinite lattice cannot.

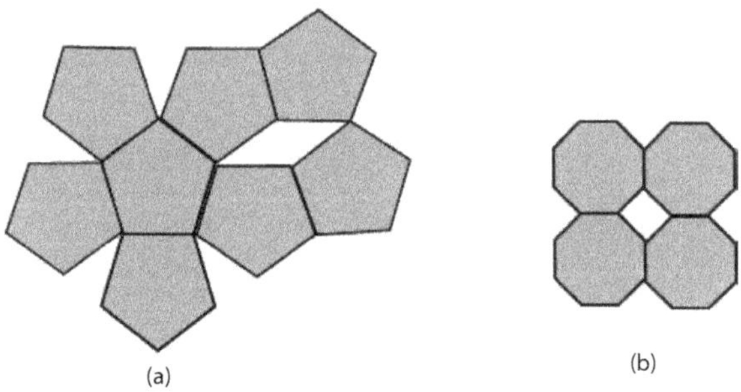

Figure 1.5 Schematic illustration of the nonexistence of (a) fivefold and (b) eightfold rotation axes in a lattice.

- **Reflection:** Reflection across a plane restores the structures.
- **Inversion:** A symmetry operation in which each point of an object is converted to an equivalent point by projecting through a common center (called center of inversion or center of symmetry) and extending an equal distance beyond this center. If the center of symmetry is at the origin of the coordinates, every point (x, y, z) becomes $(-x, -y, -z)$.

- The number of point groups (i.e., combination of certain symmetry operations, such as rotation, reflection, and inversion) in two and three dimensions is 10 and 32, respectively.
- The number of distinct space groups (i.e., group of all symmetry operations) possible in two and three dimensions is 17 and 230, respectively.

1.5 Types of Lattices

It is convenient, to classify lattices according to some basic symmetry groups. The point groups form the basis for construction of different kinds of lattice. Only a finite number of lattices is permissible which are consistent with the point group operations. Such lattices are called **Bravais lattices** [1–5]. Point group operations in two and three dimensions produce only 5 and 14 distinct Bravais lattices, as shown in Fig. 1.6 and Fig. 1.7, respectively.

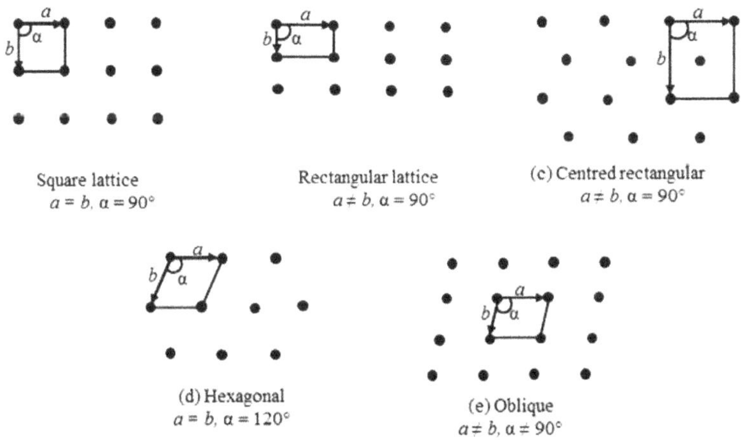

Figure 1.6 Bravais lattice in two dimensions.

To calculate the effective number of lattice points (N) belonging to a particular cell following equation is used:

$$N = N_i + \frac{N_f}{2} + \frac{N_c}{8}$$

N_i: Number of lattice points completely inside the cell

N_f: Number of lattice points occupying the face center. These lattice points are shared by two cells.

N_c: Number of lattice points occupying the corner position. These lattice points shared by eight cells.

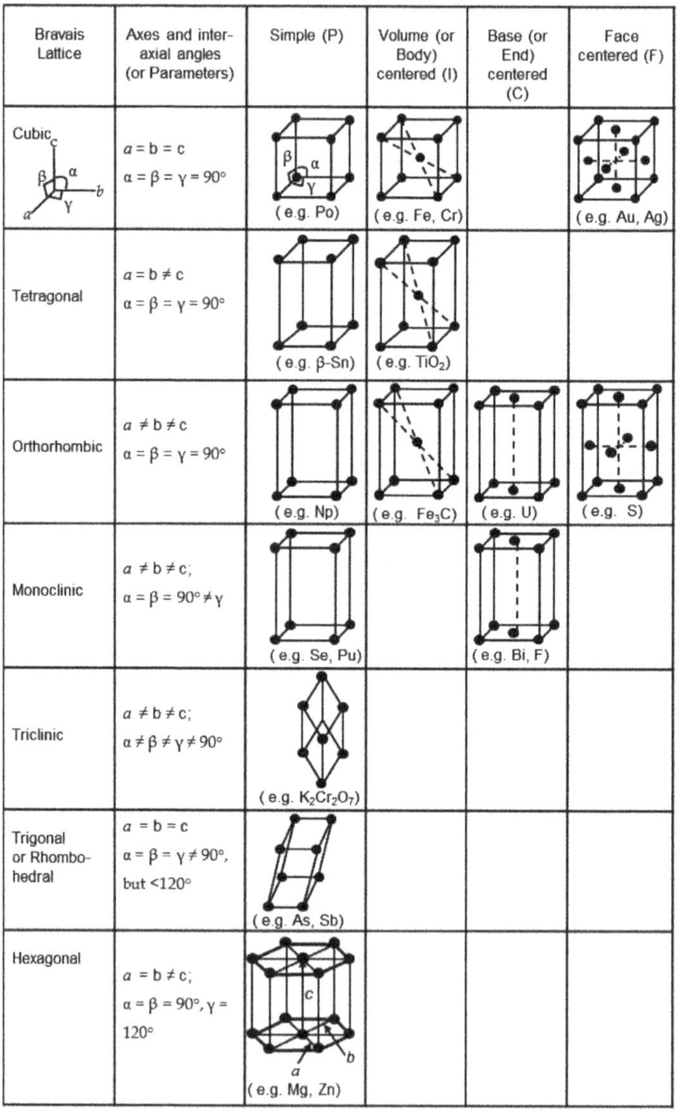

Bravais Lattice	Axes and inter-axial angles (or Parameters)	Simple (P)	Volume (or Body) centered (I)	Base (or End) centered (C)	Face centered (F)
Cubic	$a = b = c$ $\alpha = \beta = \gamma = 90°$	(e.g. Po)	(e.g. Fe, Cr)		(e.g. Au, Ag)
Tetragonal	$a = b \neq c$ $\alpha = \beta = \gamma = 90°$	(e.g. β-Sn)	(e.g. TiO_2)		
Orthorhombic	$a \neq b \neq c$ $\alpha = \beta = \gamma = 90°$	(e.g. Np)	(e.g. Fe_3C)	(e.g. U)	(e.g. S)
Monoclinic	$a \neq b \neq c$; $\alpha = \beta = 90° \neq \gamma$	(e.g. Se, Pu)		(e.g. Bi, F)	
Triclinic	$a \neq b \neq c$; $\alpha \neq \beta \neq \gamma \neq 90°$	(e.g. $K_2Cr_2O_7$)			
Trigonal or Rhombo-hedral	$a = b = c$ $\alpha = \beta = \gamma \neq 90°$, but <120°	(e.g. As, Sb)			
Hexagonal	$a = b \neq c$; $\alpha = \beta = 90°, \gamma =$ 120°	(e.g. Mg, Zn)			

Figure 1.7 Bravais lattice in three dimensions.

1.6 Index System for Crystal Planes and Directions

The orientation of a surface or a crystal plane may be defined by considering how the plane intersects the main crystallographic axes of the solid. A set of numbers which quantify the intercepts called **Miller indices** (hkl) and thus are used to uniquely identify the plane or surface [1–5]. The following steps are employed to determine the Miller indices of a plane:

(i) Find the intercepts of the axes in terms of the lattice constants a, b, and c.
(ii) Take the reciprocal of these numbers.
(iii) Simplify to remove fractions and enclose the numbers obtained in parentheses.

For example, a plane shown in Fig. 1.8a intercepts the crystallographic axes of the cubic lattice at a, $b/2$, $2c$. The reciprocal of these numbers are 1, 2, 1/2. Simplified numbers (or smallest three integers) after removal of fractions are 2, 4, 1. When these numbers are enclosed in parentheses (241) define the indices of the plane. Thus, the Miller indices are proportional to the **inverses** of the intercepts of the plane with the unit cell. If one or more of the indices is zero, it simply means that the planes do not intersect that axis (i.e., the intercept is at infinity). If a plane intercepts an axis on the negative side of the origin, the corresponding index is negative. In that case, a bar (or negative sign) is put above (or before) the corresponding number of the Miller indices, for instance, ($\bar{1}00$) or (–100). Specific planes are denoted by parentheses (or round brackets): (hkl), for example, (100). And the family of equivalent planes is denoted by braces (or curly brackets): {hkl}. For instance, {100} indicates a family of (100), ($\bar{1}00$), (010), ($0\bar{1}0$), (001), and ($00\bar{1}$) planes.

The angle θ between the two crystallographic directions $[h_1k_1l_1]$ and $[h_2k_2l_2]$ is given by

$$\cos\theta = \frac{h_1h_2 + k_1k_2 + l_1l_2}{\sqrt{(h_1^2 + k_1^2 + l_1^2)(h_2^2 + k_2^2 + l_2^2)}}$$

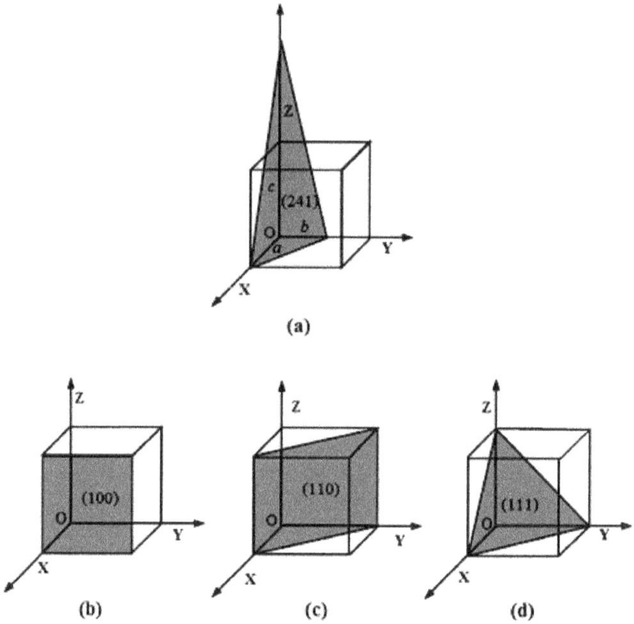

Figure 1.8 Miller indices of different planes in a cubic lattice.

The interplanar spacing (or distance) between successive parallel (hkl) planes for an orthorhombic lattice is

$$d = \frac{1}{\sqrt{\left(\frac{h^2}{a^2} + \frac{k^2}{b^2} + \frac{l^2}{c^2}\right)}}$$

where a, b, and c are the lengths of the axes.

For a cubic lattice, $a = b = c$:

$$d = \frac{a}{\sqrt{(h^2 + k^2 + l^2)}}$$

To determine the indices of a line, first the line is made to pass through the origin by shifting the origin. Thereafter, projections of the vector drawn from the origin to any point on this line on the crystallographic axes are determined. Let these projections be u, v, and w along the x, y, and z axes, respectively. Apparently, u, v, and w also represent the coordinates of that point. These coordinates are then simplified to get a set of the smallest possible integers which when enclosed in a square bracket represent the indices of the line. For instance, to determine the indices of the direction OR in a cubic crystal shown in Fig. 1.9, we may choose a point anywhere on this line, for example, at (½, ½, ½) or (¾, ¾, ¾) or (1, 1, 1), either of these points yields the indices of the direction OR as [111]. The same are the indices of any other direction parallel to OP because by shifting the origin to any appropriate position, the new direction can be made to pass through the points O and R. If one or more of the indices is zero, it simply means that the line (or direction) is perpendicular to that corresponding axes and hence do not form any projection on that axes. In the case of a direction having projection on the negative side of the origin, the corresponding index is negative which is written by putting a bar (or negative sign) over (or before) the indices. Specific directions are denoted by brackets: [*uvw*], for example, [100]. And the family of equivalent directions is denoted by angle brackets: <uvw>. For instance, <100> indicates a family of [100], [−100], [010], [0−10], [001], and [00−1] directions. In the case of a cubic lattice, the direction [*hkl*] is perpendicular to the plane (*hkl*).

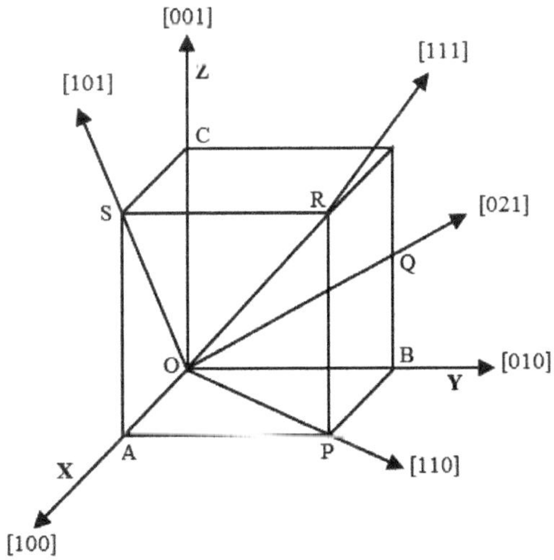

Figure 1.9 Schematic representation of indices of different directions in a cubic lattice.

1.7 Cubic Structures

There are three types of cubic structures: simple cubic (SC), body-centered cubic (BCC), and face-centered cubic (FCC) [5–7].

1.7.1 Simple Cubic or Primitive Cubic Structure

The lattice vectors of an SC crystals are orthogonal and of equal length. The unit cell of an SC structure is a cube with an atom at each corner and the same as the primitive cell. The atoms at the corner touch each other along the cube edge, as shown in Fig. 1.10. As the each atom at the corner is shared by eight unit cells, only a 1/8 part of the atom is shown at the corner. Polonium is the only metal that forms a simple cube crystal structure at room temperature. In the case of SC structure, the side a of a unit cell is related with the radius r of the corner atom as follows:

$$a = 2r$$

The coordination number (number of nearest neighbors) of each atom is 6. The atomic packing factor (APF), which is the fraction of the volume of the cell occupied by the hard spheres, is given by

$$\text{APF} = f = \frac{\text{Sum of atomic volumes}}{\text{Volume of cell}} = \frac{1(4/3)\pi r^3}{a^3} = 0.52$$

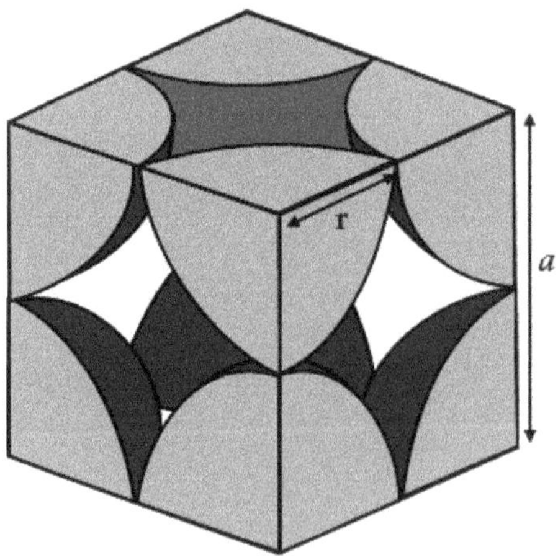

Figure 1.10 Simple cubic structure.

1.7.2 Body-Centered Cubic Structure

The unit cell of a BCC structure is a cubical shape with an identical atoms on the corners and at the center, as shown in Fig. 1.11. The effective number of atoms per unit cell is 8(1/8) + 1 = 2. Thus the unit cell of a BCC structure is not a primitive cell. The atoms touch each other along the body diagonal. The coordination number of each atom is 8. Example of a BCC structure are Li, Na, K, Mo, W, etc. In this case, the side a and the atomic radius r are related as follows:

$$\text{Cube diagonal} = \sqrt{\left(\sqrt{2}a\right)^2 + a^2} = \sqrt{3}a \qquad 4r = \sqrt{3}a$$

The packing fraction is given by

$$f = \frac{\text{Number of atoms per unit cell} \times \text{Volume per atom}}{\text{Volume of unit cell}}$$

$$= \frac{2(4/3)\pi r^3}{a^3} = 0.68$$

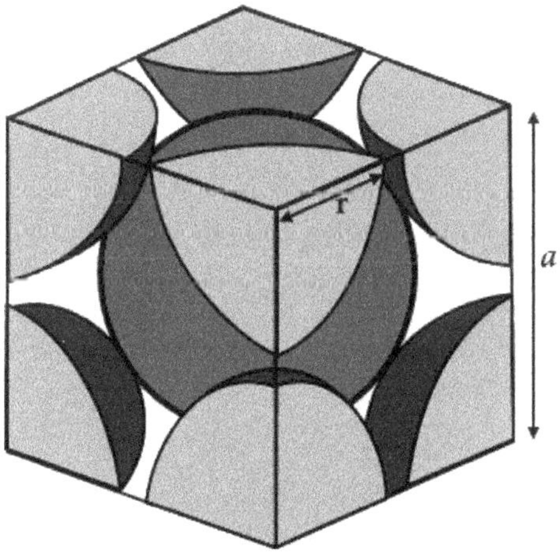

Figure 1.11 Body-centered cubic structure.

1.7.3 Face-Centered Cubic Structure

The unit cell of an FCC structure also starts with identical atoms on the eight corners of the cube and at the centers of the six faces as illustrated in Fig. 1.12. The atoms touch one another along the face diagonals. The effective number of atoms per unit cell is 8(1/8) + 6(1/2) = 4. Thus the unit cell of an FCC structure is also not a primitive cell. The coordination number of each atom is 12. Examples of the materials having an FCC structure are Cu, Ag, Au, Al, etc. The a is related to r as follows:

$$4r = \sqrt{2}a$$

The packing fraction is given by

$$f = \frac{\text{Number of atoms per unit cell} \times \text{Volume per atom}}{\text{Volume of unit cell}}$$

$$= \frac{4(4/3)\pi r^3}{a^3} = 0.74$$

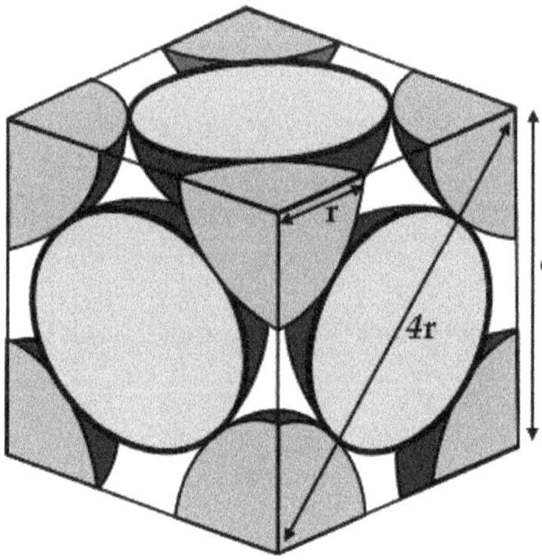

Figure 1.12 Face-centered cubic structure.

1.8 Stereographic Projection

The stereographic projection is a projection of points from the surface of a sphere onto its equatorial plane [3, 8–11]. Mathematically, it is conformal projection in which angular relationships of the faces are preserved. It is a convenient method to discuss a three-dimensional angular relationship between planes and directions and the symmetrical arrangement of crystal faces on a two-dimensional flat surface (i.e., plane). The stereographic projection is an invaluable tool in crystallography to represent the whole range of planes and directions on a two-dimensional graph. It can be said here that the stereographic projection is a two-dimensional drawing of three-dimensional data. The etch rates of different crystallographic planes, which are achieved by the study of the etching of a hemisphere in aqueous solution, are represented by stereographic projection.

In this section, the construction of stereographic projection is discussed briefly. An elaborate explanation about stereographic projection can be found in crystallography related books; few of them are listed in the References section. To construct the stereographic projection of a crystal, the following steps are used:

(i) Place the crystal at the center of the sphere which is called the sphere of projection, as shown in Fig. 1.13.

(ii) Analogous to the globe, define north and south poles, N and S, as shown in Fig. 1.13. The plane of projection is the equatorial plane of the sphere, which passes through the center of the sphere and normal to the line NS. The equatorial plane cut the sphere in a great circle (i.e., a circle of maximum diameter) is called equator or primitive circle or primitive of projection. Every plane passing through the origin of the crystal intersects the sphere of projection in a **great circle**, called the trace of the plane.

(iii) Draw the normals to each face. These normals are originated from the center of the sphere where the crystal is located. This set of face normals is representative for the crystal planes.

(iv) Extend the normals. These normals intersect the sphere in a set of points called poles. The pole of (011) plane is represented by P in Fig. 1.13.

(v) Connect the intersection point P (or the poles on the northern hemisphere) with the south pole S of the sphere.

(vi) The line PS cuts the equatorial plane at p', which is called the stereographic projection of P (or the stereographic projection of face (011). So this way, we can represent all crystallographic planes and directions on a plane piece of paper. A pole on stereographic projection represents either normal to the plane or the plane itself. The stereographic projection of poles on the sphere is usually called a stereogram.

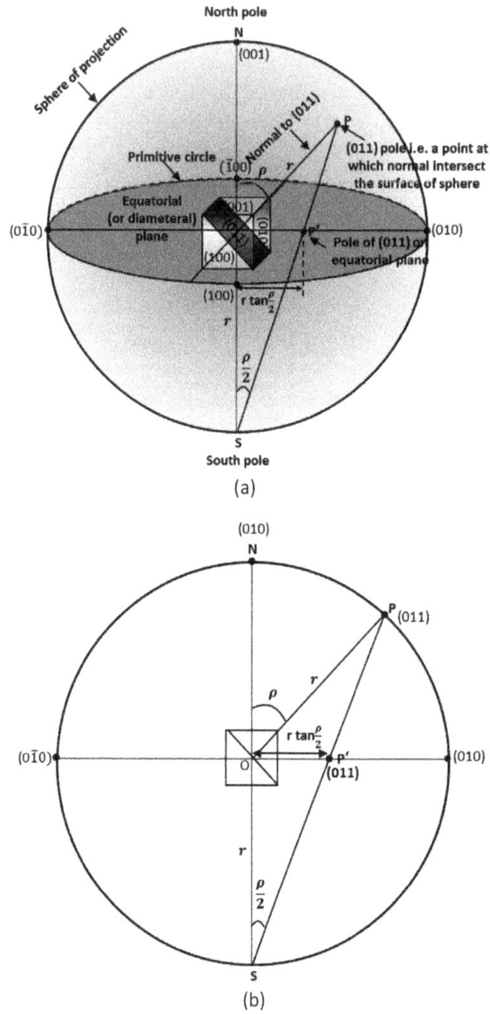

Figure 1.13 Schematic representation of (a) the pole of a plane of a cubic crystal and its projection on the equatorial plane (i.e., stereographic projection of the normal of a crystal face) and (b) the longitudinal section of the sphere of the projection.

The stereographically projected distance from the center of projection of the pole of any plane (i.e., OP’) can be understood from the longitudinal section of the sphere of projection including crystal in the plane of the zero meridian, as shown in Fig. 1.13b. The angle ρ in Fig. 1.13 is the angle between the (001) and (011) planes.

As shown in Fig. 1.14, the plane itself can be projected onto the equatorial plane instead of just the pole of the face. In this case, a plane to be projected is extended through the center of the sphere. This plane intersects the sphere of projection in a great circle. The half part of the great circle lying in the north hemisphere is projected onto equatorial plane by connecting the each point of the intersection with the south pole. This great circle is projects as the line on the equatorial plane (or on a stereogram). The great circle of a plane is the locus of the poles 90° from the pole of that plane. For instance, the poles of the planes lying on the great circle of the (011) plane, as shown in Figs. 1.14 and 1.15, make an angle of 90° with the (011) pole (i.e., pole of great circle). All planes whose poles lie on a great circle mutually intersect along a common direction

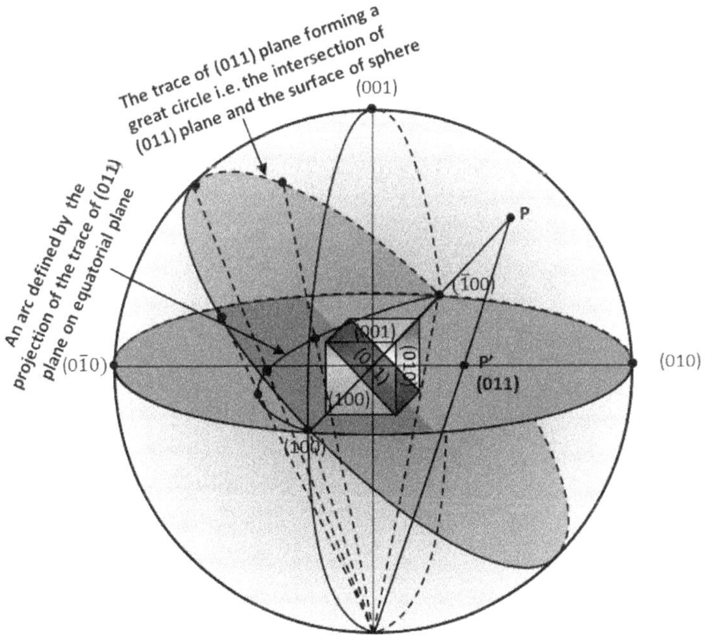

Figure 1.14 Projection of the (011) plane as an arc and a pole on the equatorial plane.

which is called zone axis. A zone is a set of planes {*hkl*} that mutually intersect along (or are parallel to) a common direction or axis [*uvw*] with respect to a given crystal. The line of intersection [*uvw*] is called a zone axis. It means, the zone of planes is represented by a great circle and the zone axis is represented by a corresponding pole. In other words, the perpendiculars to all planes that belong to one zone axis are located on one great circle. As shown in Fig. 1.16, the vertical great circles project as straight lines passes through the center of the projection. These lines become the diameter of the primitive circle. In the remaining cases, they project as circular arcs. The primitive of the projection is also a great circle. The standard stereographic projections of a cubic crystal on (001), (011), and (111) are shown in Figs. 1.17, 1.18, and 1.19, respectively [10, 11]. The center of the stereographic projection represents the orientation of the crystal.

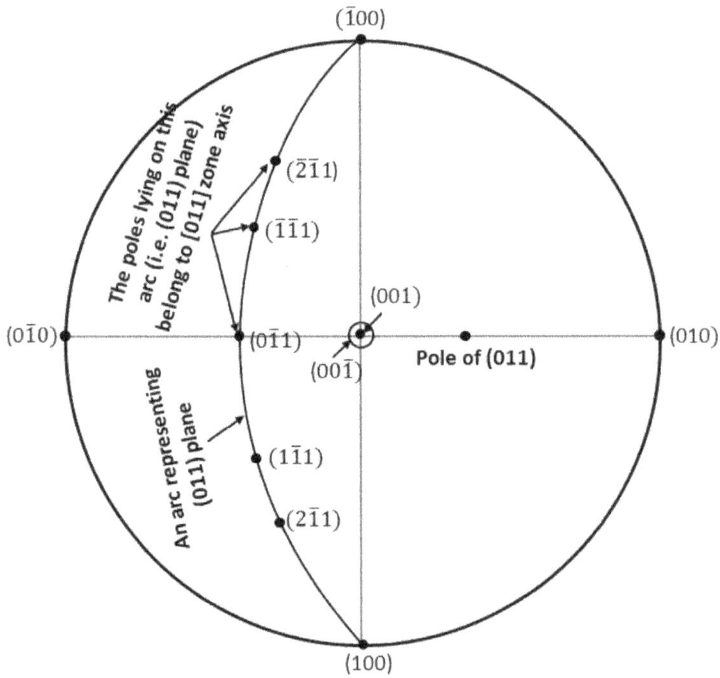

Figure 1.15 Stereogram of a cubic crystal illustrates the stereographic projection of the (011) plane as an arc and a pole. The poles of the planes lying on the circular arc belong to the same zone axis (i.e., [011]).

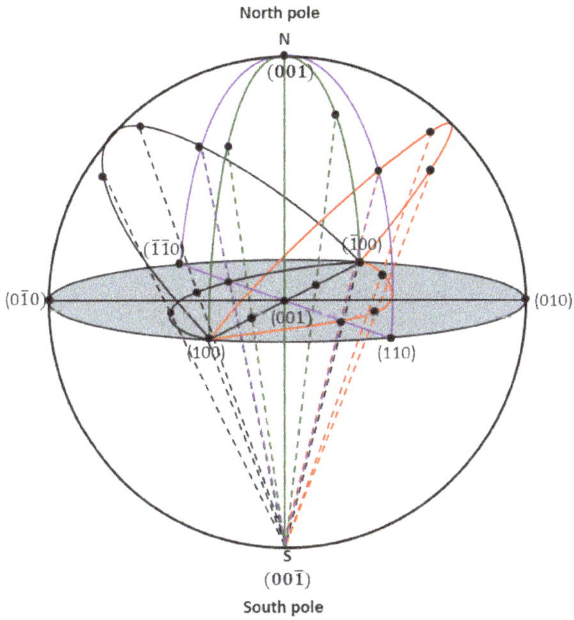

Figure 1.16 Stereographic projection of the planes and poles lying on the trace of the planes.

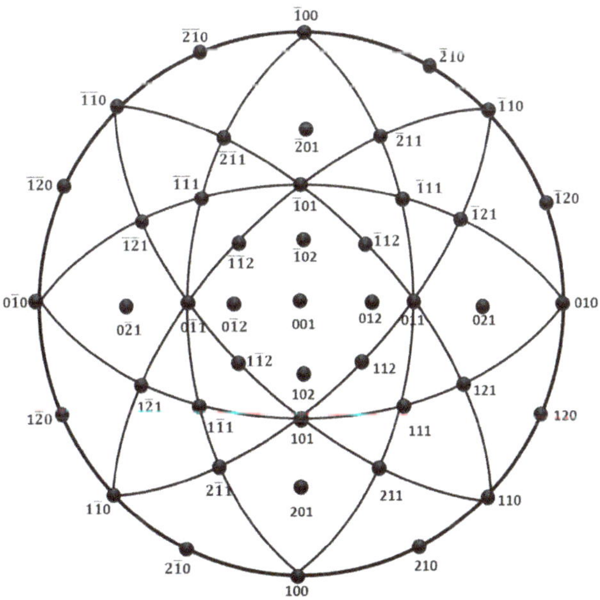

Figure 1.17 Stereographic projection of (001).

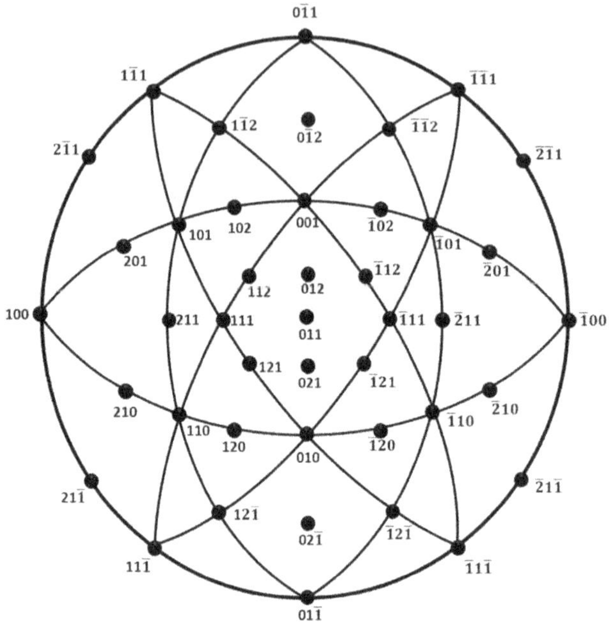

Figure 1.18 Stereographic projection of (011).

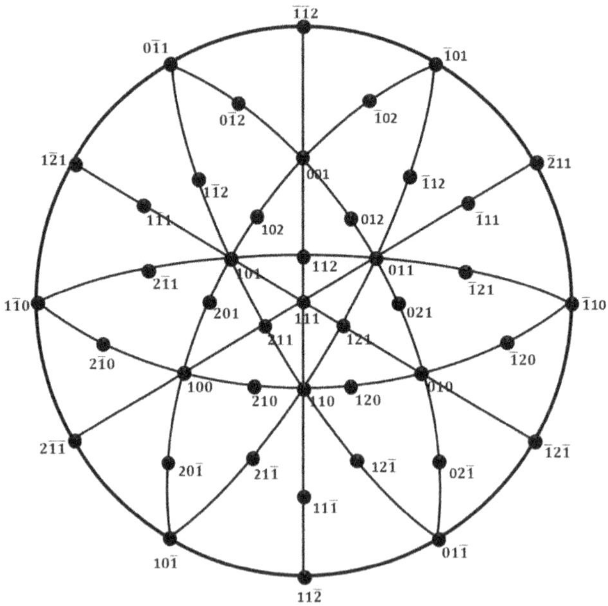

Figure 1.19 Stereographic projection of (111).

A Wulff net shown in Fig. 1.20 is used for measuring the angles between directions (or planes) and finding the pole of a great circle and vice versa. The lines (or great and small circles) on a Wulff net are drawn at intervals of 1°–2°. To measure the angle between two poles, the stereographic projection and the Wulff net are superimposed and pinned at the center. The angle between two planes is measured if, and only if, their corresponding poles lie on a same great circle, as shown in Fig. 1.20. If the poles of the planes do not lie on same great circle, the projection is rotated with respect to the Wulff net until they lie on the same great circle.

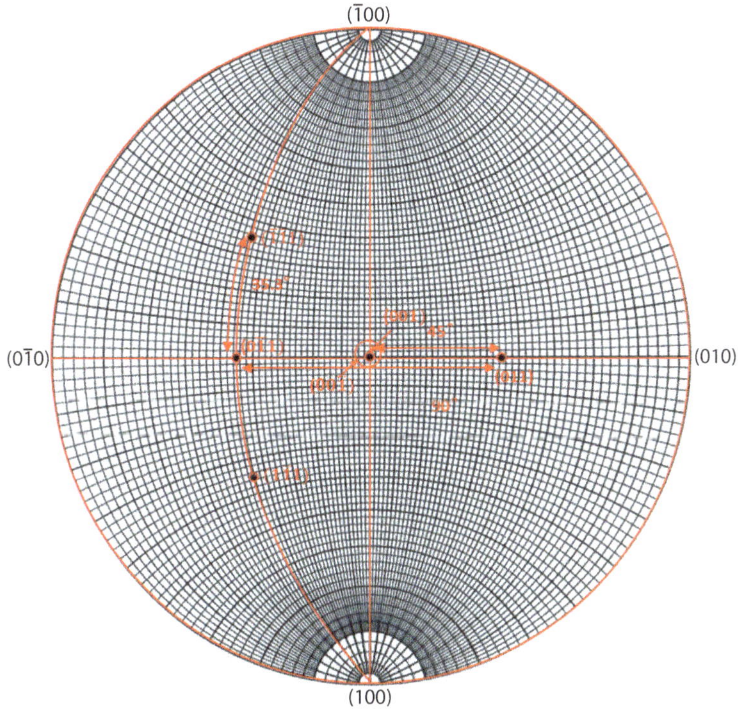

Figure 1.20 The Wulff net drawn to 2° intervals.

References

1. Dekker, A. J. (1981). *Solid State Physics* (Macmillan India, Madras).

2. Galperin, Y. M. (2009). *Introduction to Modern Solid State Physics* (free online).
3. Cullity, B. D. (1956). *Elements of X-Ray Diffraction* (Addison-Wesley, Massachusetts).
4. Kittel, C. (2009). *Introduction to Solid State Physics*, 7th ed. (Wiley India, New Delhi).
5. Puri, R. K., and Babbar, V. K. (2010). *Solid State Physics* (S. Chand, New Delhi).
6. http://lessons.chemistnate.com/simple-cubic-fcc-and-bcc.html
7. http://spaceflight.esa.int/impress/text/education/Glossary/Glossary_B.html
8. Klein, C., and Dutrow, B. (2002). *Manual of Mineral Science*, 22nd ed. (John Wiley & Sons, New York).
9. Kelly, A. A., and Knowles, K. M. (2012). *Crystallography and Crystal Defects*, 2nd ed. (John Wiley & Sons, Chichester).
10. http://www.jmtilli.iki.fi/~juhis/stereographic/ (free online software to draw stereographic projection).
11. http://www.jcrystal.com/products/winwulff/ (free demo version to draw stereographic projection).

Chapter 2

Brief Overview of Silicon Wafer Manufacturing and Microfabrication Techniques

2.1 What Is Silicon?

Silicon is a semiconductor material which has the symbol Si and atomic number 14. It has a diamond lattice structure where each atom forms four identical covalent bonds with its neighboring atoms in a tetrahedral arrangement. The unit cell is constituted by two face-centered cubic (FCC) cells interpenetrating along 25% of the body diagonal (i.e., <111> direction), as shown in Fig. 2.1. It is the second most abundant material in earth's crust after oxygen and exists as compound. It is most widely distributed in dusts, sands, planetoids and planets in various forms of silicon dioxide (silica) or silicates. Silicon is the most extensively used material for the fabrication of integrated circuits (ICs) and microelectromechanical systems (MEMS). Main properties of silicon are given below:

- Unit cell: FCC
- Total number of atoms per unit cell = (4 atoms completely inside cell) + (Each of the 8 atoms on corners are shared among 8 cells) + (Each of the 6 atoms on the faces shared among 2 cells) = 4 + 1 + 3 = 8
- Lattice constant = 5.431 Å
- Cell volume = $(5.431\ \text{Å})^3 = 1.6 \times 10^{-22}\ \text{cm}^3$

Silicon Wet Bulk Micromachining for MEMS
Prem Pal and Kazuo Sato

ISBN 978-981-4613-72-9 (Hardcover), 978-1-315-36492-6 (eBook)
www.panstanford.com

- Density of silicon atoms = (8 atoms)/(cell volume) = 5×10^{22} atoms/cm^3
- Energy gap (E_g) at 300 K = 1.12 eV
- Young's modulus (E) = 130–180 GPa (it varies with crystallographic directions) [1]
- Shear modulus = 64.1 GPa
- Thermal conductivity = 1.412 W/cm-K
- Melting point = 1414°C

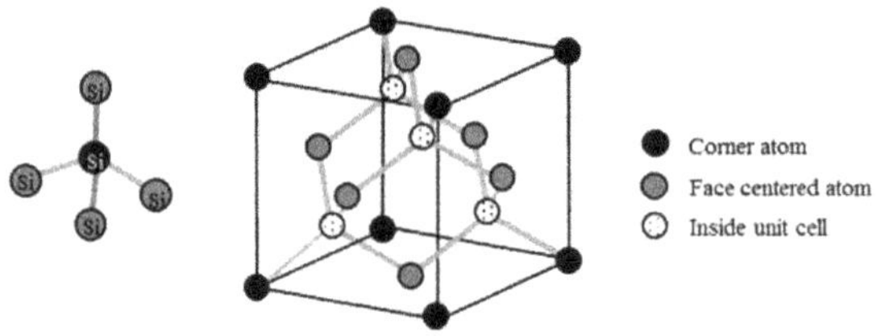

Figure 2.1 Crystal structure of silicon.

2.2 Why Is Silicon Used as a Material for ICs and MEMS?

Silicon is a semiconductor material which has an electrical conductivity intermediate in magnitude between that of a conductor and an insulator. Silicon is used for IC devices due to the following reasons:

- It is a second-most abundant element in the earth's crust, which makes it a low-cost material.
- Its electrical conductivity can be tailored and controlled over a very wide range by the introduction of dopants.
- The study of the silicon was mainly focused for the application in IC industry and most Fab-lines were developed for silicon processing. When the concept of MEMS was introduced, silicon was the first choice for the development of MEMS-based devices due to the following causes:
    - The same equipment, which was developed for IC fabrication, being utilized
    - Better scalability

- Well-known material properties
- Readily available in pure form
- Electronic circuitry integrated easily on the same chip

2.3 What Are Microelectromechanical Systems?

The integration of electrical and mechanical components, as shown in Fig. 2.2, on the same silicon chip gave birth to a new technology, appropriately named MEMS or microsystem technology (MST). When the dimensions of the MEMS structures are scaled down to the nanoregime, they are named nanoelectromechanical systems (NEMS). MEMS/NEMS has found applications in wide-ranging areas such as chemical sensors, biosensors, aerospace, radio-frequency (RF) circuits, consumer electronics, etc. and it has been identified as one of the most promising technologies of the 21st century. MEMS-based devices can range in size from a few micrometers to millimeters and have the ability to sense, control and actuate on the micro-/nanoscale. MEMS promises to revolutionize nearly every product category as it allows the integration of mechanical and electrical components and therefore making the realization of complete system-on-a-chip. Over the years, several new materials, concepts, and technologies have been incorporated to cater the needs of new applications.

Electronic components are fabricated by well-developed IC technology (e.g., bipolar complementary metal-oxide-semiconductor [BiCMOS] processes). The mechanical structures (movable or nonmovable) are fabricated using techniques such as micromachining, LIGA (German acronym of *LIthographe, Galvanoformung, Abforming*), microstereolithography (μSL), etc. Microstereolithography is used for polymer MEMS. Silicon micromachining is dominant method for the formation of mechanical components. This is briefly discussed in the microfabrication section. In this method, most of the process equipment such as thermal oxidation furnace, RF sputtering, thermal evaporator, chemical vapor deposition (CVD), spin coater, etc., are the same as those utilized for IC fabrication.

In this chapter, silicon wafer manufacturing and the common microfabrication methods which are employed for silicon-based ICs and MEMS fabrication are briefly summarized.

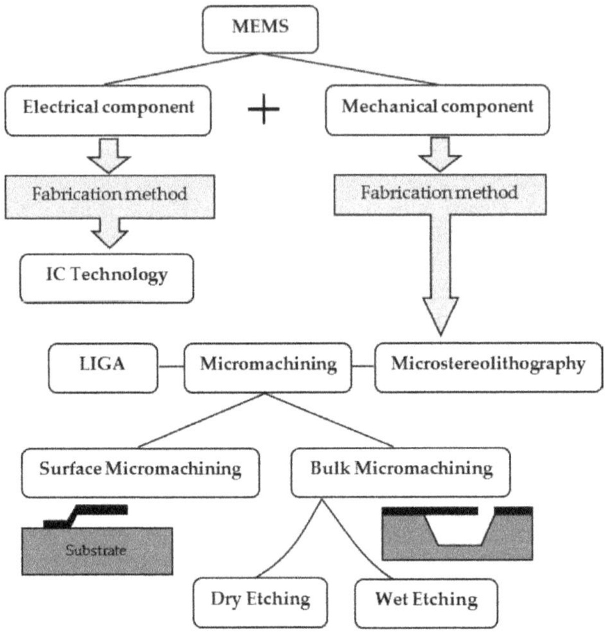

Figure 2.2 MEMS and its fabrication techniques.

2.4 Wafer Manufacturing

The manufacturing of silicon wafers of different orientations and diameters involves several process steps. These are briefly discussed in the following sections [2–7].

2.4.1 Crystal Growth and Doping

The raw material for silicon manufacturing is sand (SiO_2) called quartzite. The sand is heated in a furnace containing a source of carbon (coal, coke, and wood chips). A number of reactions take place in the furnace, which eventually produces metallurgical-grade silicon (MGS). Although MGS is of relatively high purity (98%), it still contains a number of contaminants (such as iron and aluminum). MGS is further purified to electronic-grade silicon (EGS) using the process of distillation. EGS is a polycrystalline material and is used to prepare device-quality, single-crystal silicon wafers. The Czochralski

(CZ) and float-zone (FZ) techniques are employed to grow cylindrical ingots from EGS for the production of wafers.

In the CZ method, EGS is broken into small pieces and placed in a fused-silicon (SiO_2) crucible. In an argon ambient, the crucible is heated above the melting temperature of silicon (above 1414°C) in order to keep EGS in a molten state. A single-crystal seed is then lowered into the melt, as shown in Fig. 2.3. Crystal orientation is determined by seed orientation. Pull rate, melt temperature, and rotation rate (typical values: pull rate 50–100 mm/h, crystal rotation 10–30 rpm, crucible rotation 2–20 rpm) are all important control parameters. Progressive freezing at the solid–liquid interface yields a large single crystal. Dopants are added to the melt to provide a controlled p-type (or n-type) doping level in the wafers. For this purpose, a known amount of dopant is added to the melt during the crystal growth process in order to achieve the desired doping concentration in the grown crystal. Boron and phosphorus are the most common dopants for p- and n-type silicon material, respectively. Figure 2.4 shows the picture of a silicon ingot grown by the CZ method.

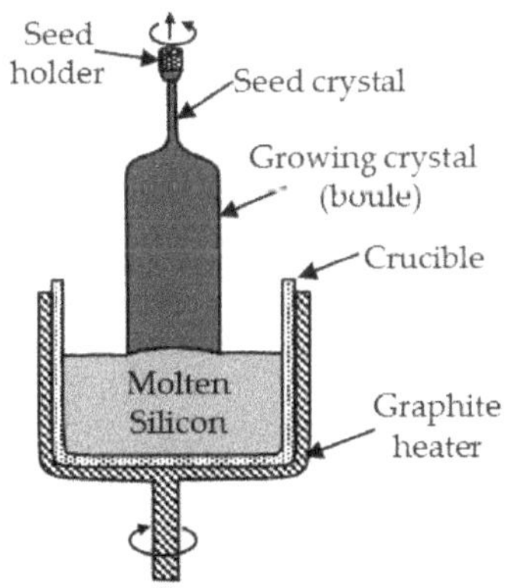

Figure 2.3 The CZ crystal-growing puller.

The FZ growth method is less common, and is reserved for situations where oxygen and carbon impurities cannot be tolerated. A schematic setup of this method is shown in Fig. 2.5. In this method, a rod of EGS polysilicon with a seed crystal at the bottom is held in vertical position and rotated. The rod is enclosed in a quartz envelop within which an inert atmosphere is maintained. During

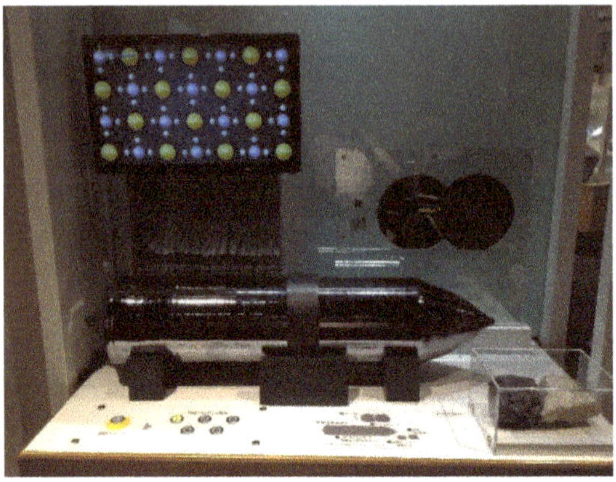

Figure 2.4 CZ-grown silicon ingot. Photograph provided by the Nagoya City Science Museum, Japan.

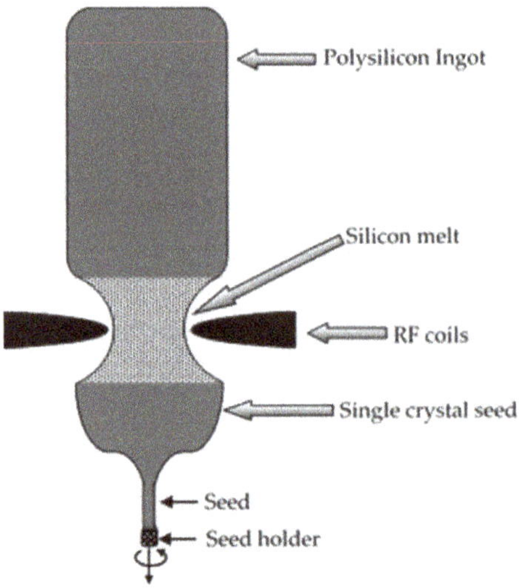

Figure 2.5 Schematic setup of the FZ process.

the operation, localized heating is done using a high-power RF coil. The heater is moved along the length of the polysilicon rod, where eddy current heating causes localized melting and crystallization of the silicon. Surface tension and levitation due to the RF field keeps the system stable. Since no crucible is used in the FZ process, there is no contamination from the crucible. In this method, the crystal orientation is set by a seed crystal and the doping concentration is accomplished either by starting with a doped polysilicon rod or by adding the dopant gas (e.g., diborane [B_2H_6] for p-type and phosphine [PH_3] for n-type) to the inert gas (Ar) during the FZ process.

2.4.2 Ingot Trimming, Grinding, and Slicing

After a crystal is grown by the CZ (or FZ) method, the seed and tang ends are removed from the ingot. After this step, it is trimmed to obtain ingot with uniform and desired diameter, as shown in Fig. 2.6. Thereafter, one or more **flats** are normally ground along the length of the ingot to indicate crystal orientation and doping type. The largest flat is called **primary flat** and work as reference plane for automatic processing equipment and the reference direction for aligning the mask pattern along specific crystallographic direction, particularly in wet chemical–based bulk micromachining. In the case of {100} wafers, it is conventionally oriented along <110> direction. However, it can be ground according to the requirement of customers. Other smaller flat, called **secondary flat** indicates the orientation and conductivity type of the wafer. These days secondary flat is not common to determine the orientation and conductivity type. Generally manufactures supply the wafers with only primary flat and the data sheet provides the information about various parameters such as orientation, conductivity type, resistivity, thickness, diameter, taper, bow, flat location, impurity, etc. For crystals with diameters equal or larger than 200 mm, no flats are ground; instead, a groove is made on the edge of the wafer for positioning and orientation purpose. After trimming and grinding, the ingot is sliced into individual wafers using inner diameter (ID) sawing or wire sawing, as shown in Fig. 2.7. Slicing determines four wafer parameters: surface orientation, thickness, total thickness variation (TTV), and bow (i.e., wafer curvature). The thickness of the sliced wafers depends upon the size of the wafer in order to provide adequate mechanical support during device fabrication. The standard thickness versus wafer diameter is given in Table 2.1. Figure 2.8 shows the use of secondary flat to indicate the orientation and conductivity type.

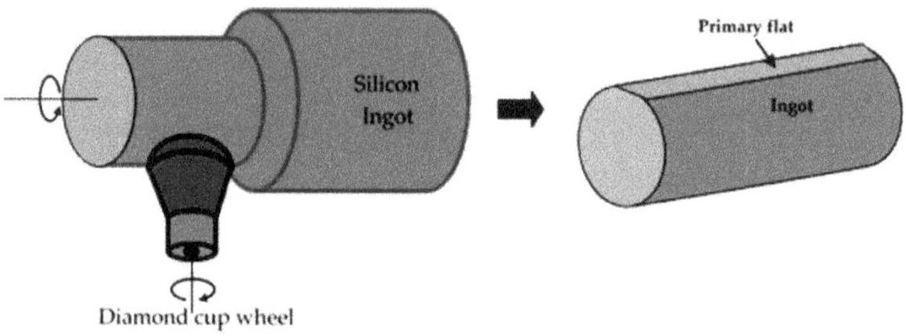

Figure 2.6 The trimming of an ingot to a desired diameter and grinding a flat along the length.

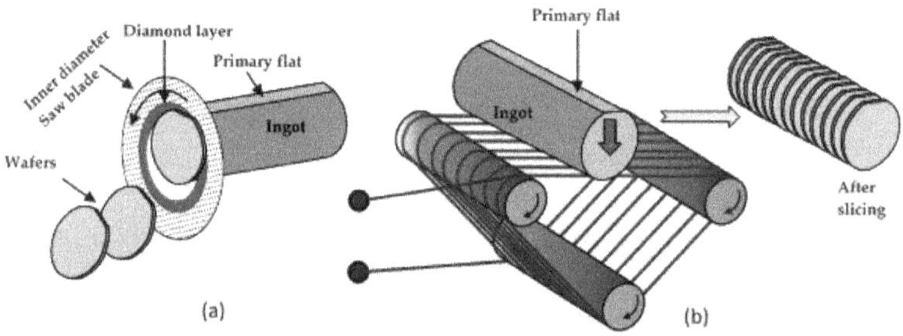

Figure 2.7 Wafer slicing from ingot. (a) Inner diameter sawing and (b) wire sawing.

Table 2.1 Standard silicon wafer size versus thickness

Wafer size	Standard thickness
51 mm (2 inch)	275 μm
76 mm (3 inch)	375 μm
100 mm (4 inch)	525 μm
125 mm (5 inch)	625 μm
150 mm (6 inch)	675 μm
200 mm (8 inch)	725 μm
300 mm (12 inch)	775 μm
450 mm (18 inch) (proposed for future)	925 μm

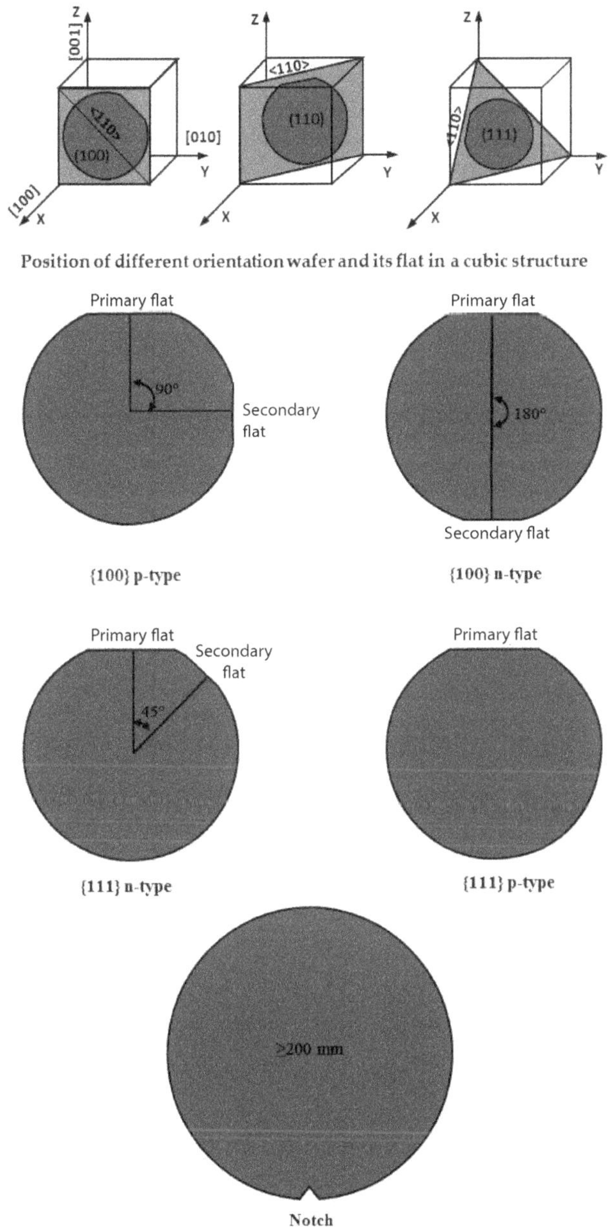

Figure 2.8 Identification flats on silicon wafers and the reference notch for 200 mm and larger wafers.

2.4.3 Polishing, Cleaning, and Packaging

Fabrication processes for the realization of ICs and MEMS require highly flat and smooth surface for achieving submicron resolution. To achieve highly flat and smooth surface, several process steps are performed on the sliced wafers. Firstly, both sides of the wafers are lapped using a mixture of Al_2O_3 water and glycerin. This step serves to improve the flatness of the wafers to about ±2 μm, removing most of the taper, saw damage on the surface, and bow that results from the sawing process. The edges of the lapped wafers are then rounded (i.e., edge contouring), as shown in Fig. 2.9, in order to avoid chipping, dislocation and other defects during device fabrication process. The lapping operation usually damages and contaminates the wafer surface. The damaged (or microscopic cracks) and contaminated regions are removed by chemical etching in a mixture of hydrofluoric, nitric, and acetic acids (or HNA). The acetic acid serves to dilute the mixture and control the etch rate. The nitric acid oxidizes the silicon surface and the hydrofluoric acid dissolves the resulting oxide in this process. In this process, few tens of microns of silicon are etched off on each side of the wafer and hence remove the surface layer containing microscopic cracks or damages. The acid solution is removed by a series of de-ionized (DI) water baths.

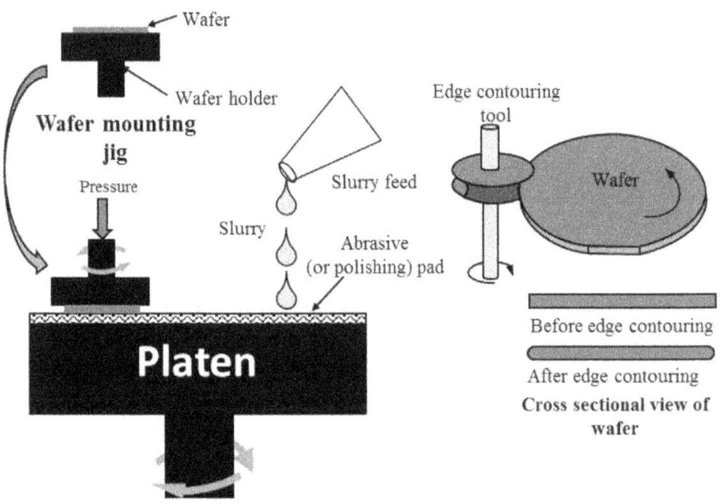

Figure 2.9 Schematic view of wafer lapping/polishing and edge contouring.

The final step of wafer shaping is polishing in order to achieve a highly smooth mirror finished surface. In this step, wafers are polished in a series of combination of chemical and mechanical polish processes which is usually called chemical-mechanical polishing (CMP), as illustrated in Fig. 2.9. The wafers are held in a hard ceramic chuck using either a wax bond or vacuum and buffed with a slurry of silica powder and DI water. The wafers are then cleaned to remove contamination caused by CMP. This is usually done by the RCA cleaning method developed by Werner Kern at the Radio Corporation of America (RCA) laboratories in 1965 and comprises mainly three process steps. The wafers are thoroughly rinsed in DI water after each step. This cleaning method is briefly discussed in the next section (Section 2.5.1). At last, the wafers are packaged in plastic boxes containing cassettes with 25 slots (standard number).

2.5 Silicon Microfabrication Processes

In this section, various fabrication methods which are used for both silicon-based IC and MEMS fabrications are briefly summarized [3–9]. Microfabrication processes are usually carried out in **cleanrooms**. Cleanrooms are defined as a specially constructed enclosed area, environmentally controlled with respect to airborne particulates, temperature (68°F–72°F), humidity (40%–46% RH), air pressure (positive in order to blow dust out), airflow patterns, air motion, vibration, noise, viable (living) organisms, and lighting. Cleanrooms are classified by the cleanliness of the air. Class X simply means that in each cubic foot of air in the room, there are less than *X* particles greater than or equal to 0.5 μm in size. The most common classifications of cleanrooms are classes 1, 10, 100, 1000, and 10,000. A typical office building is about class 100,000. The following precautions must be taken care of in order to preserve the cleanliness of a cleanroom:

- Wear a bunny suit (or cleanroom suit). The suit covers the wearer to prevent skin and hair being shed into a cleanroom environment.
- Use air showers when entering into a cleanroom. An air shower blows filtered air on and around the person to remove any surface contaminants before entering the clean space.

- Do not use paper, pencils, or markers that leave dust or lint.
- Change gloves whenever they get dirty or torn.
- Use cleanroom paper and dust-free ballpoint pens.
- The cleanroom should be held at a positive pressure in order to blow dust out.
- Doors should open inward so that the room pressure closes them shut.

2.5.1 Wafer Cleaning

Cleaning involves removing of particles, organics (photoresist [PR]), and any other trace contaminants from wafer surfaces. Silicon wafer cleaning processes before and after metallization are called **front-end** and **back-end** processes, respectively. Particles are largely removed by ultrasonic (20–50 kHz) agitation during cleaning. The sound waves in an appropriate liquid cause the generation, expansion, and violent collapse of tiny vapor bubbles under the alternating tensile and compressive stresses of the ultrasound. Megasonic agitation (1 MHz) is also quite effective for wafer cleaning. Organics like PRs are removed in an O_2 plasma or in H_2SO_4:H_2O_2 :: 1:1 (called **Piranha bath**). This mixture is a strong oxidizer and therefore removes most organic matter. Moreover it hydroxylates (adds OH groups) most surfaces and makes them extremely hydrophilic. Other mixture ratios of H_2SO_4 and H_2O_2 (e.g., 1:3, 1:4, etc.) are also called piranha solutions. A piranha solution must be prepared with great care. It is highly corrosive and oxidizing. Make sure surfaces are reasonably clean and completely free of organic solvents from previous wash steps before coming into contact with a piranha solution.

RCA cleaning is a standard process used to remove organics, heavy metals, and alkali ions from the wafer surface. RCA cleaning is performed mainly in three steps shown in Fig. 2.10:

(i) The first step of cleaning SC-1 (standard clean-1) uses a solution (NH_4OH:H_2O_2:DI water :: 1:1:5) to remove organic impurities and particles from the wafer surface. This step of cleaning results in the formation of a thin silicon dioxide layer

(~1 nm) on the silicon surface in which certain degrees of metallic contaminations are accumulated.

(ii) In the second step, the thin oxide layer is removed using a diluted hydrofluoric acid (e.g., H_2O:HF :: 50:1) solution.

(iii) The final step is carried out in SC-2 (standard clean-2) solution (HCl:H_2O_2:DI water :: 1:1:6) to remove the remaining traces of metallic (ionic) contaminants. This treatment results in the formation of a superclean thin oxide layer on the wafer surface, which acts as a passivation layer to protect the surface from subsequent contamination as the bare silicon is contaminated immediately if exposed to the atmosphere. All three steps are followed by thorough rinsing in DI water.

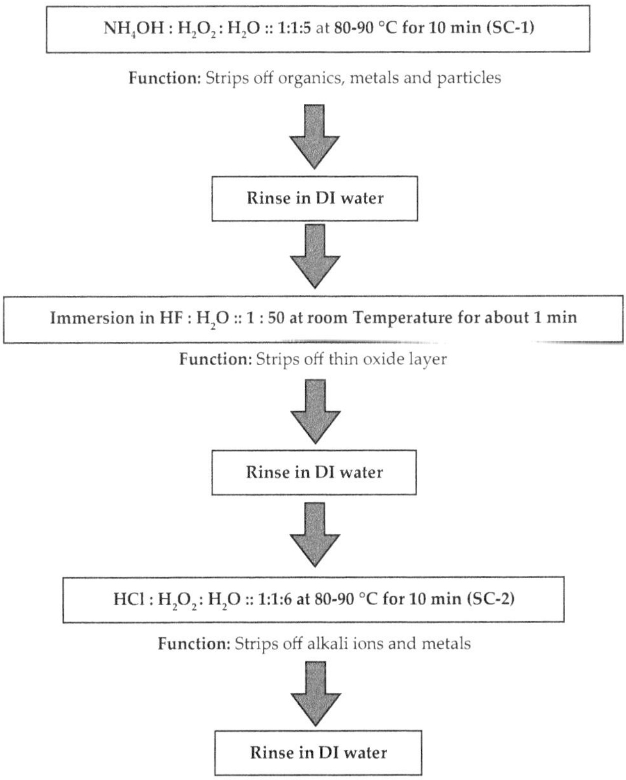

Figure 2.10 Process steps employed in the RCA cleaning method.

2.5.2 Thin-Film Formation

Thin films are essential building materials in ICs and MEMS. A thin film may be deposited on a substrate by both physical and chemical means. These films include dielectrics, semiconductors, and metals. The deposited films are divided into three categories, namely amorphous, polycrystalline, and single crystal. The most commonly used deposition techniques in MEMS and IC Technology are discussed in the following sections.

2.5.2.1 Thermal oxidation (or reactive growth)

In the thermal oxidation process, silicon dioxide is grown by exposing the silicon wafer to oxygen (i.e., dry oxidation) or water vapor (i.e., wet oxidation) at high temperature (900°C–1200°C) and therefore the process is referred to as **thermal oxidation**. In this deposition method, the chemically reactive species combines with the substrate to form a new thin film. Hence it is called reactive growth. The basic setup of a thermal oxidation system is shown in Fig. 2.11. The system consists of a resistance-heated furnace, a cylindrical fused quartz tube containing the silicon wafers held vertically in a slotted quartz boat, gas cylinders, and a quartz water bubbler to produce water vapor. Oxidation temperature is generally in the range of 900°C–1200°C and the typical gas flow rate is about 1 L/min. Thermal oxidation is classified into three categories: dry, wet, and pyrogenic. In the dry oxidation process, oxygen is directly flown into the tube which is set at the desired oxidation temperature, while in wet oxidation, oxygen passes through boiled DI water.

$$\text{Dry oxidation: Si (solid)} + O_2 \text{ (gas)} = SiO_2 \text{ (solid)}$$

$$\text{Wet oxidation: Si (solid)} + H_2O \text{ vapor (gas)} = SiO_2 \text{ (solid)} + 2H_2$$

In the case of pyrogenic oxidation, the mixture of hydrogen (H_2) and oxygen (O_2) gases is converted into pure steam (H_2O) by autoignition in internal or external torch. The oxidation reaction is same as that of wet oxidation.

To create an inert atmosphere, nitrogen gas (N_2) is flown during the temperature ramp-up and ramp-down. A silica rod is used for pushing/pulling into/from the furnace. The Si–SiO_2 interface moves into the silicon during the oxidation process. To grow the oxide layer of thickness x by thermal oxidation, $0.44x$ (0.44 times x) thickness of silicon is consumed, as shown in Fig. 2.12. The volume changes in oxide and silicon in thermal oxidation process causes compressive stress in the grown oxide layer. The basic kinetics of the reaction involves two limiting factors, (i) the reaction rate at the sample surface and (ii) the diffusion of new species through the already formed films. Oxygen is diffusing faster than Si in SiO_2 and hence reaction takes place at the interface between SiO_2 and Si.

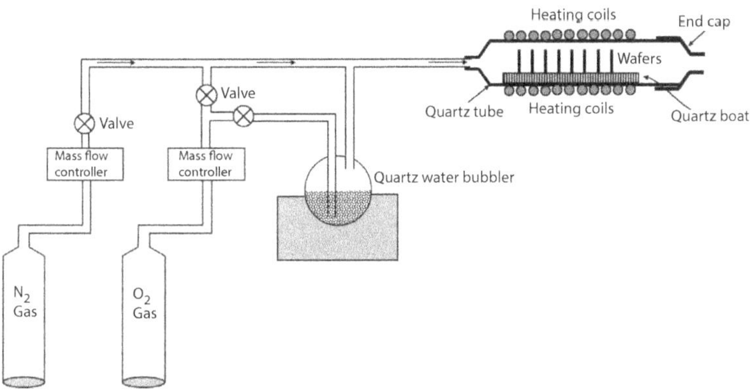

Figure 2.11 Basic setup of the thermal oxidation system.

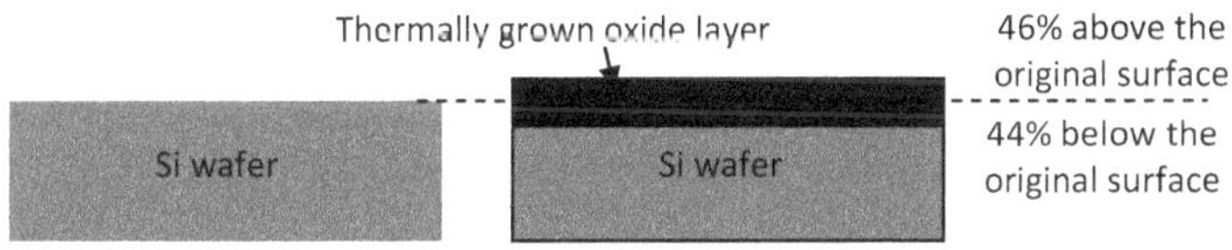

Figure 2.12 Growth of a thermal oxide layer on a silicon surface.

The Deal–Grove model is a simple and popular method to understand the SiO_2 growth. The details of this model can be found in many publications related to microfabrication. According to this model, oxidation takes place in two stages, linear and parabolic. The linear stage (reaction limited) is applicable during the early stage of oxide growth when the oxide thickness varies linearly with time. In this case, chemical reaction occurs between silicon and oxidants at wafer surface and growth is called reaction limited. The reaction is limited by the number of silicon atoms available to react with oxidants. During the first 500 Å of oxide growth, the oxide grows linearly with time (t) and the thickness (x) of grown oxide in time t is given by the following simplified relation:

$$x = \frac{B}{A}(t + \tau)$$

B/A is called linear rate constant and τ represents the time coordinate shift to account for the initial oxide layer.

As the oxide layer becomes thicker (more than 1000 Å) the oxidant must diffuse through the oxide to react at Si–SiO_2 interface and the reaction becomes diffusion limited. This stage of oxidation is called the parabolic stage (i.e., diffusion-limited case). In this case, oxide thickness x in time t is given by the following expression:

$$x^2 = B(t + \tau)$$

where B is the parabolic rate constant.

SiO_2 grows much faster in an H_2O ambient than it does in dry O_2. The principal reason for this is that the oxidant solubility in SiO_2 is much higher for H_2O than for O_2. As a result, both linear (B/A) and parabolic (B) rate constants are much larger for H_2O than for O_2. Due to fast growth rate in wet oxidation process, τ is negligible.

Advantages of a thermally grown oxide layer:

- It works as an excellent electrical insulator (resistivity $>10^{20}$ ohm-cm, energy gap = 9 eV).
- It exhibits a high breakdown electric field $>10^6$ V/cm.
- It serves as a mask against implantation and diffusion. The diffusivity of impurities in SiO_2 is very much less than that of in Si, that is, $D_{SiO2} << D_{si}$.
- It acts as a component in metal-oxide-semiconductor (MOS) structures.
- It can be used as structural (or sacrificial) layer in the fabrication of MEMS.
- It works as surface passivation.
- It provides electrical isolation between conducting layers.
- It is used for isolating one device from another.

2.5.2.2 Chemical vapor deposition

CVD is a technique for the formation of a nonvolatile solid film on a substrate by reaction of a vapor-phase chemical. This technique of thin-film deposition is extensively used to deposit a wide range of materials (e.g., Si_3N_4, poly-Si, W, SiC, etc.) for microelectronics and MEMS applications. Materials are deposited in various forms, including monocrystalline, polycrystalline, amorphous, and epitaxial. This technique offers conformal coating with good step coverage and can coat a large number of wafers at a time. A typical CVD process is performed in the following steps:

(i) Gases (or reactants) are introduced into a reaction chamber, which is kept at a temperature ranging from 100°C to 1000°C.
(ii) Gas species move to the substrate (e.g., wafer) surface.
(iii) Reactants are adsorbed on the substrate surface.
(iv) Film-forming chemical reactions, including decomposition, surface mitigation to attachment sites, site incorporation, and other surface reactions, take place.
(v) Desorption and removal of gaseous by-products by forced convection take place.

For example, silicon nitride can be formed as a result of a reaction between dichlorosilane (SiH_2Cl_2) and ammonia (NH_3). Figure 2.13 shows the schematic diagram of horizontal CVD reactor.

There are various forms of CVD techniques referenced in the literature. Three types of CVD methods are commonly employed: atmospheric pressure CVD (APCVD), plasma-enhanced CVD (PECVD), and low-pressure CVD (LPCVD). CVD films may have large tensile/compressive stress, which is a problem for some applications. The stress in the deposited film can be reduced by careful choice of the deposition conditions.

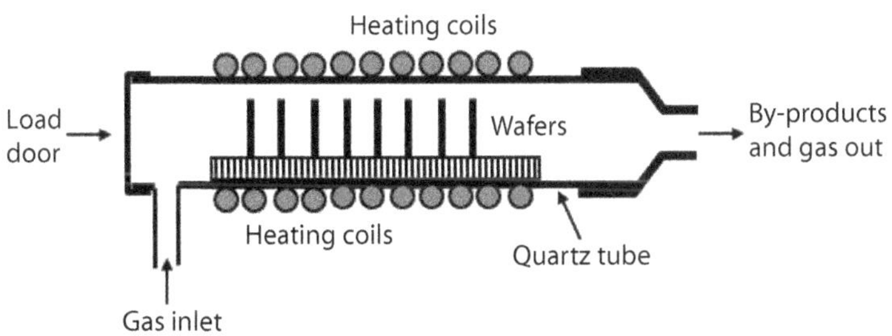

Figure 2.13 Schematic diagram of a horizontal CVD reactor.

2.5.2.3 Physical vapor deposition

Physical vapor deposition (PVD) techniques are widely used in micro-/nanoelectronics for the deposition of thin conducting, semiconducting, and insulating materials. PVD involves the direct impingement of particles on the substrate surface, as shown in Fig. 2.14. In comparison to CVD, PVD is a low-temperature deposition technique. The most common methods of PVD are evaporation and sputtering. In the evaporation method, the source material is heated above its melting point in an evacuated chamber. The heating process can be performed by several means such as resistive (i.e., thermal), focused electron beam (e-beam), etc. The evaporated atoms travel at high velocity and get deposited on the substrate. Thermal evaporation is suitable for metals with sufficiently low melting temperature, while e-beam evaporation can be employed for metals with both low and high melting points.

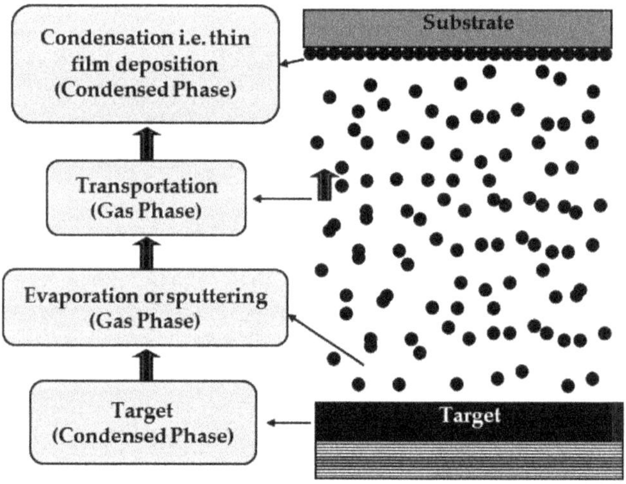

Figure 2.14 Schematic representation of PVD.

In the sputtering method, plasma is formed by way of electrical discharge to throw ionized atoms from a target material onto the surface of the substrate. This works on a wide range of metals, semiconductors, and insulators. DC sputtering is used for the deposition of conducting layers, while RF sputtering is employed for both conducting and insulating layers. To generate plasma using

RF (or DC) voltage, argon gas is most commonly used. The plasma usually consists of argon ions and the pressure of argon gas affects both the deposition rate and the film quality.

2.5.3 Photolithography

Photolithography is a process of transferring a geometric pattern from a photomask to a thin layer of radiation-sensitive material (called resist) covering the surface of the substrate (e.g., silicon wafer). A photomask is a transparent plate of varying sizes depending on the wafer size. It is made of soda lime glass (or quartz), onto which the computer-generated design is formed in a thin chrome (or emulsion) layer.

Various types of lithographic methods such as optical lithography, e-beam lithography, X-ray lithography, ion beam lithography, etc., are employed in microfabrication process. Each method has its own advantages and disadvantages. Mainly three parameters are used to determine the performance of an exposure tool, namely resolution, registration, and throughput [5]. **Resolution** is the minimum feature dimension that can be transferred with high fidelity to a resist film. **Registration** is a measure of how accurately patterns on successive masks can be aligned with respect to previously defined patterns on the same wafer. **Throughput** is the number of wafers that can be exposed per unit time for a given mask level. Optical lithography (or ultraviolet [UV] lithography or photolithography) is most widely used as it has high throughput, good resolution, low cost and ease in operation. However, optical lithography is not suitable for deep-submicron resolution. To pattern at submicron level, e-beam lithography, X-ray lithography, ion beam lithography, etc., are employed.

There are three classes of optical exposure methods, as shown in Fig. 2.15: contact, proximity, and projection printing methods. In contact printing, which is the oldest technique, the mask is placed (chrome or emulsion side down) in direct contact with the resist layer of the wafer. In this case, aligning of the mask to the patterns already on the wafer takes place prior to exposure by observing both the mask and wafer patterns through a microscope, with the mask slightly separated from the wafer. Contact printing offers high resolution (around 1 μm, depending on the PR thickness and the wavelength of UV light). The hard contact between the mask and the wafer results in damage to both the mask and the resist layer. It happens due to the sandwiching of dust particles (or fleck) during the contacting process. Proximity printing largely solves the damage/defect problem associated with contact printing, because the mask and the wafer are kept at a small gap (10–50 μm). The separation of the mask and wafer degrades the resolution of the printing pattern due to diffraction effect. The resolution falls into the range of 2–5 μm. Projection printing avoids the problem of mask damage associated with shadow printing and is dominantly employed in production line. In this method, optical system reduces the mask image by 4X to 5X and scanned (or stepped) over the wafer to cover the entire wafer surface. It is capable of printing 1 μm (or smaller) features and have throughput of about 25–50 wafers per hour.

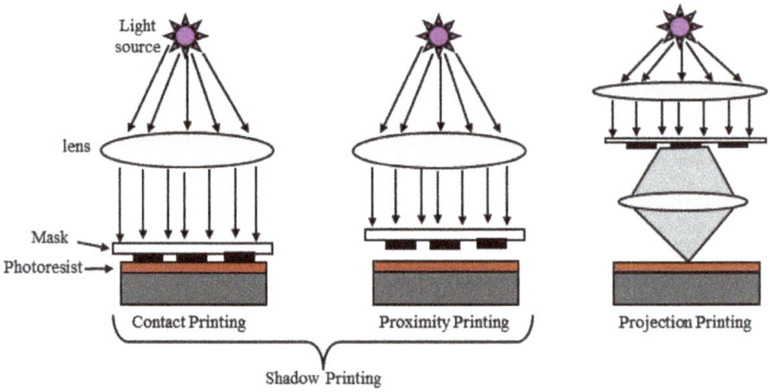

Figure 2.15 Three basic exposure methods in photolithography.

The most popular light source for photolithography is high-pressure mercury arc lamp. This lamp source provides a wavelength spectrum from 310 to 440 nm with several peaks termed "g-line" (436 nm), "h-line" (405 nm), and "i-line" (365 nm), as shown in Fig. 2.16. The i-line offers higher resolution and is dominantly used. Advanced exposure tools such as KrF excimer laser (248 nm), ArF excimer laser (183 nm), and F_2 excimer laser (157 nm), as shown in Fig. 2.16, have been developed for mass production for resolution up to 70 nm.

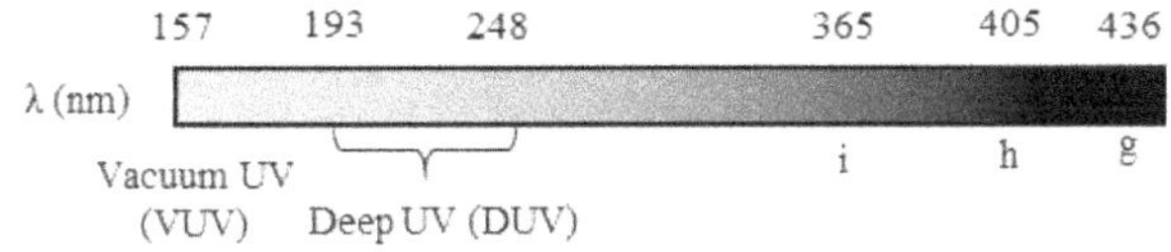

Figure 2.16 Common UV wavelengths used in optical lithography.

Figure 2.17 shows the process steps employed for patterning of an oxide layer on a silicon wafer. Photolithography mainly includes four steps: (i) PR coating, (ii) alignment, (iii) UV exposure, and (iv) development. A PR is a photosensitive polymer that acts as a photographic film, allowing the desired pattern to be transferred from the mask to the substrate. There are two types of PRs, **positive** and **negative**, each with a different chemical reaction to UV light. A positive PR becomes more soluble, while a negative resist crosslinks and becomes less soluble in developing solution when exposed to UV light. Lithography is performed in a cleanroom, which is illuminated with yellow light as a PR is insensitive to wavelengths greater than 500 nm. Firstly the substrate (or wafer) is cleaned to make sure that it is free from dust and dirt, followed by dehydration bake to evaporate the moisture. Thereafter adhesion promoter (usually hexamethyldisilazane [HMDS]) is spin-coated, which is immediately followed by spin coating of the PR. Spin coating produces a constant thickness of the PR across the sample. The thickness of the spin-coated PR (T_{PR}) is inversely proportional to the square root of the speed of the spin coater ($T_{PR} \propto 1/\sqrt{}$ spin speed). A PR-coated wafer is prebaked at 90°C–100°C for about 30 min in an oven or for about 3–4 min on a hot plate to evaporate the solvent and to improve the adhesion of PR. After this step, the mask is aligned over the substrate. UV light is then directed over the mask/substrate assembly. This step is called UV exposure, wherein the UV light passes only through the transparent areas of the mask to the substrate below. After UV exposure, the substrate is placed in a developing solution to remove the soluble areas of the PR. The choice of the PR depends on the requirement as the positive and negative PRs provide the reverse image to each other for the same mask. Now the wafer is postbaked at 120°C–130°C for about 30 min in an oven or for about 2–3 min on a hot plate to harden the resist and to improve its etch resistance. After transferring the pattern to the PR, the exposed layer (SiO_2 in

Fig. 2.17) is etched out in an etchant (e.g., buffered hydrofluoric acid [BHF] for oxide) which should not attack the resist. Etching may be wet or dry etching. Finally, the PR is stripped off using solvent (e.g., acetone for positive PR) or oxygen plasma. The sequence of these steps is summarized in Fig. 2.18 by a block diagram.

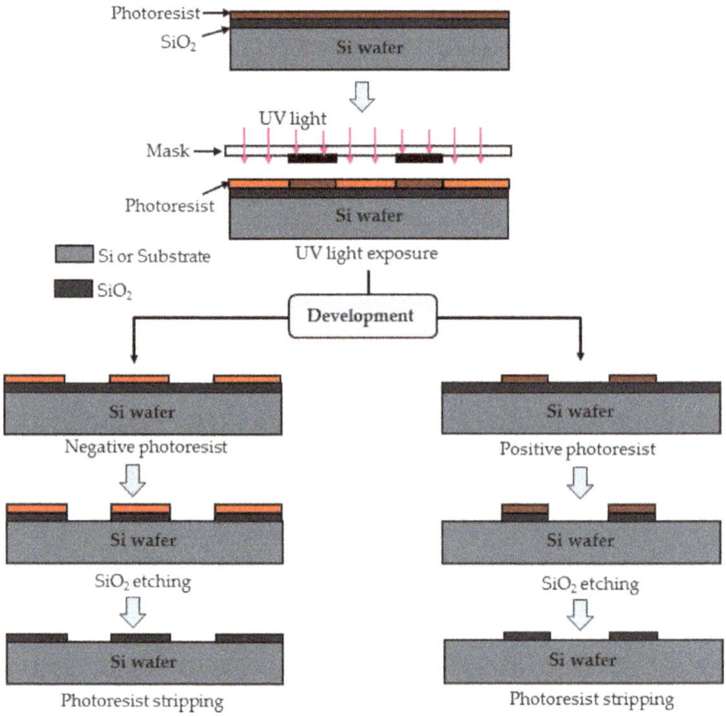

Figure 2.17 Pattern transfer by photolithography using positive and negative photoresists.

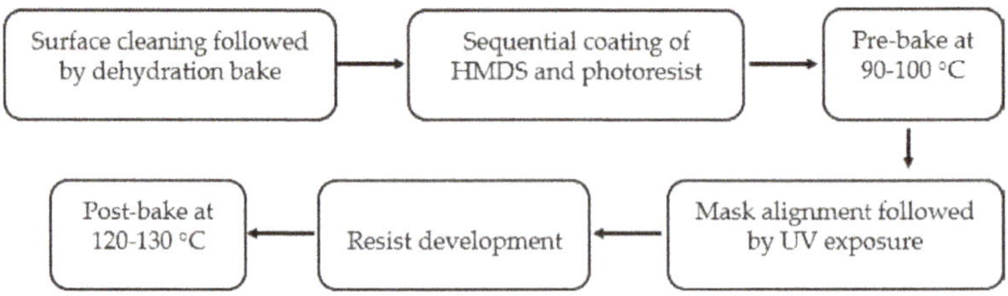

Figure 2.18 Summary of the process steps employed for pattern transfer by photolithography.

2.5.4 Etching

Etching is the removal of material from unprotected areas. The lithography process is employed for patterning to etch the desired portions on the substrate surface, as shown in Fig. 2.19, where the oxide layer is patterned by lithography and etching steps. A PR is used as mask to etch the oxide layer selectively. Etching is employed at many points during the fabrication of devices. Etching is divided into two categories, **wet** and **dry**. Wet and dry etching techniques are further classified into isotropic and anisotropic types. Wet etching is performed in liquid etchants. The wafers are immersed in the etching solution and the exposed part of the material is etched by chemical processes. The etching process takes place in four steps: (i) diffusion (or transport) of the reactant molecules to the silicon surface, (ii) adsorption of the reactant molecules on the surface, (iii) chemical reaction between etchant and surface atoms, and (iv) desorption of the reaction products and dissolution of desorbed products into the bulk etchant. These steps are schematically illustrated in Fig. 2.19. Dry etching involves the use of gas-phase etchants in plasma. Due to the involvement of plasma in dry etching, it is often called plasma etching. Ion beam- and laser beam-based etching also fall into the category of dry etching.

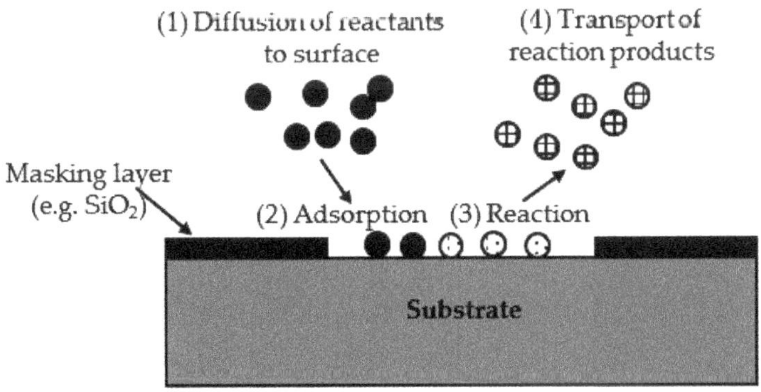

Figure 2.19 Basic process steps involved in wet chemical etching.

In the etching process, the following parameters are very commonly used and they must be known for the etchant before the etching experiment:

- **Mask layer:** It is used to protect the regions of the surface where etching is not required, for instance, a PR in Fig. 2.17. The mask layer should have the high selectivity with material to be etched out.
- **Aspect ratio:** The etch-depth-to-width ratio is called aspect ratio.
- **Isotropy:** The same etch rate in all directions, that is, the lateral etch rate is about the same as the vertical etch rate (Fig. 2.20a). In the case of a single crystal, the etch rate does not depend upon the crystallographic direction in isotropic etchants.
- **Anisotropy:** The etch rate is not equal in all directions (Fig. 2.20b). In the case of etching of a single crystal by wet chemical anisotropic etchants, the etch rate depends upon the crystallographic directions.

$$\text{Degree of anisotropy } (A) = 1 - \frac{\text{Lateral etch rate } (R_L)}{\text{Vertical etch rate } (R_V)}$$

$$= 1 - \frac{b}{d}$$

- **Bias:** The amount of undercutting (or lateral etching) at the mask edges, as shown in Fig. 2.20, is known as bias.

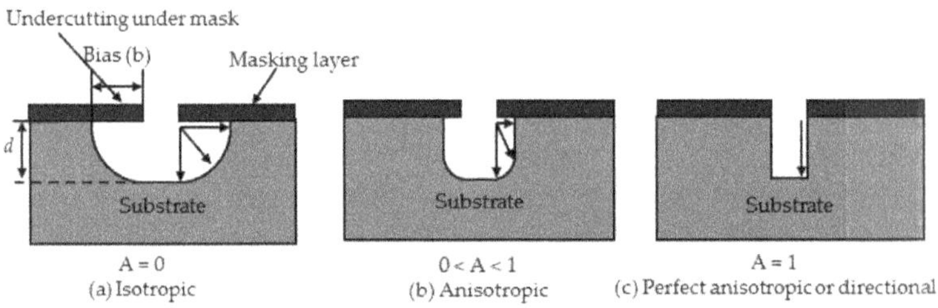

Figure 2.20 Schematic illustration of isotropic, anisotropic, and perfect anisotropic etching.

- **Overetch:** The etching after the required etch depth (or film thickness) is called overetching. The overetching results in more lateral etching (i.e., bias) as illustrated in Fig. 2.21.
- **Selectivity:** The ratio of the etch rate of the material to be etched to that of the material not to be etched (e.g., masking

layer), as shown in Fig. 2.22, is defined as selectivity. In terms of the patterning of the thin film deposited on the substrate, selectivity is the ability to etch the desired film but not etch the underlying substrate and the mask layer. For instance, BHF can be used to etch SiO_2 layer grown onto a Si substrate with excellent selectivity.

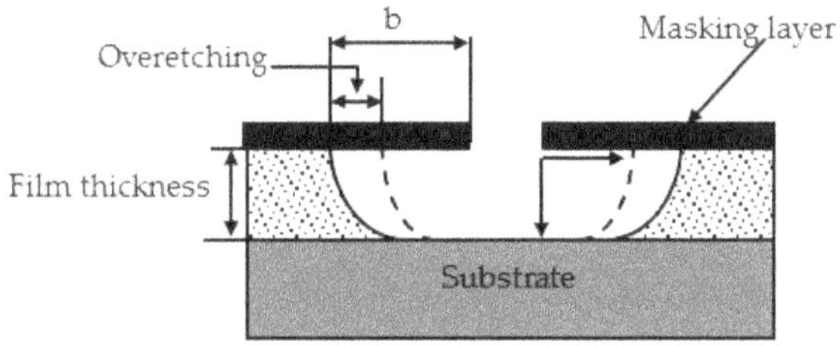

Figure 2.21 Schematic representation of overetching.

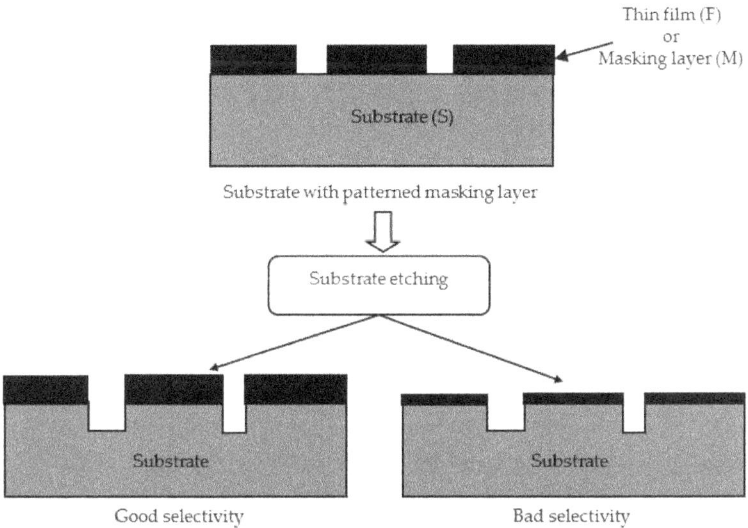

Figure 2.22 Schematic illustration of selectivity between masking layer and substrate.

$$\text{Selectivity between thin film (or mask layer) and substrate}$$
$$(S_{FS} \text{ or } S_{MS}) = \frac{\text{Etch rate of film } (R_F) \text{ or mask layer } (R_m)}{\text{Etch rate of substrate } (R_S)}$$

- **Uniformity:** The variation of the etch rate across surface to be etched out is termed "uniformity."

$$\text{Etch rate uniformity (\%)} = \frac{\text{Maximum etch rate } (R_{\text{max}}) - \text{Minimum etch rate } (R_{\text{min}})}{\text{Maximum etch rate } (R_{\text{max}}) + \text{Minimum etch rate } (R_{\text{min}})} \times 100\%$$

In general, highly anisotropic, selective, and uniform etching is desirable in order to scale down devices. However, these can be difficult to achieve simultaneously.

Table 2.2 Commonly used materials in microfabrication and their etchants

S. no.	Material	Etchant
1	Single-crystal silicon	Alkaline solution (e.g., KOH, TMAH, etc.) (for anisotropic etching)
2	Amorphous, polycrystalline, and crystalline silicon	HNO_3 (69%) H_2O:NH_4F (40%) :: 64%:33%:3% (for isotropic etching)
3	Gold (Au)	40 g potassium iodide (KI) + 10 g iodide + 400 mL DI water
4	Chromium (Cr)	16.45 g ammonium ceric nitrate + 4.3 perchloric acid + 100 mL DI water
5	SiO_2	HF (49% in water):NH_4F :: 1:6 (buffered HF [BHF or BOE])
6	Al	H_3PO_4:H_2O + HNO_3:CH_3COOH :: 16:2:1:1
7	Si_3N_4	HF (49% in water) or boiled H_3PO_4 (temp. 130°C–150°C)
8	Ti	NH_4OH:H_2O_2:H_2O :: 1:1:5
9	Positive photoresist	Acetone

In all cases, the etch rate depends upon etching conditions such as concentration, temperature, agitation during etching, etc.

2.5.4.1 Silicon wet isotropic etching

Isotropic etching exhibits equal etch rates in all directions. The isotropic etchants dissolve the material being etched at the same rate in all directions. For silicon wet isotropic etching, "HNA," a mixture of hydrofluoric acid (HF), nitric acid (HNO_3), and acetic acid (CH_3COOH) is commonly used. The HNO_3 drives oxidation of the silicon, while fluoride ions from HF form the soluble silicon compound H_2SiF_6. The etch rates depend on the ratio of the chemicals used and silicon doping. A schematic diagram of isotropically etched profiles is shown in Fig. 2.23. The details of isotropic etching of silicon are discussed in Chapter 3.

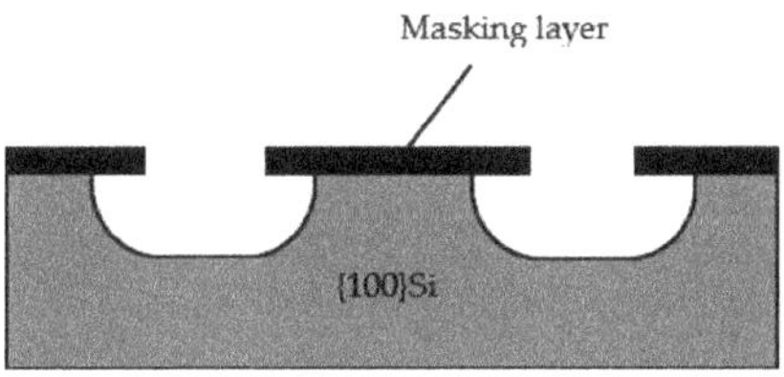

Figure 2.23 Schematic view of a wet isotropically etched profile.

2.5.4.2 Silicon wet anisotropic etching

Wet anisotropic etching is a process of orientation-dependent etching of material. The anisotropic etchants etch much faster in one direction than in another, exposing the slowest-etching crystal planes over time. Silicon wet anisotropic etching is performed in alkaline solution in which KOH and tetramethylammonium hydroxide (TMAH) are most commonly used. In the alkaline solutions, {111} are the slowest-etching planes, while {110} and {100} are the fast-etching planes. The etch rate and etched surface morphology depend on several parameters of the etchant such as concentration, etching temperature, agitation during etching, additives, etc. The effect of the slowest etching rate of {111} planes compared to other planes leads to specific profiles on different types of silicon substrates. Owing to this slow etching nature of {111} planes, the final etched structure comprises of different {111} planes on differently oriented wafers. The {111} silicon planes are inclined at an angle of 54.7° to the surface of a Si{100} wafer as illustrated in Fig. 2.24. This leads to the formation of a V-shaped profile during wet anisotropic etching. The anisotropic etching in a Si{110} wafer is schematically represented in Fig. 2.25. As can be seen in figure, the structure is bounded by six {111} planes, four of which are vertical, while the remaining two {111} planes are inclined at an angle of 35.5° to the Si{110} surface. KOH- and TMAH-based wet anisotropic etching is discussed in detail in Chapter 4 and Chapter 5, respectively.

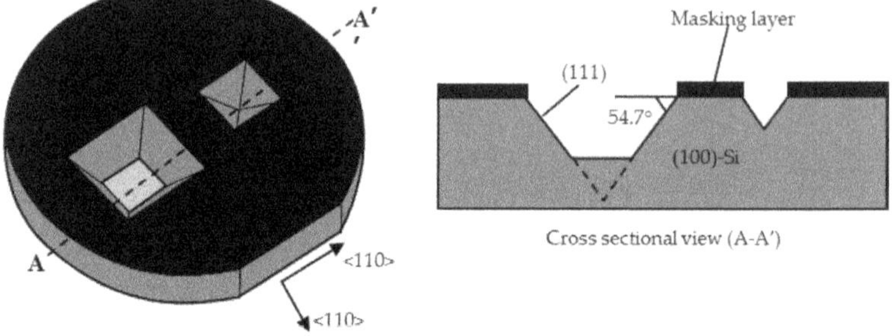

Figure 2.24 Schematic illustration of anisotropic etching in a Si{100} wafer.

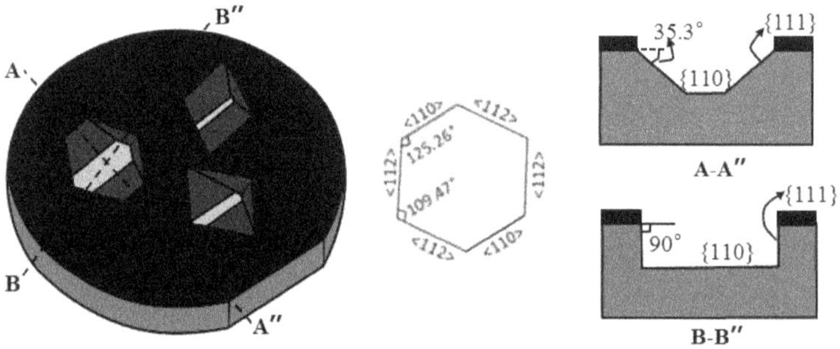

Figure 2.25 Schematic representation of anisotropic etching in a Si{110} wafer.

2.5.4.3 Silicon dry etching

Dry etching involves the use of gas-phase etchants in plasma. This method of etching is employed to avoid undercutting, which is a common phenomenon in wet etching, for better control over the dimensions of the etched patterns to achieve ultra-large-scale integration (ULSI). Due to the involvement of plasma in dry etching, it is often called plasma etching. The terms "dry etching" and "plasma etching" are used synonymously. Ion beam– and laser beam–based etching also falls into the category of dry etching. Plasma etching classified into two categories, physical and chemical. The former includes sputter etching and the latter includes pure chemical etching. Physical etching, however, is rather unselective. On the other hand, pure chemical etching is very selective but isotropic. The combination of physical and chemical etching, as shown in

Fig. 2.26, is called reactive ion etching (RIE), which is the most popular type of plasma-based etching. RIE is a good compromise between anisotropy and selectivity and has become the backbone of etch technology in microfabrication. According to electrode configuration, modes of operation are classified into two types, RIE mode and plasma mode. In the case of RIE mode, wafers are loaded onto the powered electrode, while in plasma mode wafers are loaded onto the grounded electrode. Analogous to wet etching (Fig. 2.19), chemical dry etching takes place in five steps: (i) etchant species are generated in the plasma, (ii) etchant species are transported by diffusion to the surface, (iii) the reactants are adsorbed on the surface, (iv) a chemical reaction takes place between reactants and surface atoms that form volatile compounds (or by-products), and (v) the volatile by-products are desorbed from the surface, diffused into the bulk gas, and pumped out by the vacuum system. These etching steps are schematically illustrated in Fig. 2.27.

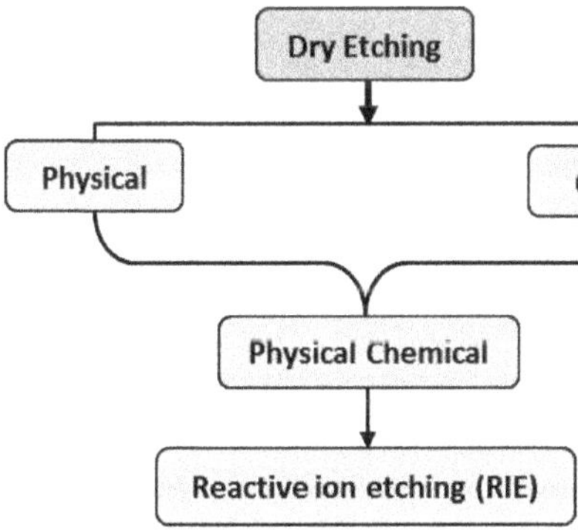

Figure 2.26 Combination of physical and chemical dry etching (i.e., RIE).

A reactor for an RIE system is schematically shown in Fig. 2.28. It contains a vacuum chamber, pump system, power supply generator, gas flow control units, etc. In this method appropriate reactive etch gases, depending on the material to be etched out, are introduced into the chamber continuously at reduced pressure (1–100 mTorr). Sulfur hexafluoride (SF_6) is commonly used for etching silicon. By applying an RF (13.56 MHz) voltage across two electrodes (cathode and anode), some of the gas atoms are ionized, producing positive ions and free electrons and creating plasma. An RF generator is used as energy source for generating plasma. The wafers are held on the RF-powered cathode, while the grounded chamber walls act as the anode. The RF voltage accelerates the light electrons in the system to much higher kinetic energies (~10 eV) than the heavier ions. The high-energy electrons collide with neutral gas molecules and cause a variety of reactions, including electron induced ionization, dissociation, recombination and excitation. Hence the plasma contains free electrons, ionized molecules, neutral molecules, and free radicals. The bombarded ions and neutrals react with the substrate, and therefore the etching process involves both physical and chemical etching components. A chemical reaction occurs at the substrate surface and forms volatile by-products, which desorb from the surface and diffuses into the bulk gas. Finally, the by-products are pumped out by a vacuum pump. During the etching process, normal surface is bombarded by both neutrals and ions, while the sidewalls are bombarded only by neutrals. Consequently the etching front moves more rapidly in the vertical direction than in the lateral direction, which results in high anisotropy (i.e., vertical sidewalls).

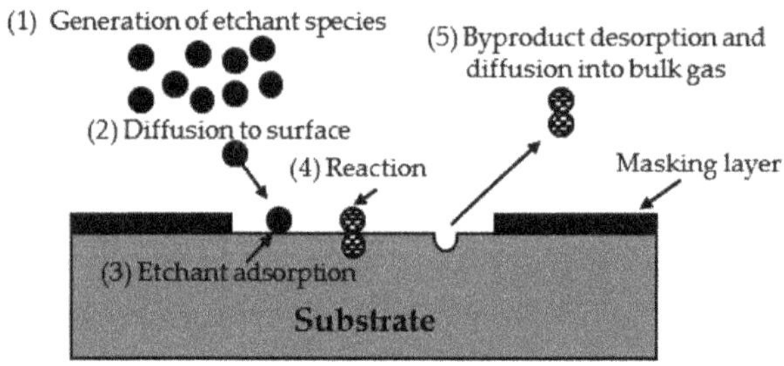

Figure 2.27 Basic steps in chemical dry etching.

Although plasma etching provides an etch rate higher than wet etching, achieving deep grooves/trenches with high anisotropy (or vertical sidewalls) is a major concern for applications mainly in bulk micromachining for the realization of MEMS components. To realize high-aspect-ratio structures and deep grooves with vertical sidewalls, a high-density plasma-based dry etching method, called **deep reactive ion etching (DRIE)**, was developed. In this technique, an etch-resistant protective layer is deposited on the sidewalls in between the etching process. During the etching process alternating processes of plasma etching of the substrate and the deposition of etching protective layer on the sidewalls are performed as illustrated in Fig. 2.29. This process was first developed by the German company Robert Bosch GmbH, and thus it is also called the Bosch process. The most commonly used method for the deposition of a protective layer is polymerization during the etching process. Silicon etching is carried out using SF_6 chemistry, while the deposition of an etch-resistant polymer layer on the sidewalls uses C_4F_8 chemistry. Recent developments have substantially improved the performance of DRIE with better sidewall-protecting materials (e.g., SiO_2).

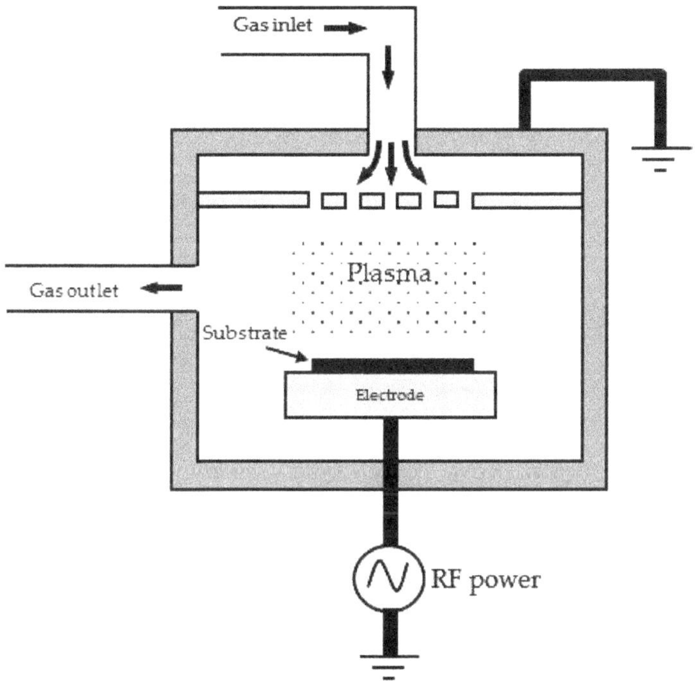

Figure 2.28 Schematic diagram of the RIE system.

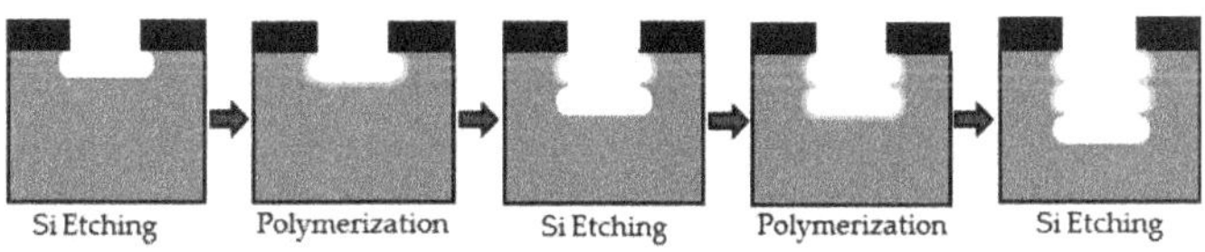

Figure 2.29 Schematic representation of deep reactive ion etching (DRIE) of silicon.

2.5.5 Local Oxidation of Silicon

In the microfabrication process, local oxidation of silicon (LOCOS) is a technique to grow the silicon dioxide layer on selected areas on a silicon wafer. This technique was developed to isolate neighboring MOS transistors. To grow the oxide layer on selected regions, the area which is not to be oxidized should be coated by a material that does not permit the diffusion of oxygen at high temperature (i.e., thermal oxidation). Silicon nitride (Si_3N_4) is the best-suited masking material for thermal oxidation process in order to oxidize the silicon wafer selectively.

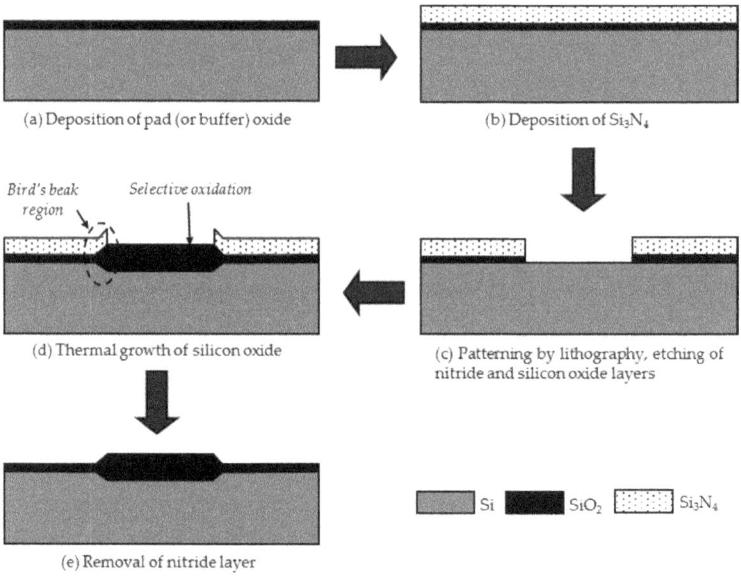

Figure 2.30 Process sequence for LOCOS.

The process steps employed during LOCOS are illustrated in Fig. 2.30. At first, a very thin silicon dioxide layer (20–50 nm) is grown on the wafer. This oxide layer is called pad oxide (or buffer oxide layer). Then a layer of silicon nitride (about 100–150 nm) is deposited. The major role of the oxide layer under the nitride is to buffer (or relieve) the strong tensile stress of the nitride layer induced during the CVD process, eventually reducing the stress in silicon substrate. The deposited layers are patterned by photolithography and etching process (Fig. 2.30c) to define the active areas for oxidation process. Thereafter, thermal oxidation is carried out to grow the oxide layer on exposed silicon. This step is the main part of the LOCOS process. Finally the nitride layer is etched out. The major advantage of the LOCOS process is the deposition of a high-quality oxide layer on selective areas. However, oxygen diffusion inside the silicon oxide is an isotropic process. When the oxide layer is grown, the oxidant diffuses under the nitride layer, bending it up, and forming a structure called the bird's beak, as shown in Fig. 2.30d.

2.5.6 Micromachining

Micromachining is the ability to make mechanical components of dimensions ranging from a few microns to thousands of micron (i.e., a few millimeters). The fabrication of submicron (or nanoscale) structures is termed as "nanomachining". Micromachining is a basic technology of microfabrication and is more popular to realize micromechanical structures. This technology can further be divided into two categories, bulk micromachining and surface micromachining.

2.5.6.1 Bulk micromachining

As the name suggests, in bulk micromachining technology, the bulk material of the substrate is involved in the fabrication process. The micromechanical structures are fabricated by selective removal of the bulk material using either a wet etch process or a dry etch process or both, depending on the requirements. If the etchant is a chemical solution, the etching process is called wet chemical etching. The cavities shown in Figs. 2.24 and 2.25 are the examples of wet bulk micromachining. The process steps for the formation of a suspended microcantilever beam by wet anisotropic etching are illustrated in Fig. 2.31. Wet etching can either be isotropic or anisotropic depending on the structure of the materials and the etchants used. If the material is amorphous or polycrystalline, the wet etching is always isotropic in nature. This book is focused on the basic and advanced wet chemical–based silicon micromachining for application in MEMS.

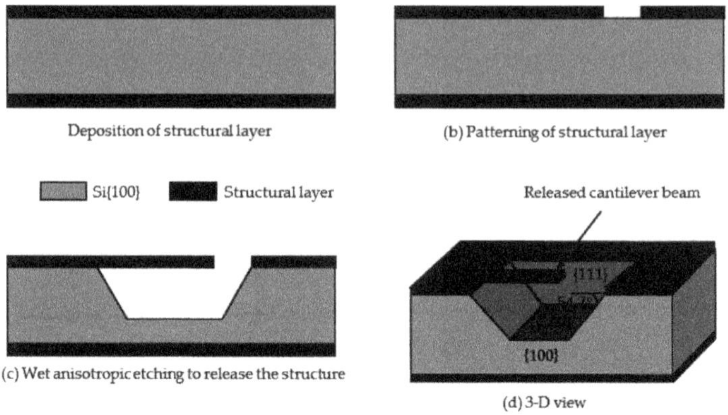

Figure 2.31 Process sequence for the formation of a cantilever beam on a Si{100} wafer using wet bulk micromachining.

2.5.6.2 Surface micromachining

In surface-micromachining technology, the microstructures are synthesized using deposited layers on the substrate. These deposited materials are used as structural and sacrificial layer. The surface-micromachining approach is attractive because smaller structures can be produced with better dimensional control compared to bulk micromachining. The prime advantage of surface micromachining is the ease of integration with IC components, since the same wafer is also used for realizing signal-conditioning circuits. Figure 2.32 illustrates a typical surface-micromachining process. The micromechanical structures are formed by first depositing a sacrificial layer onto the substrate. The sacrificial layer is then patterned by photolithography steps. Next, a structural layer of required thickness is deposited and then patterned to form a desired mechanical structure. Finally, the sacrificial layer is removed by wet or dry etching to obtain freestanding microstructures.

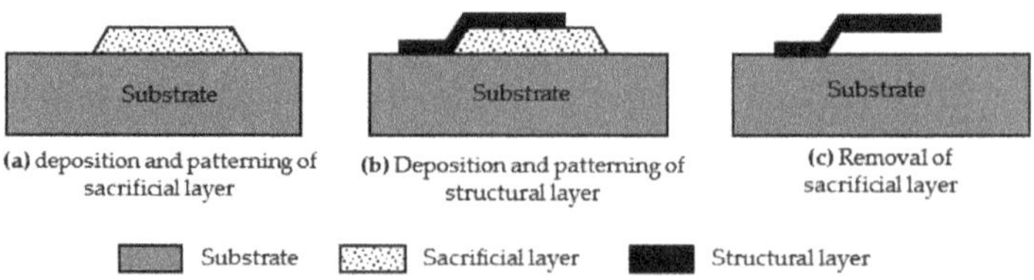

Figure 2.32 Schematic representation of process steps employed for the fabrication of surface-micromachined structures.

2.5.7 Doping

Impurity doping is the introduction of a controlled amount of impurity dopants into a substrate in order to modulate its electrical properties. Semiconductors doped with accepter impurities are called p-type, while those doped with donor impurities are known as n-type. In the case of silicon, trivalent (e.g., aluminum, indium, boron, gallium) and pentavalent (e.g., antimony, arsenic, phosphorous) impurities are used for p-type and n-type doping, respectively. Some dopants are added when the crystal (i.e., boule or ingot) is grown. In this case, it provides almost uniform doping in silicon wafers. To define circuit elements, selected areas, typically defined by photolithography, are further doped, for instance, to form bases, emitters, and resistors in bipolar transistors, to define the source and drain regions in field-effect transistors (FETs). Heavily doped p-type silicon (P^+-Si) is used as an etch stop in the alkaline-based wet etching process in order to make micromechanical structures for MEMS. There are two methods commonly used to add impurities (i.e., dope) into the substrate, **diffusion** and **implantation**, the latter method being more popular in large production runs because of increased controllability.

2.5.7.1 Diffusion

Diffusion describes the movement of particles through random motion from an area of high concentration to an area of low concentration. Diffusion takes place at elevated temperature (800°C–1000°C). It is done by placing silicon wafers in a high-temperature quartz tube and supplying a gas mixture that contains the desired dopant. The atoms of dopant gas diffuse into the silicon substrate. The maximum concentration of a dopant that can be dissolved under equilibrium conditions, without forming separate phase, is termed the solid solubility. Boron is the most popular dopant for p-type impurities, whereas for n-type impurities phosphorous is extensively employed. There are three types of dopant sources which are commonly used: (i) a solid source (e.g., boron trioxide [B_2O_3] or boron disc for p-type, phosphorous pentaoxide [P_2O_5] or phosphorous disc for n-type; (ii) a liquid source (e.g., boron tribromide [BBr_3] for boron and phosphorus oxy chloride [$POCl_3$] for phosphorous); and (iii) a gas source (e.g., diborane [B_2H_6] for boron and phosphine [PH_3] for phosphorous). Solid sources are placed in the quartz tube upstream from the substrate. During the diffusion process, the carrier gas transports vapor from this source and deposits them on the silicon wafers (or substrate). Disc-shaped solid planar sources consisting of dopant species and a binder are commonly used to provide vapor transport over large-size wafers. Figure 2.33 shows a schematic view of a diffusion system where solid discs are used as dopant source. The discs of the same diameter as the silicon wafer are stacked with the silicon wafer. In the case of diffusion using liquid sources, an inert carrier gas is bubbled through the liquid before being introduced into the furnace tube, whereas in the diffusion using gaseous sources, dopant species are directly supplied to the furnace tube in the gaseous state.

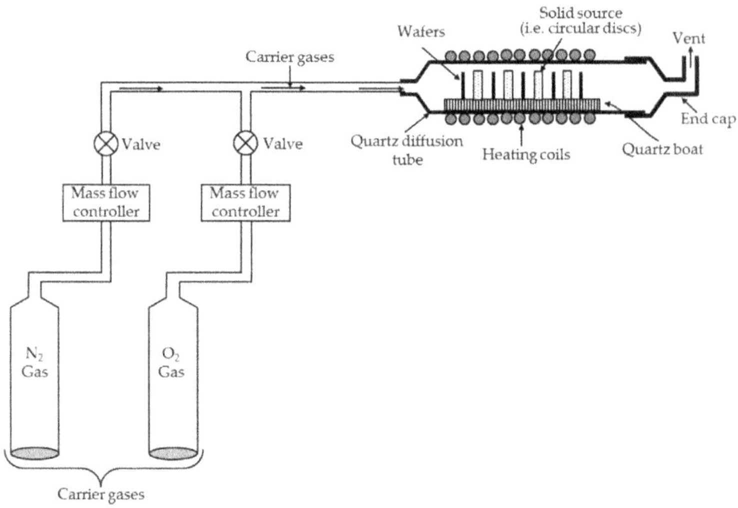

Figure 2.33 Schematic diagram of a solid-source diffusion system.

A two-step diffusion process, as shown in Fig. 2.34, is commonly used. Silicon nitride and thermal oxide are commonly used as diffusion mask material. However thermally grown SiO_2 layer is most widely employed as diffusion mask layer. To diffuse the selective regions, oxide layer is patterned using the photolithography and etching steps. Thereafter dopants are introduced into these patterned wafers in a high-temperature (800°C–1000°C) diffusion furnace. The dopants are gradually transported from the high-concentration region near the surface into the substrate through diffusion. This step is called predeposition. It is followed by a drive-in diffusion (or redistribution) at high temperature (800°C–1000°C) under total constant dopant condition. This step is employed to achieve the desired junction depth. The dopants are effectively blocked or masked by the oxide layer because their diffusivity in oxide is very low. The distribution of impurities in the substrate is calculated by solving diffusion equations with appropriate boundary conditions. If the source of dopant atoms at the substrate surface is limited, a Gaussian distribution is obtained. On the other hand, if the dopant atoms are supplied continuously, the doping profile is given by a complementary error function.

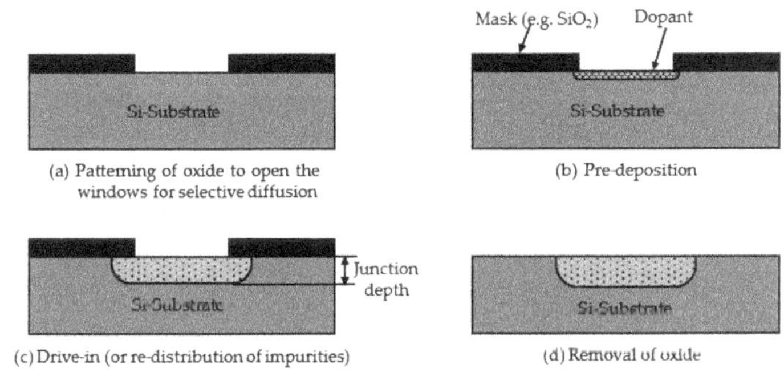

Figure 2.34 Schematic illustration of a two-step diffusion process.

The main advantages and disadvantages of diffusion are as follows:

Advantages:

- It requires a simple system.
- It is suitable for batch processing.

Disadvantages:

- It is a high-temperature process.
- It exhibits high lateral diffusion (75%–80% of vertical diffusion).
- It has poor doping control.
- It is not suitable for thin junctions.

2.5.7.2 Ion implantation

Ion implantation is a low-temperature technique for the introduction of impurities (dopants) into semiconductors (e.g., silicon). This method offers more flexibility than diffusion. For example, in MOS transistors (e.g., metal-oxide-semiconductor field-effect transistor [MOSFET]), ion implantation can be used to accurately adjust the threshold voltage. Dopant ions such as boron, phosphorus, and arsenic are generally created from a gas source (e.g., BF_3, AsH_3, PH_3). These gases tend to be very hazardous. If the desired implant species is not available in gaseous form, a solid charge can be heated and the resultant vapor is used as the source. Implantation energies of charged particles (i.e., dopant ions) are between 1keV and 1 MeV, resulting in ion distributions with average depth ranging from 10 nm to 10 μm. A typical schematic diagram of an ion implantation system is shown Fig. 2.35.

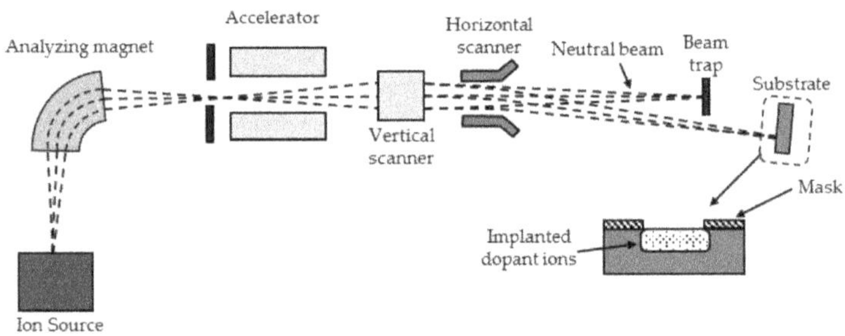

Figure 2.35 Schematic diagram of an ion implanter.

The ions to be implanted are created by breaking up the source gas into charged ions (B^+ or As^+) using a heat filament or plasma discharge. The ions are extracted by a voltage bias to a mass analyzer. The magnetic field of the analyzer is chosen such that only ions with the desired charge-to-mass (q/m) ratio can travel through it without being filtered. The selected ions then enter the acceleration tube where they are accelerated to the final energy of implantation. Some beam neutralization may occur during this acceleration phase and this is a problem because neutral atoms cannot be electrostatically scanned or counted as ion current. They implant an uncontrolled dose at the center of the wafer. For this reason, the ion path typically undergoes an electrostatic deviation from the linear path just before the final implantation, which acts to trap the neutrals which continue undeflected. The ion beam is then focused onto the substrate (i.e., wafer) and scanned over the surface using electrostatic deflection plates in order to achieve uniform doping. Either the wafer or the beam could be stationary. The highly energized ions enter the substrate, collide with the electrons and nuclei of the substrate, and finally come to rest at a certain depth within the solid. The average penetration depth is determined by the dopant, substrate materials, and acceleration energy and can be controlled by adjusting the acceleration energy. The dopant dose (i.e., total implant ions per unit area) can be controlled by monitoring the ion current during implantation. To implant the selected areas, a masking layer, which is patterned by lithography and etching processes, is used. The major side effect of ion implantation is the damage of the semiconductor lattice due to ion collisions. The damaged crystal thus needs to be restored. Furthermore, if the implanted species are intended to act as dopants, they must occupy the lattice site (i.e., substitutional site). Most of the ions as implanted are not located in substitutional sites. The process of moving a large fraction of the implanted impurities onto lattice sites is known as impurity activation. To activate the implanted ions and to repair the lattice damage, the substrate is annealed at an appropriate combination of time and temperature. The annealing step accomplishes both tasks simultaneously.

The main advantages and disadvantages of ion implantation are as follows:

Advantages:

- It is a low-temperature process.
- It exhibits less transverse straggle (standard deviation of the distribution of implanted ions in the direction normal to the direction of the ions' motion). It is analogous to lateral diffusion.
- In this technique, impurity depth and dose can be controlled independently.
- Complex profiles can be achieved by multienergy implants.

Disadvantages:

- It is an expensive technique.
- It causes damage to the substrate.
- In this method, all the damage cannot be repaired by annealing.

- Dopant redistribution takes place during annealing.
- It often uses extremely toxic gas sources such as arsine (AsH_3) and phosphine (PH_3).

2.5.8 Silicon Direct Wafer Bonding

Direct wafer bonding (DWB) is direct contact between two surfaces without using any specific bonding agent (e.g., glue, wax, epoxy, etc.) [10–12]. It is mostly referred to as bonding with silicon. DWB is exploited for wide applications such to create a silicon-on-insulator (SOI) wafers, silicon-to-silicon bonding can be used as substitute for the epitaxial layer for bipolar devices, two silicon wafers of opposite conductivity can be joined to form a p-n diode etc. In MEMS, it is explored for manufacturing different micromechanical structures that find wide use in sensors. In the DWB method two silicon wafers are fused by heating and therefore it is called **fusion bonding** also. The bonding process is based on chemical bonds between two surfaces. Two wafers can also have SiO_2 or some other thin-film layer over the surfaces that are bonded together. The successful bonding of two silicon wafers requires clean, flat, and ultrasmooth surface totally contamination (or particle) free otherwise unbonded areas so called voids (i.e., interface bubbles) can occur. Initially the direct bonding is induced by the forces of attraction between two bonding surfaces (e.g., van der Waals forces). These forces of attraction become stronger as the distance between the surfaces reduces. They also depend on the nature of the surfaces, that is, the chemical structure of the surfaces. Consequently, bonding method is divided into two categories, **hydrophilic bonding** and **hydrophobic bonding**. The main difference between these two surface types is the presence of films of water of just a few molecular layers that are adsorbed on hydrophilic surfaces. The surface state of a silicon wafer can be measured by the contact angle a water drop adapts as a result of its placement on the wafer surface. Surfaces with contact angles larger than 90° are categorized as hydrophobic, whereas surfaces with contact angles smaller than 90° are considered as hydrophilic. It means that the lower contact angle results in higher hydrophilicity (i.e., higher wettability)

Silicon DWB is performed in three steps: (i) surface preparation (i.e., polishing, cleaning, surface activation, etc.), (ii) prebonding at room temperature, and (iii) annealing at high temperature in order to improve the bond strength. Hence, in this method, two flat, clean and smooth wafer surfaces are brought into contact and form a weak bonding based on physical forces. The physical forces can be van der Waals forces, capillary forces, or electrostatic forces. The wafer pair is then annealed at high temperature (>600°C) and the physical forces are converted to chemical bonds. DWB is schematically presented in Fig. 2.36. Prior to bonding, the wafers are usually cleaned using RCA clean method as discussed in Section 2.5.1.

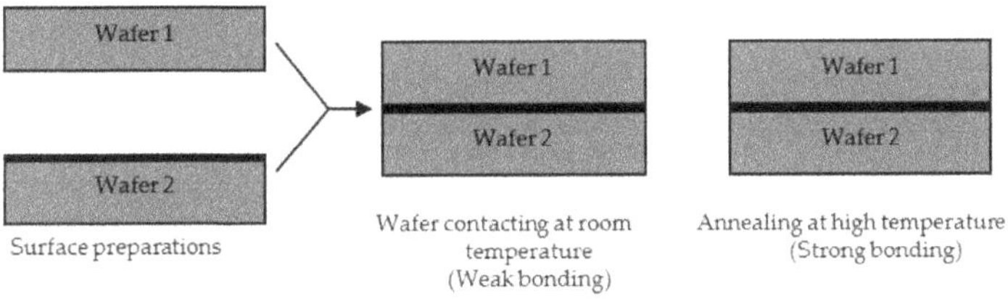

Figure 2.36 Schematic view of direct wafer bonding between oxidized and bare silicon wafers.

2.5.8.1 Hydrophilic bonding

In hydrophilic bonding of silicon, the silicon wafer surface is covered with an oxide layer. The oxide can be a thin native oxide, thermally grown oxide or deposited oxide. The surface of oxidized silicon normally contains silicon hydroxyl groups (i.e., Si-OH). The amount of Si-OH groups on the surface determines the hydrophilicity of the surface due to the polarization of hydroxyl (OH) groups. The SC-1 of RCA clean method (1:1:5 NH_3:H_2O_2:H_2O) is mostly used to enhance the hydrophilicity of the surface. When the wafers are contacted together and heated, the two opposite OH groups interact, forming H_2O and the Si-O-Si siloxane bond. The reaction between two hydrophilic silicon surfaces during bonding can be described by the following equation:

$$\text{Si-OH} + \text{HO-Si} \rightarrow H_2O + \text{Si-O-Si}$$

After the prebonding in air (or vacuum), the bonded pair is annealed (800°C–1100°C) in a special gaseous atmosphere or vacuum for increasing the bonding strength. Thus the annealing provides a certain amount of thermal energy which forces more silanol (Si-OH) groups to react among each other and new, highly stable chemical bindings are formed. During this reaction between silanol groups, more water molecules are released and Si-O-Si bonding is formed. The water molecules diffuse into the silicon dioxide on the surfaces. If the water reaches the silicon, it reacts with it to form silicon dioxide and hydrogen according to the following reaction:

$$Si + 2H_2O \rightarrow SiO_2 + 2H_2$$

Hydrogen has a high solubility into SiO_2 and by having an oxide layer of thickness >50 nm on at least one wafer surface, hydrogen-induced voids can be avoided. The bonding strength rises with increasing annealing temperature as the strong Si-O-Si covalent bonds form at high temperature.

2.5.8.2 Hydrophobic bonding

In hydrophobic bonding, the wafers to be bonded have a bare silicon surface, which either is hydrogen and fluorine terminated or in some cases has dangling bonds on the surface. A hydrophobic surface is generated if the native oxide layer is removed by either plasma treatment or by fluoride containing etching solutions (e.g., hydrogen fluoride [HF] or ammonium fluoride [NH_4F]). A hydrogen- and fluorine-terminated surface results from hydrofluoric acid etching, which is a standard process for removing silicon dioxide. A few hydroxyl groups (OH) also exist on the hydrophobic surface due to the water rinse. The hydrophobic surface is quickly contaminated with hydrocarbons; therefore the wafers should be quickly contacted or stored in a vacuum after removal of the silicon dioxide layer. The aim is to form a crystalline grain boundary, while avoiding the native oxide at the silicon wafer surfaces. The bonding at room temperature is mostly based on van der Waals forces between those hydrogen and fluorine atoms. Similar to bonding of hydrophilic surfaces, the room temperature bonded wafers are annealed at high temperature (600°C–1000°C) in order to achieve stronger DWB through Si-Si covalent bonds. During annealing, hydrogen and fluoride desorbs from the surfaces and Si-Si bonds are formed according to the following reaction:

$$\text{Si-H} + \text{Si-H} \rightarrow \text{Si-Si} + H_2$$

During annealing, surface diffusion of silicon takes place and closes the microgaps between the surfaces. The bonding energy reaches cohesive strengths of bulk silicon. The problem with the bonding process is again the presence of hydrogen at the bonded interface, which may cause voids.

References

1. Hopcroft, M. A., Nix, W. D., and Kenny, T. W. (2010). What is the Young's modulus of silicon?, *J. Microelectromech. Syst.*, **19**, pp. 229–238.
2. Barron, A. R. (2009). Formation of silicon and gallium arsenide wafers, *Connexions*, http://cnx.org/content/m16627/1.5/.
3. Madou, M. J. (2002). *Fundamentals of Microfabrication: The Science of Miniaturization*, 2nd ed. (CRC Press, Boca Raton).
4. Plummer, J. D., Deal, M. D., and Griffin P. B. (2000). *Silicon VLSI Technology: Fundamentals, Practice, and Modeling* (Pearson Higher Education & Professional Group).
5. Sze, S. M. (2008). *Semiconductor Devices: Physics and Technology*, 2nd ed. (Wiley India, New Delhi).
6. Sze, S. M. (1988). *VLSI Technology*, 2nd ed. (McGraw-Hill, New Delhi).
7. Streetman, B. G., and Banerjee, S. K. (2009). *Solid State Electronic Devices*, 6th ed. (PHI Learning, New Delhi).
8. Hsu, T. R. (2003). *MEMS & Microsystems: Design and Manufacture*, 1st ed. (McGraw-Hill, New Delhi).
9. Gandhi, S. K. (2009). *VLSI Fabrication Principles: Silicon and Gallium Arsenide* (Wiley India, New Delhi).
10. Moriceau, H. (2005). *A Bright Future of Direct Wafer Bonding*, *Clefs CEA, No. 52,* Summer 2005.
11. Tong, Q. Y., and Gosele, U. (1998). *Semiconductor Wafer Bonding: Science and Technology* (John Wiley & Sons, New York).
12. Suni. T. (2006). *Direct Wafer Bonding for MEMS and Microelectronics*, VTT publications 609, dissertation for the degree of doctor of science in technology, Helsinki University of Technology, Finland.

Chapter 3

Isotropic Etching of Silicon and Related Materials

3.1 Isotropic Etching Properties

Isotropic etching is a process in which etching advances normal to solid surfaces with a uniform etching rate at any place on the solid surface except on masking materials. Etching advances perpendicular to the wafer surface and, at the same time, in any directions from the mask edge with the same etching rate. This results in rounded etch profiles both in cross-sectional and in top views.

A cross-sectional view of isotropically etched recesses on a silicon wafer is schematically shown in Fig. 3.1. The etching advanced under the mask is called **mask undercut**, or **side etching** (see Fig. 3.1a). The mask opening size should be designed considering the amount of mask undercut.

When etching is applied to mask patterns whose apertures are closely spaced (Fig. 3.1b), side etching from the separated apertures meets in between two apertures. When the etching advances further, the masking material is lost from the wafer surface. This is the reason why isotropic etching is not used for deep etching of the wafer, aiming at a high aspect ratio of etched profiles.

When the wafer is etched in ideally isotropic manner from a pinhole of the mask, the etched profile becomes hemispherical concave, as shown in Fig. 3.1c.

Silicon Wet Bulk Micromachining for MEMS
Prem Pal and Kazuo Sato

ISBN 978-981-4613-72-9 (Hardcover), 978-1-315-36492-6 (eBook)
www.panstanford.com

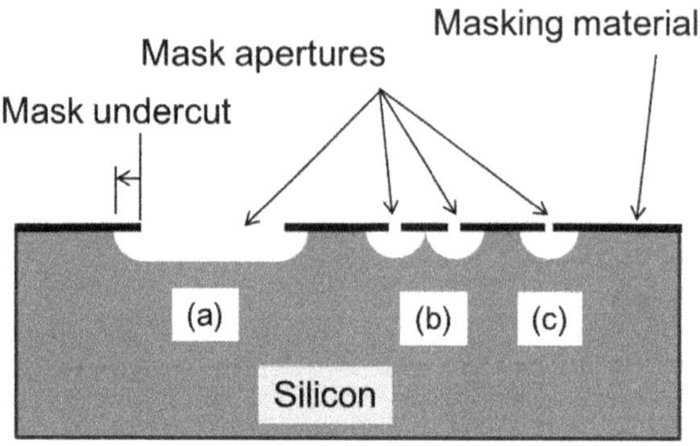

Figure 3.1 Cross section of isotropically wet-etched recesses for different types of mask apertures. (a) Large-sized single aperture, (b) closely separated mask apertures, and (c) pinhole mask aperture.

3.2 Isotropic Etching Solutions

Isotropic etching solutions are aggressive acidic etchants. Examples of isotropic etching solutions for silicon and related materials are listed in Table 3.1.

Table 3.1 Isotropic wet etching systems for silicon and silicon compounds

Etched material	Masking materials	Etchant
Si (crystalline, polycrystalline, and amorphous)	Au/Cr, LPCVD Si_3N_4	49% HF + 69% HNO_3 + CH_3COOH (typical volume fraction 2:3:3)
SiO_2 (amorphous)	Photoresist	concentrated 49% HF solution 49% HF + 50% NH_4F (typical volume fraction 1:6) 49% HF + 40% NH_4F (typical volume fraction 1:7)
Glass (amorphous)	Photoresist, Cr	HF water solution
Si_3N_4 (amorphous)	Negative photoresist, SiO_2	H_3PO_4 water solution boiling at 180°C

When etched material is amorphous or polycrystalline, such as SiO_2, glass, and Si_3N_4 in Table 3.1, the wet etching property is always isotropic. On the other hand, single crystals such as Si and quartz (crystalline SiO_2) tend to show an orientation-dependent etching rate reflecting the anisotropy in chemical activeness of each crystal orientation. Single-crystal silicon shows a strong anisotropic nature against alkaline etching solutions, as mentioned in subsequent chapters of this book.

Single-crystal silicon is etched in an isotropic manner, when the etching reaction is highly aggressive. Examples of such etchants are nitric acid, H_2O_2, and Br_2. Though they etch silicon very fast, etching results are incontrollable. As a milder isotropic etchant, a mixture of HF, HNO_3, and CH_3COOH solutions (so-called HNA solution) is widely used [1–4]. A typical formulation of HNA solution is shown in Table 3.1. Etching advances according to the series of reactions as follows.

$$HNO_2 + HNO_3 \rightarrow 2HNO_2 + 2OH^- + 2H^+$$

$$Si + 2H_2O + 4h^+ \rightarrow SiO_2 + 4H^+$$

$$SiO_2 + 6HF \rightarrow H_2SiF_6 + 2H_2O$$

The first formula means that HNO_3 generates OH^- and holes. The holes are injected into the silicon valence band. The HNO_2 works as a catalyzer in this system. The second formula is the oxidization of silicon by the holes. The third is the reaction where the silicon dioxide is dissolved by HF. The CH_3COOH added in the mixture is used to moderate etching reactions.

Etching properties significantly change, due to the fractional change in the three components, as shown in Fig. 3.2. In the figure, photographs are attached comparing top views of etched cavities, and both are etched from same-sized pinhole mask apertures, but under different mixture conditions A and C in the three-component system. The concentration of CH_3COOH as a buffer is the same between A and C. However, the concentration ratio of HNO_3/HF is higher with condition A than that with condition C. The cavity etched under condition A is slightly square shaped, representing an anisotropic nature, while the one etched under condition C is quite isotropic.

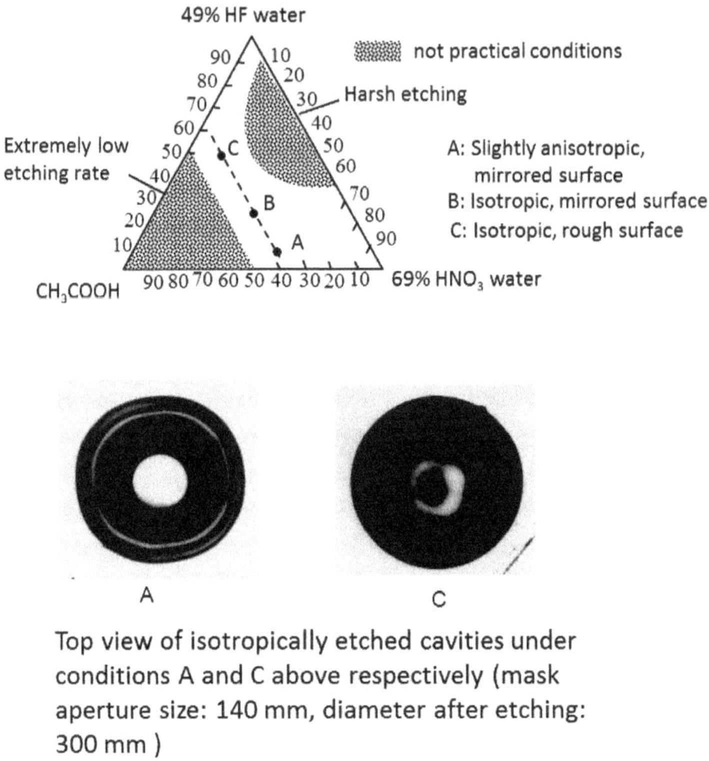

Figure 3.2 Etching properties change due to the fractional component change in the HF-HNO_3-CH_3COOH system. Reprinted from Ref. [5] with permission of the Japan Society of Applied Physics.

On the other hand, smoothness of the etched cavity is better with condition A, showing a clear reflection of the illumination of optical microscope. In such manners, the isotropy and surface smoothness are in a trade-off relationship. For compromising the isotropy and surface smoothness, it is recommended to choose condition B (fraction of HF:HNO_3:CH_3COOH is 2:3:3, as in Table 3.1, and the etching temperature 50°C [5]).

The isotropic etching solution for SiO_2 is a HF water solution. A concentrated 49% HF solution rapidly etches SiO_2 with an etch rate of 1.8 μm/min at room temperature. Instead of such a strong etchant, buffered hydrogen fluoride (BHF) is usually used to etch SiO_2 films, which is deposited on silicon wafers by thermal oxidization or chemical vapor deposition (CVD). Typical components are 49%

HF + 50% NH_4F with a volume fraction of 1:6 or 49% HF + 40% NH_4F with a volume fraction of 1:7. The etching rate is about 0.1 μm/min at room temperature. This is a mild solution in contrast to a concentrated 49% HF solution.

The isotropic etching solution for Si_3N_4 is a H_3PO_4 water solution at elevated temperature of 160°C–180°C. Positive-type photoresist materials are not bearable under such a harsh condition. Negative photoresists are used for patterning Si_3N_4 films. Alternatively, a SiO_2 film deposited on the nitride is also used as a mask material against a boiling H_3PO_4 solution.

3.3 Conditions for Isotropic Etching

Single-crystal silicon is an anisotropic material in terms of physical/chemical properties. We are going to discuss etching mechanisms that make etching results isotropic or anisotropic for single-crystal materials in this section. Amorphous and polycrystalline materials are excluded from the following discussion.

In a chemical isotropic etching system, etching rates are strongly enhanced by stirring motion of etching solutions. This fact contrasts with chemical anisotropic etching of silicon using alkaline solutions, where the stirring motion of the etching solution little influences the etching rates. This is due to the difference in the mechanism of the two etching systems.

The isotropic etching system is based on the fact that the chemical reaction between the etchant and the silicon surface is so aggressive that the etching rate is not dominated by the reaction but by material transfer via diffusion in the solution. Materials to diffuse are reactant components in the etching solution moving toward the silicon surface and etched products leaving the silicon surface. Such a reaction system is called a **diffusion-limited process**. The time for the reactant approaching to the solid surface and/or the time for the reaction products leaving the solid surface dominate the etching rate, while the time for the etching reaction is much faster than that for diffusion.

The etching species in the isotropic etching system is quite reactive to the work material. In the case of the HNA system, HNO_3 is highly reactive to silicon, and so with HF to SiO_2. It is easy to enhance the etching rate by stirring the etching solution, that is, enhancing the material transfer by the turbulence of the solution.

On contrary, in anisotropic etching system like silicon etching using alkaline solutions, the etching species are less reactive. The activeness of crystalline orientation strongly influences the etch rate. This makes the etched profiles anisotropic polygonal shapes. The etching rate can be significantly increased by raising the etching temperature that enhances chemical etching reactions. The main purpose of stirring the alkaline etching solution is not for enhancing diffusion in the solution but for minimizing temperature differences in the solution.

3.4 Applications

Isotropic wet etching is applied in micromachining for different purposes as follows:

- Selectively removing a film on a wafer surface, in particular removing the sacrificial layer in surface micromachining
- Fabrication of sharp needle tips on a wafer
- Fabrication of concave hemispherical profiles on a wafer
- Removing sharp corners of surface profiles for decreasing stress concentration
- Removing a work-damaged surface layer caused by physical processes such as mechanical machining or ion bombardment

Firstly, the isotropic etching system is usually utilized for the purpose of selectively removing materials from a wafer surface by the difference in the properties of the materials coexisting on the wafer. For example, mixture of 49% HF and 50% NH_4F (ammonium fluoride) in a volume ratio of 1:6 (Table 3.1) dissolves a SiO_2 film at a rate of 0.1 μm/min at room temperature, but it attacks Si or poly-Si to quite a small extent. Such selectivity is utilized for removing a native oxide film on a silicon wafer.

A good example of utilizing etching selectivity between silicon and oxide is seen in surface micromachining. The same etching solution is widely used for removing the sacrificial layer of SiO_2 in surface micromachining, as shown in Fig. 3.3. Undercutting of isotropic etching is essential for the removal of the sacrificial layer under a poly-Si structural material to become a cantilever released from the substrate surface. However, undercut etching for a long distance takes time. The etching rate of SiO_2 can be enhanced by doping P, and in some case also doping B, both in the oxide film. Doped oxide films such as phosphosilicate glass (PSG), and boron-doped phosphosilicate glass (BPSG) are easily removed by diluted HF solutions [6].

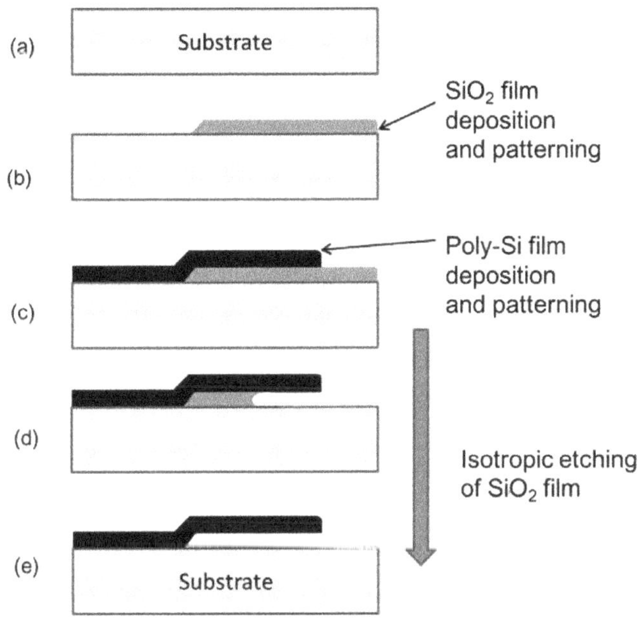

Figure 3.3 Surface-micromachining process including isotropic etching of a SiO_2 film as a sacrificial layer for releasing the polycrystalline silicon cantilever from the substrate.

Secondly, the isotropic etching system is utilized for the fabrication of sharp needle tips on a wafer, as shown in Fig. 3.1b. In this case, a circular disk-shaped mask is patterned in a position of the needle tip on a wafer, as schematically shown in Fig. 3.4. When the mask undercut reaches the center of the circular disk, the mask falls apart from the wafer surface. Etching has to be stopped at this moment because the needle height quickly decreases by further etching, as shown in Fig. 3.4c. Etching time control is quite critical for fabricating sharp needle tips.

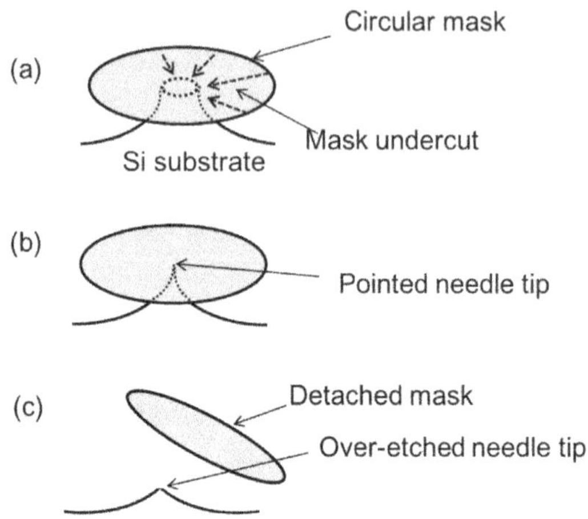

Figure 3.4 Schematic process sequence of pointed needle tip formation (a–c) by using isotropic etching of silicon.

This type of fabrication process is often found in the fabrication of arrayed cold cathodes for field emission devices and tall needle arrays for medical purposes such as for penetrating human skin. Both process sequences are shown in Fig. 3.5.

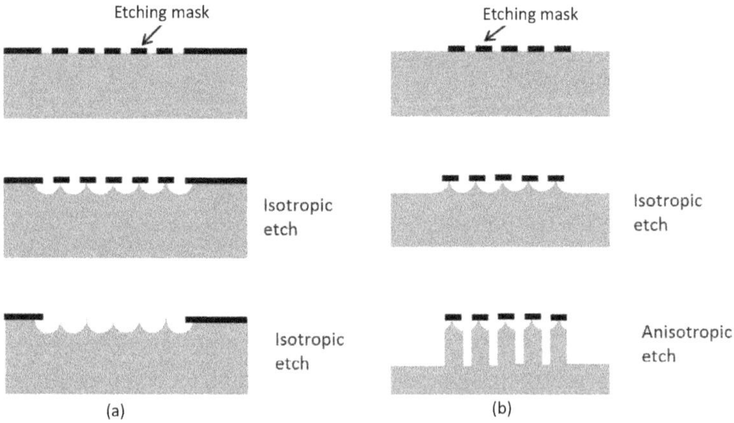

Figure 3.5 Application of isotropic etching process steps for silicon needle arrays for field emitters (a) and tall needles for medical applications fabricated in combination with anisotropic etching (b).

Isotropic etching for fabricating sharp needle tips in Fig. 3.5a was first investigated by using an HNA solution [7] already in the

1970s when silicon micromachining had just started. Due to an increasing demand for tip uniformity in fabricating silicon cold emitter arrays, the HNA isotropic system was used in combination with anisotropic etching [8]. A sharp and tall needle array was also demanded in medical applications for transdermal drug delivery and for electrodes to be attached to the human skin. Figure 3.6 is an example of a sharp needle tip having a through-hole fabricated by a combination of deep reactive ion etching (DRIE) for through-hole connected to rear side of the wafer and wet isotropic etching for tip shaping [9]. In this case, wet etching was done by a HNO_3:H_2O:HN_4F solution at room temperature. Though wet isotropic etching used to be applied to needle tip forming, isotropic plasma etching followed by anisotropic DRIE is also used for needle fabrication currently.

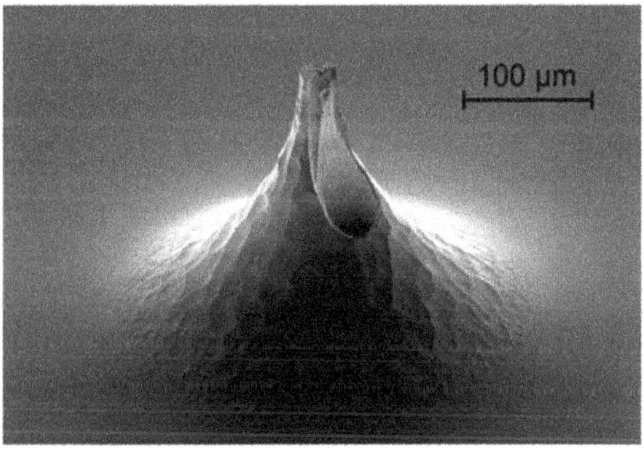

Figure 3.6 Isotropically wet-etched needle tip having a through-hole DRIE etched from rear side of the wafer. From Ref. [9]. Copyright @ 2005, IEEE.

Thirdly, HNA wet etching was applied for fabrication of an acoustic lens on a silicon substrate [5]. The photographs of the silicon acoustic lens are shown in Fig. 3.7. It has a concave profile at the tip, 55 μm in radius, with an excellent sphericity of 0.2 μm with a mirror-finished surface. It was mounted on a scanning acoustic microscopy system and used for focusing and receiving ultrasonic vibration at 1 GHz, allowing an acoustic image resolution of 1.8 μm. The etchant used was a $HF+HNO_3+CH_3COOH$ water solution, mentioned as condition B in Fig. 3.2 at a temperature of 50°C. The periphery of the cavity was mechanically ground in a tapered shape using an abrasive grinding machine.

Figure 3.7 Isotropically etched silicon lens applied to an acoustic microscopy system. Reprinted from Ref. [10], Copyright (2008), with permission from Elsevier.

The etching of silicon by using a HF+HNO_3+CH_3COOH water solution is a diffusion-limited process, as mentioned in the previous section. A small amount of fluctuation in the flow of the etching solution on the wafer surface results in asymmetric etching profiles. To achieve repeatability and uniformity in the etched profiles, the wafer was placed horizontally on the bottom of an etching bath, with the wafer surface facing up, and etched without stirring or agitation.

References

1. Robbins, H., and Schwartz, B. (1959). Chemical etching of silicon: 1. The system HF, HNO_3, and H_2O, *J. Electrochem. Soc.*, **106**, pp. 505–508.
2. Robbins, H., and Schwartz, B. (1960). Chemical etching of silicon: 2. The system HF, HNO_3, and H_2O, and $HC_2H_3O_2$, *J. Electrochem. Soc.*, **107**, pp. 108–111.
3. Schwartz, B., and Robbins, H. (1961). Chemical etching of silicon: 3. A temperature study in the acid system, *J. Electrochem. Soc.*, **108**, pp. 365–372.
4. Schwartz, B., and Robbins, H. (1976). Chemical etching of silicon: 4. Etching technology, *J. Electrochem. Soc.*, **123**, pp. 1903–1909.

5. Hashimoto, H., Tanaka, S., Sato, K., and Ishikawa, I. (1993). Chemical isotropic etching of single-crystal silicon for acoustic lens of scanning acoustic microscope, *Jpn. J. Appl. Phys.*, **32**, pp. 2543–2546.

6. Monk, D. J., Soane, D. S., and Howe, R. T. (1994). Hydrofluoric acid etching of silicon dioxide sacrificial layers. I. Experimental observations, *J. Electrochem. Soc.*, **141**, pp. 264–269.

7. Thomas, R. N., Wickstrom, R. A., Schroder, D. K., and Nathanson, H. C. (1974). Fabrication and some application of large-area silicon field emission arrays, *Solid State Electron.*, **17**, pp. 155–163.

8. Lee, H. C., and Huang, R. S. (1992). A study on field-emission array pressure sensors, *Sens. Actuators A*, **34**, pp. 137–154.

9. Stoeber, B., and Liepmann, D. (2005). Arrays of hollow out-of-plane microneedles for drug delivery, *J. Microelectromech. Syst.*, **14**, pp. 472–479.

10. Gianchandani, Y. B., Tabata, O., and Zappe, H. (2008). *Comprehensive Microsystems*, Vol. 1, *1.07 Wet Etching* (Elsevier, Oxford).

Chapter 4

KOH-Based Anisotropic Etching

4.1 Etching in Pure KOH Solutions

4.1.1 The Etching Mechanism

The KOH-based solution has been used for decades for the purpose of anisotropic etching of single-crystal silicon [1, 2]. Many alternative solutions such as an ethylenediamine pyrocatechol (EDP)–water solution [3], N_2H_2-based solution [4], NaOH-based solution [5], and tetramethylammonium hydroxide (TMAH)-based solution [6] show similar anisotropic effects in silicon etching. The chemical reaction is the same for all of the above solutions [7]:

$$Si + 2H_2O + 2OH \rightarrow Si(OH)_2O_2^{2-} + 2H_2 \text{ (gas)} \quad (4.1)$$

When the etching reaction proceeds, silicon atoms are hydroxylated to become water-soluble hydroxide ions. Hydrogen gas bubbles are generated as a by-product. It should be noted that the chemical reaction needs both the OH^- ion and water as reactants.

The anisotropic nature of etching comes from the silicon crystal lattice structure. A silicon crystal has a diamond lattice structure, as shown in Fig. 4.1. Every silicon atom has four bonds connected to four neighboring atoms. When a single-crystal silicon wafer is cut out of an ingot, atoms on the solid surface have a few number

Silicon Wet Bulk Micromachining for MEMS
Prem Pal and Kazuo Sato

ISBN 978-981-4613-72-9 (Hardcover), 978-1-315-36492-6 (eBook)
www.panstanford.com

of bonds connected to their neighbors. The rest bonds become so-called dangling bonds, being less stable and chemically reactive.

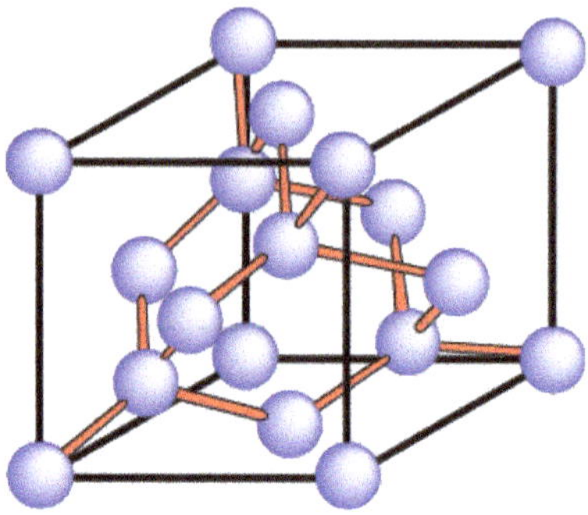

Figure 4.1 Diamond lattice structure of a Si crystal. The atoms' spatial locations are schematically shown as balls.

Those dangling bonds are schematically illustrated in Fig. 4.2 and visually photographed using a plastic atomic model in Fig. 4.3 for three principal surface orientations: (100), (110), and (111). The number of dangling bonds for a surface atom on the Si(100) surface is two. They are the origin of hydroxylation. Comparing the surface atoms on Si(110) and Si(111), the number of dangling bonds per surface atom is the same "one." A difference is found in the back-bond locations. In the case of a Si(110) surface atom, neighboring atoms are located parallel to the surface, and thus the back bonds are exposed to the surface, being easily hydroxylated than atoms on Si(111). In the case of a Si(111) surface atom, all the back bonds are connected to neighboring atoms having no dangling bonds. This makes the surface atoms stable on the ideal (111) surface. This is the explanation why Si(111) is hardly etched compared to other orientations.

As dangling bonds are quite active, they are quickly terminated with hydrogen atoms in water or etching solutions, as shown in Fig. 4.3. Those hydrogen-terminated bonds on the crystal surface are the source of the etching reaction. In the case of the Si(100) surface, as an example, each of the two dangling bonds of a surface atom is terminated with a hydrogen atom. This is the initial condition of the Si(100) etching. Etching reactions are sequentially took place in the following steps from Eqs. 4.2 to 4.5 [1, 8]. In the formulae, the number of back bonds of silicon atom are indicated by the number of short bars, such as = or ≡, attached to the symbol Si.

$$= SiH_2 + OH^- + H_2O \rightarrow = SiHOH + H_2 + OH^- \quad (4.2)$$

$$= SiHOH + OH^- + H_2O \rightarrow = Si(OH)_2 + H_2 + OH^- \quad (4.3)$$

$$(\equiv Si)_2 = Si(OH)_2 + 2H_2O \rightarrow 2(=Si\text{-}H) + Si(OH)_4 \quad (4.4)$$

$$Si(OH)_4 + 2OH^- \rightarrow Si(OH)_2O_2^{2-} + 2H_2O \quad (4.5)$$

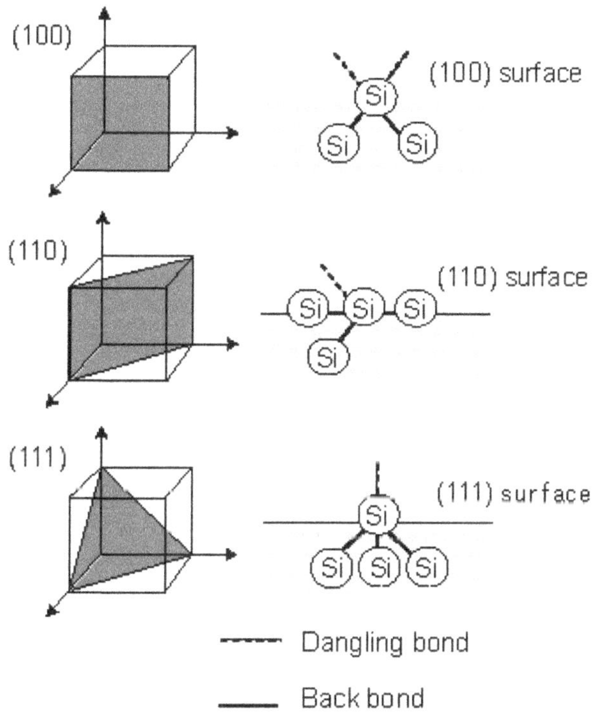

Figure 4.2 Number of dangling bonds exposed on the surface orientations (100), (110), and (111).

The first step of the reaction is described with Eq. 4.2, where hydrogen-terminated silicon atoms on the crystal surface are attached with an OH^- ion by the replacement of the hydrogen atom. Then the second step advances in the reaction in Eq. 4.3, and the silicon surface atom is gradually hydroxylated by an increase in the number of attached OH as Eq. 4.4, finally forming a water-soluble hydroxide ion (Eq. 4.5). Hydrogen gas bubbles are generated as a by-product. By summing the above four steps from Eqs. 4.2 to Eq. 4.5, the total reaction, Eq. 4.1, is obtained.

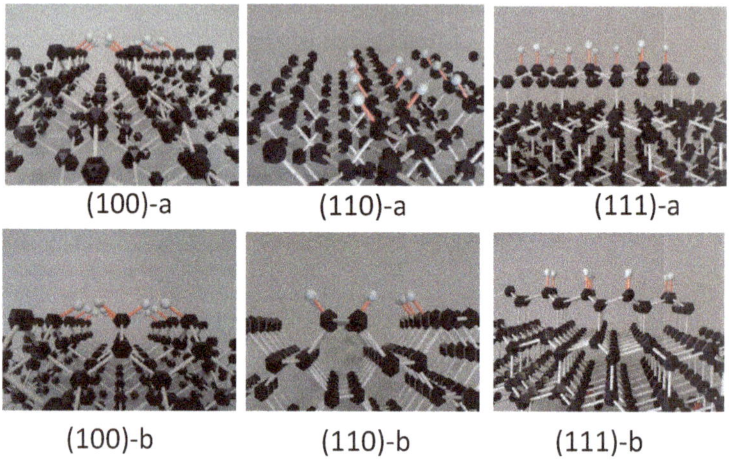

Figure 4.3 Atom arrangements near the surface of silicon (100), (110), and (111), visualized using a plastic atom model. Dangling bonds (pink-colored bars) are terminated by hydrogen atoms (white balls).

The situation is the same with other silicon surfaces having different crystal orientations. The difference in the surface orientation is the difference in the number of hydrogen-terminated silicon bonds exposed to etching solution.

It is interesting that cations such as K^+ do not participate to the reactions above. This is the reason why many alkaline solutions show somehow similar anisotropic natures in silicon etching. In an actual process, different cations influence the etching anisotropy by weakly adsorbing to the silicon surface in different manners. This will be discussed later in Section 4.1.8 as a factor causing a difference in anisotropy between KOH and TMAH solutions. Also some nonionic surfactants strongly influence the etching anisotropy. This will be discussed later in Section 4.2 with TMAH solutions with a surfactant.

4.1.2 Advantages and Disadvantages of KOH

Many alkaline solutions were reported to show strong anisotropy in Si etching suitable for fabricating 3D microstructures on a single-crystal silicon wafer. However, two of them, that is, KOH and TMAH solutions, are currently used in industry. They both are selected due to easiness of operation from viewpoints of toxicity of chemicals, low material cost, and less expensive fabrication facility due to less toxicity than EDP.

The KOH–water solution has the following advantages over other etching solutions.

- KOH is a less expensive chemical than TMAH, even when KOH is highly graded, containing an extremely low level of contamination.
- Etch rate anisotropy is stronger than that of TMAH. This is advantageous for fabricating 3D structures such like V grooves, diaphragms, and vertical grooves.
- The etched surface is flat and smooth, especially with high concentrations of more than 30% KOH.
- The etch rate for Si(100) is moderate, being typically 1 μm/min or less at 80°C, though it depends on the KOH concentration, as mentioned in the next section.
- On the other hand, the disadvantages of a KOH solution against a TMAH solution are as follows:
- A KOH solution noticeably dissolves the etching mask of the SiO_2 film. The etch rate selectivity of Si to the SiO_2 film is about 150 under normal etching conditions. If one needs to etch through a silicon wafer 500 μm thick from one side of the wafer, a thermally grown oxide film 2–3 μm thick is not enough for the etching mask. Instead of the oxide, Si_3N_4 or silicon oxynitride is needed as a masking material. This will demand a problem of patterning the Si_3N_4 mask using a H_3PO_4 water solution boiling at 180°C, which is not an easy process. On the other hand, the TMAH solution does not attack the SiO_2 film noticeably. A submicron-thick SiO_2 film, easy to make by thermal oxidation of Si and easy to pattern using buffered hydrofluoric acid (BHF) solution, can work as an etch mask for deep anisotropic etching.
- A KOH solution contains potassium (K^+) ions, which are harmful to electronic circuits. Process engineers tend to avoid using a KOH solution for fabricating microelectromechanical system (MEMS) devices having integrated electronics inside.

It seems reasonable for avoiding possible contamination of the silicon wafer by the potassium ion during KOH etching. However, this may be too conservative, since some foundries are successfully fabricating integrated MEMS using KOH under a condition of complete removal of potassium ions from wafer by immersing it in boiling water after KOH etching.

Although both KOH and TMAH are useful for bulk micromachining of silicon, they should be handled with care for protection of human health. Strong alkaline solutions easily attack the human skin and the surface of the eyeball. Protections such as gloves and face masks are needed to handle those solutions. Also they have to be handled in a draft bench in order to avoid inhaling vapor from the solutions. TMAH has been categorized as a toxic chemical since 2014 (see "Material Safety Data Sheet" [MSDS]). It should be handled with care during use and also in the processing as a waste.

4.1.3 Characterization of Etch Rate Anisotropy

Etching of single-crystal materials such as quartz and silicon has been investigated for decades for industrial purposes. One of the common research interests is how to characterize the etching rates which strongly depend on crystal orientations. Ueda [9] measured the thickness change of a number of quartz wafers having different surface orientations. This was a direct method of evaluating anisotropy. However, a dense net of data is hard to accumulate because of the numerous experiments required for a single etching condition. Weirauch [10] measured the side etch around the periphery of a circular mesa on a silicon wafer whose top was masked. Kendall [11] and Seidel [1] measured the side etch but along straight mask edges that were aligned with a number of different directions on a silicon wafer. This was so-called wagon-wheel method due to its mask pattern shape looking like a radial spoke arrangement schematically shown in Fig. 4.4. Though these side-etch measurements were qualitatively effective in visualizing anisotropy, the obtained data did not necessarily show correct etch rates, because the side-etched planes were not always perpendicular to the wafer surface. Measurement of sidewall angle is demanded for every orientation to calculate etch rates.

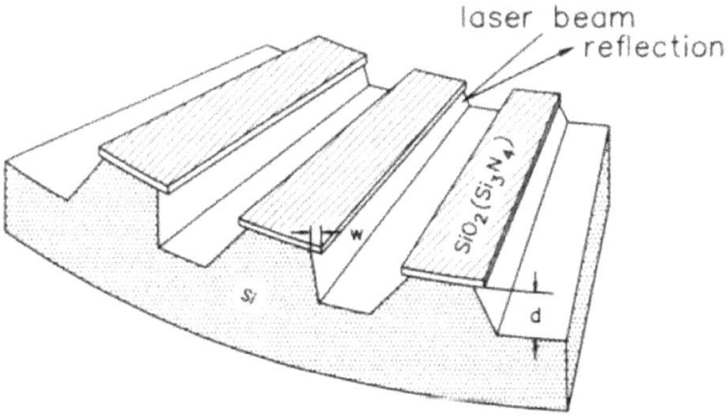

Figure 4.4 Wagon-wheel mask pattern (a part) used for characterization of etch anisotropy. From Ref. [1]. Reproduced by permission of The Electrochemical Society.

Sato [12] proposed the use of a concave hemispherical silicon specimen for the measurement of orientation dependence. The etch rates were calculated for some orientations from the dimensional change of the hemisphere, by comparing the surface profiles before and after etching. They recommended later a use of convex hemisphere with a large specimen diameter, typically of 40 mm, and a relatively small etch depth of 100–200 μm [13] to minimize the risk that some slow-etching orientations disappeared invaded by a fast-etching orientation on the same spherical surface. Etch rates of any orientations are available with this method.

Figure 4.5 shows the photos of the single-crystal Si specimen before and after KOH etching. The figure also shows how the surface profile of the hemisphere was measured by the 3D profiler UPMC-CARAT (a product of Carl-Zeiss Co.) having a contact probe.

Figure 4.6 shows an example of an etch rate contour map of Si etched in 40% KOH at 70°C. High-etch-rate regions are colored in yellow and low-etch-rate regions in blue. The following features were found:

- Si(100) and Si(111) show local minima in the contour map.
- Si(111) shows an extremely small etch rate, being a minimum among all orientations.
- Si(110) belongs to a high-etch-rate orientation. However, (110) is not the one showing a maximum. The etch rate maximum is deviated from (110) on the map.

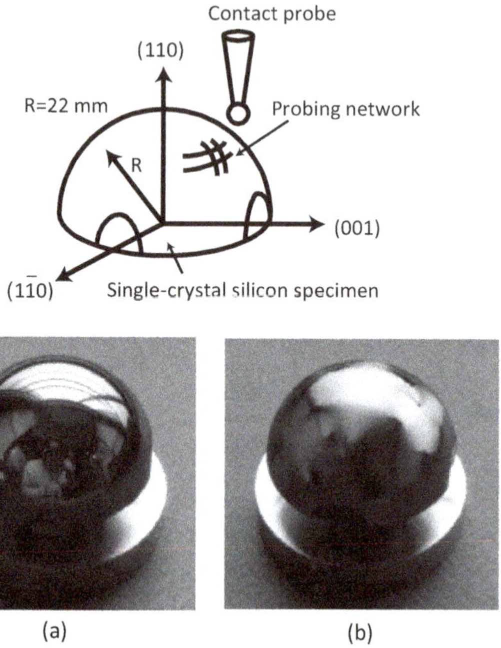

Figure 4.5 Hemispherical Si specimen for characterizing etch anisotropy. Photos are specimens before (a) and after (b) etching. Etch depths are calculated for arbitrary orientations by measuring specimen profiles (a) and (b). Reprinted from Ref. [13], Copyright (1998), with permission from Elsevier.

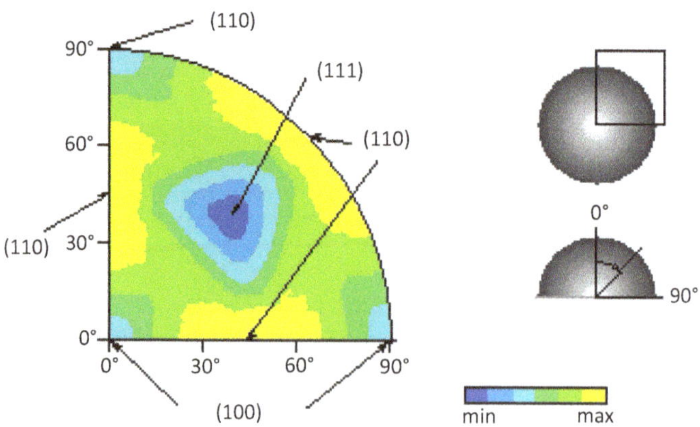

Figure 4.6 Example of etch rate contour map as a function of orientation for an etching condition 40% KOH solution at 70°C. Reprinted from Ref. [13], Copyright (1998), with permission from Elsevier.

The etch rate anisotropy varies depending on KOH concentration, temperature, and some additives, including surfactant and metallic ions. The etched profile also varies reflecting the change in the etch rate anisotropy. An example will be given in the later section showing different etch profiles due to variation of KOH concentration. Etch rates for total orientation are necessary for etch profile analysis, in particular in the case of fabricating 3D complicated structures by utilizing a higher order of Miller indices as etched profiles. It is necessary to know the etch rate anisotropy for every different etching condition. The etch rate database for complete orientations ODETTE was constructed for a wide range of etching conditions, including KOH concentrations of 30% to 50% and temperatures of 40°C to 90°C [13]. Examples of measured etch rates for 12 different orientations are listed up for different KOH concentrations at 70°C in Table 4.1.

There are etch profile simulation systems commercially available, such as IntelliEtch by IntelliSense Co., USA, and Fab Meister ES, whose original version was MicroCAD [14], by Mizuho Information and Research Co., Japan. They are all supported by the above-mentioned database. It also includes the Si etch rate data for TMAH solutions.

Table 4.1 Measured etch rates of 12 different orientations as a function of KOH concentration at 70°C. Etch rates are in μm/min

	KOH concentration in wt%		
Orientation	**30%**	**40%**	**50%**
(100)	0.797	0.599	0.539
(110)	1.455	1.293	0.870
(210)	1.561	1.233	0.959
(211)	1.319	0.950	0.631
(221)	0.714	0.544	0.322
(310)	1.456	1.088	0.757
(311)	1.436	1.067	0.746
(320)	1.543	1.287	1.013
(331)	1.160	0.800	0.489
(530)	1.556	1.280	1.033
(540)	1.512	1.287	0.914
(111)	0.005	0.009	0.009

Source: Reprinted from Ref. [33], Copyright (1998), with permission from Elsevier.

4.1.4 Effects of KOH Concentration

The etch rate of silicon depends on the KOH concentration and etching temperature. Figure 4.7 shows the effects of KOH concentration on the etch rates of silicon for five different crystallographic orientations [15]. The etch rate decreases according to an increase of KOH concentration in a concentration range more than 25 wt%. One may think it strange that the diluted solution shows fast-etching characteristics. However, the measured result is reasonable considering that H_2O is a reactant and KOH is a catalyst in the chemical reaction described in Eq. 4.1. The reaction needs a sufficient amount of water as a reactant. The concentration dependence looks different between two orientation groups. The first group including (100) and (221) showed almost the same etching rates throughout the concentration range from 17 to 51 wt%. The second group including (110), (210), and (320) showed higher etch rates than the first group when the concentration was more than 25 wt%. The rate of the second group decreased when the concentration decreased to 17 wt%.

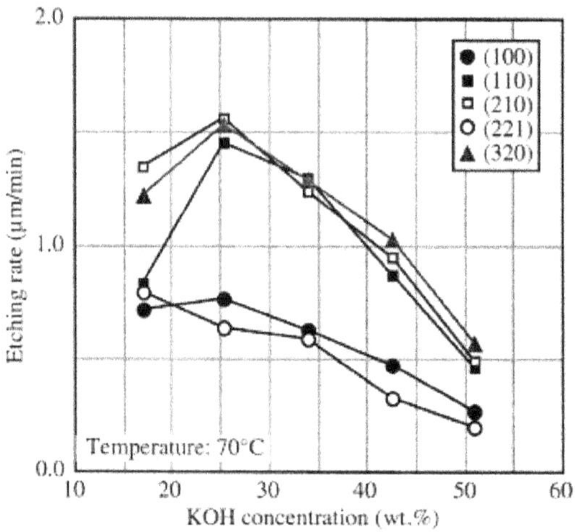

Figure 4.7 Effects of KOH concentration on etch rates of five different orientations at an etching temperature of 70°C. Reprinted from Ref. [15], Copyright (1000), with permission from Elsevier.

Though diluted solutions show high etch rates, solutions less than 10% KOH often results in a generation of gel-like white-colored products, for example, oxyhydrates, left on an etched surface. This sometimes causes surface roughening that is not desirable from a production point of view. On the other hand, a very low concentration range of 1–10 wt% KOH is recently investigated from an environmental point of view aiming at less consumption of chemicals [16].

As the KOH concentration dependence on the etch rate is different among crystal orientations observed in Fig. 4.7, it is easily understood that the anisotropy in the etch rate varies due to the change in KOH concentration. This means that the etched profiles can change due to the change in KOH concentration.

An example is shown in Fig. 4.8. The photos are cross-sectional scanning electron microscopy (SEM) images of the deeply etched vertical grooves on a Si(110) substrate, comparing two different etching conditions, 30% KOH (25.5 wt% KOH) and 40% KOH(34 wt% KOH). It is apparent, for both cases, that deep grooves were etched whose sidewalls were of (111), which has the minimum etch rate among any other orientations. However, when comparing the bottom surfaces, one is flat and the other is wedge shaped. This is the effect of KOH concentration. It was analytically proved that this is due to a small change in etch anisotropy in the cross-sectional plane [13]. The two etch profiles were both drawn by using the Wulff–Jaccodine graphical method [17] and compared.

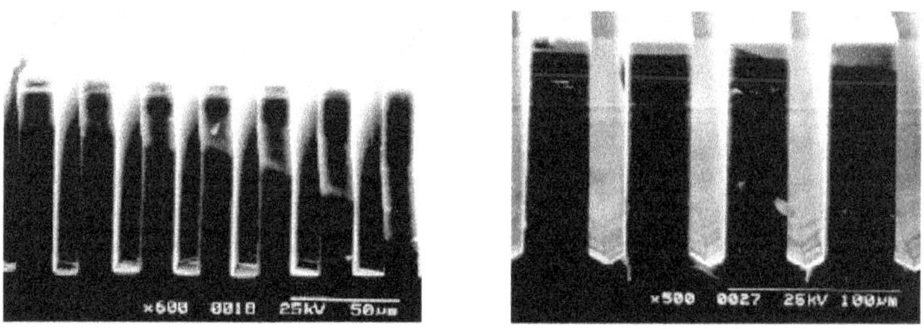

(a) 30% KOH (25.5 wt% KOH) (b) 40% KOH (34 wt% KOH)

Figure 4.8 Difference in bottom shapes of grooves on Si(110), deeply-etched using KOH solutions of different concentrations. Flat (a) and wedged (b) bottom shapes are due to a change in etch rate anisotropy, depending on the KOH concentration. Reprinted from Ref. [13], Copyright (1998), with permission from Elsevier.

In the case of concentrated KOH solutions, it is common that a wedge-shaped bottom appears in the groove. Due to a small change in anisotropy, that is, a decrease in the etch rate of (110) relative to the other orientations, a flat-shaped bottom appears with diluted solutions.

In summary, etch rate anisotropy can vary by a change in KOH concentration, and the etch profile can be controlled by KOH concentration, even such a simple case of groove fabrication.

4.1.5 Effects of Etching Temperature

In anisotropic wet etching using KOH as well as TMAH solutions, the etch rate of silicon strongly depends on etching temperature. Figure 4.9 shows etch rates of four different orientations including Si(100) and Si(110) as a function of temperature for the same concentration of a 40% KOH solution [18]. The vertical axis is the etch rate in logarithmic scale, and the horizontal is the inverse of absolute temperature. When etching temperature increases from 40°C to 80°C, the etch rate increases 10 times. The slope of the linear relationship is proportional to the activation energy of the reaction as expressed in a formula of the Arrhenius equation:

$$R = R_0 \exp[-Q/(kT)],$$

where R is the etch rate, T is the absolute temperature, Q is the activation energy of the process, k is Boltzmann's constant, and R_0 is a constant.

Each crystallographic orientation seems to have a slightly different value of activation energy. The activation energy of Si(100), for example, was about 0.60 eV. Si(111) showed the smallest activation energy among the four orientations. This suggests that anisotropy can change according to the change in etching temperature.

As mentioned in Chapter 3, the isotropic etching process is a diffusion-limited process. In contrast to this, the anisotropic etching process using a KOH solution is reaction limited. Here "reaction" means attachment of OH^- ions to silicon atoms on the solid surface by replacing the hydrogens which are already terminating the dangling bonds of silicon atoms, as mentioned in Section 4.1.1. This is a slow process dominating the total etching reaction. It demands a large activation energy by means of thermal energy. In such cases, the etching process, that is, the etch rate, is strongly dependent on the etching temperature.

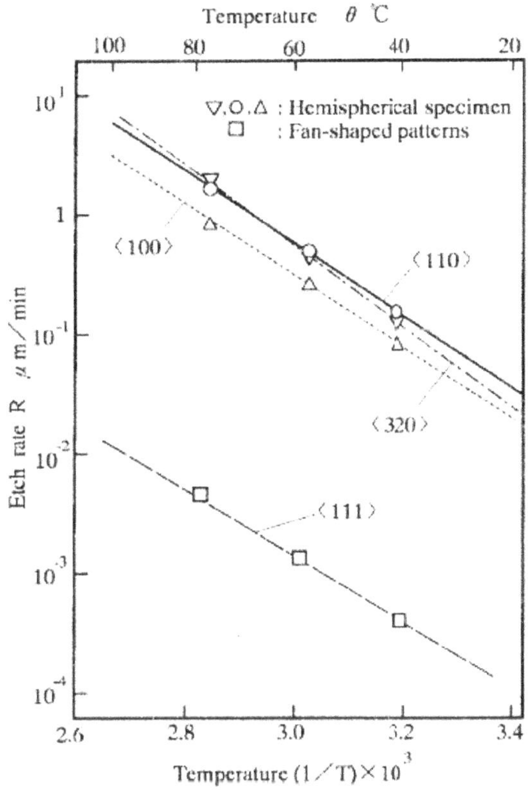

Figure 4.9 Temperature dependence of the Si etch rate in a 10% KOH–water solution for four different orientations. Reprinted from Ref. [18] with permission from JSPE.

In industrial processes, fast processing time is required as long as the etched surface quality is maintained or even becomes better. It is effective to raise the temperature when one needs to enhance the etch rate. A limitation of achievable temperature is the boiling point of the etching solution. Tanaka et al. [19] proposed the use of a highly concentrated KOH–water solution, typically 45–50 wt% KOH, aiming at increased the boiling point. When the KOH concentration is 50 wt%, the boiling point reaches 150°C due to its high ion concentration. A temperature range of 100°C–145°C allows a high etch rate of typically several micrometers per minute for Si(100), as shown in Fig. 4.10a. It is 1 order higher than that of etching below 80°C. The etched surface roughness was excellent on Si(100), as shown in Fig. 4.10b.

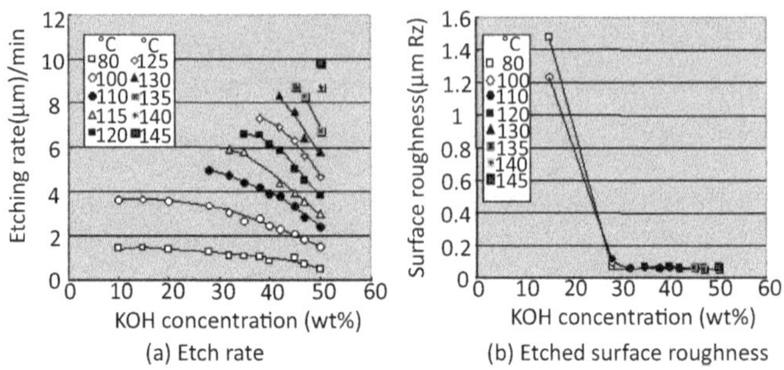

Figure 4.10 Improved etching characteristics by using highly concentrated KOH solutions at higher temperatures than 100°C. (a) Etch rate and (b) etched surface roughness both for Si(100). Reprinted from Ref. [19], Copyright (2004), with permission from Elsevier.

In the case of Si(110) etching, the etch rate increases to 10 μm/min even higher than that for Si(100). Though the surface roughness of Si(110) also decreases according to the increase of KOH concentration, hillocks start to emerge on the etched surface. It is reported that the hillocks disappeared at higher temperature than 135°C with 45 wt% KOH, and a smooth surface was obtained [19].

One drawback of using a highly concentrated KOH solution is that the etch selectivity of SiO_2 to Si decreases compared to conventional diluted solutions. In addition, it decreased also due to an increase of etching temperature. It is reported that the etch selectivity of thermally grown SiO_2 to Si is 40 at 110°C by using 45 wt% KOH and further goes down to 25 at 135°C with the same solution. It is recommended to use a plasma-enhanced chemical vapor deposition (CVD)-grown SiN film as a mask material, which shows a selectivity of 4000 and 2500 under the conditions mentioned above, respectively [19].

4.1.6 Etch-Stop Technologies

There are some etching techniques aiming at precise control of etched dimensions such as etched depth and resultant membrane thickness. Three techniques are introduced below for stopping the etching at a certain depth of the silicon wafer: (i) etch stop by heavy boron doping, (ii) electrochemical etch stop, and (iii) use of a silicon-on-insulator (SOI) wafer.

4.1.6.1 Etch stop by heavy boron doping

Impurity doping of silicon on a normal doping level does not change etching characteristics such as etch rates. Boron doping on an extremely high-concentration level of more than 1×10^{19} cm^{-3} results in a significant decrease of the etch rate. Figure 4.11 shows the etch rates of Si(100), depending on the boron-doping concentration [2]. Etch rates can be reduced by one hundreds. When applying this type of etch-stop technique, one can deeply etch a silicon wafer, while letting a doped silicon layer remain on the wafer. In the fabrication of micromechanical structures such as cantilevers, diaphragms were so far demonstrated. Figure 4.12 illustrates the process scheme fabricating a thin diaphragm on a (100) wafer. It should be noted that the membrane thickness is limited to the boron diffusion depth practically up to 2–3 microns in case that doping is done by thermal diffusion. It is known that impurities other than boron, that is, phosphor and germanium, do not show such a strong reduction in etch rates by the heavy doping.

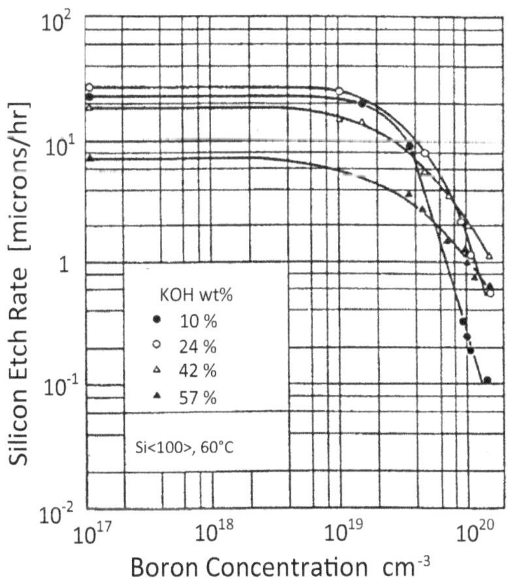

Figure 4.11 Etch rate of Si(100) in a KOH solution, depending on boron-doping concentration of the wafer. From Ref. [2]. Reproduced by permission of The Electrochemical Society.

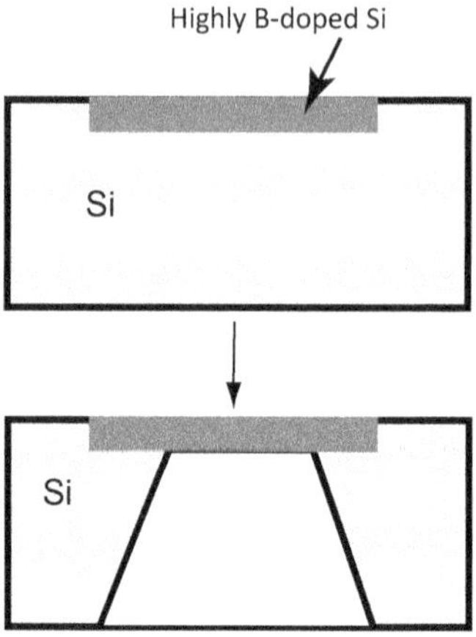

Figure 4.12 Application of the boron-doping etch-stop technique fabricating a thin diaphragm on a silicon wafer surface.

Heavy doping of boron, on the other hand, has drawbacks for the properties of fabricated microstructures. The first is mechanical internal stresses induced by the doping. A high level of tensile stress is induced by the boron doping. This results in a crystalline lattice slippage during high-temperature processing of the wafer. The induced tensile stress is due to the difference in atomic size between Si and B, which is smaller than Si. To ease the stress, codoping of Ge and B is proposed. As the atomic size of Ge is larger than Si, reduction of stress is obtained, while maintaining the etch-stop characteristics.

The second drawback of using heavy boron doping is an electrical property change in the boron-doped region. Silicon becomes a conductor with such a high-concentration range of doping. One cannot make any semiconductor circuits on the heavily doped membrane. The use of the membrane is limited for mechanical purposes such as deflectable diaphragms and cantilever beams.

4.1.6.2 Electrochemical etch stop

Silicon wafers during alkaline etching are kept in a negative potential passively in the etching solution. This potential is called open-circuit potential (OCP). The OCP differs by the impurity type, impurity concentration, and light illumination on the wafer surface. Usually, the OCP is in a range of −1.5 to −1.1 V. Under such a condition, there is no current flowing macroscopically between the wafer and the etching solution. As observed in Eq. 4.1, the OH^- ion comes from the solution to the wafer surface and hydrogen bubbles are generated. Thus electrons are injected to the wafer. However, at the same time, a hydroxyl complex as a reaction product leaves from the wafer surface. Thus equilibrium of charge is maintained between the wafer and the etching solution.

When the electrical potential of the wafer is intentionally raised against the etching solution, the etched product becomes silicon oxide, not a hydroxyl complex ion. The silicon oxide is not soluble in the solution but forms a passivation film against etching. When the electrical potential is kept +2 to +3 V against the etching solution, etching stops due to the passivation film of oxide formed on the silicon surface.

Based on this fact, an etch-stop technique has been developed, as explained below [20]. This technique is aiming to stop etching at a desired depth of the wafer remaining at a certain thickness of microstructures of a low doping level of silicon unlike the heavy boron-doping etch stop mentioned before. Figure 4.13 shows a schematic cross section of an experimental setup. A p-type silicon wafer is used as a substrate. An n-type epitaxial silicon layer of a desired thickness is grown on the substrate by using an epitaxy process. The thickness of the n-layer corresponds to that of microstructure to be fabricated on the substrate. An electrical contact is made on the n-layer connected to a direct electric source, keeping the n-layer at a positive potential level against the etching solution. An etching mask is patterned on the rear side of the p-substrate. All the wafer surfaces, including the n-layer, except etching mask openings on the p-substrate, are protected with masking materials and electrically passivated from the solution. Etching starts from the mask opening on the p-substrate. Until the etching front reaches the p–n interface of the wafer, current does not flow through the p–n junction. Thus the OCP is maintained between the p-layer and the etching solution. When the etching front meets the n-layer, current starts to flow from the n-layer to the solution, and the electrochemical etch stop works.

The thickness of the epitaxial n-layer can reach several tens of microns being enough as mechanical elements, while it is controlled within an error of 0.1 microns over the entire wafer surface. The doping concentration of the n-layer can be moderate suitable for MEMS functions. These are advantages of an electrochemical etch stop over the heavy boron-doping technique.

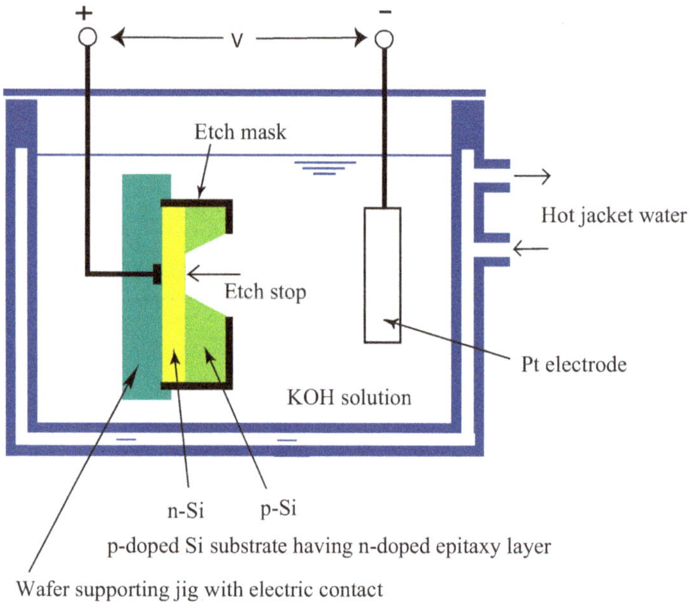

Figure 4.13 Schematic setup of an electrochemical etch-stop technique using a p-n junction as an etch-stop surface.

On the other hand, a drawback of the electrochemical technique is the need of electrical contact to the n-layer, while avoiding electric leakage over the n-layer and the p–n junction along the wafer periphery. Also uniformity in the electrochemical etch stop has to be maintained when current starts to flow from the electrical contact. The potential drop in relation to the distance from the electrical contact must be compensated over the wafer surface. This care demands additional costs for production, affecting the total production cost of the final devices.

4.1.6.3 Use of a SOI wafer

SIO is composed of a silicon substrate and a top silicon layer separated by a buried silicon oxide layer. When the top layer of silicon or the substrate is anisotropically etched by a wet process, etching stops when the etching front meets the silicon oxide layer. A large-sized thin diaphragm having the same thickness to the top Si layer of the SOI wafer is obtained by using this technique. When microstructures are fabricated on the top layer of the SOI, and the silicon substrate is sequentially etched using the buried oxide as an etch-stop layer, thin microstructures, for example, cantilever beams like an atomic force microscope (AFM) probe, can be easily fabricated, as schematically shown in Fig. 4.14.

As the etching front of anisotropic KOH etching is smooth and flat, and etching advances quite uniformly over a Si(100) top layer of SOI, the etch-stop technique using a SOI wafer is quite easy to control and etch results are excellent. A drawback of the use of a SOI wafer is the expensiveness of the wafer, which costs 10 times higher than a normal wafer.

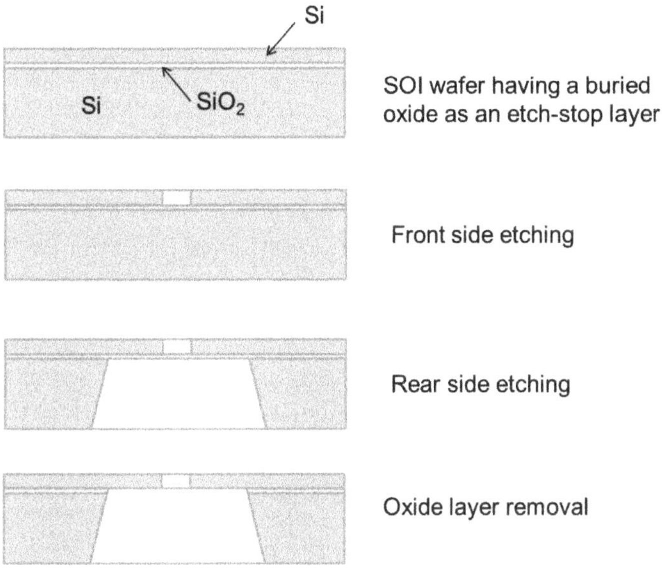

Figure 4.14 Fabrication of a thin membrane/cantilever structure by using a SOI wafer having a buried oxide as an etch-stop layer.

4.1.7 Etched Surface Roughness

Anisotropically a wet-etched silicon surface shows a unique surface morphology reflecting crystal orientation. When using a hemispherical Si specimen described in Section 4.1.3, one can see the morphology of the etched surface as a function of crystal orientation. Figure 4.15 compares optical microscopy images of different surface orientations on the same hemispherical specimen etched in a pure KOH–water solution. The specimen diameter was initially 44 mm and the surface roughness was 0.005–0.007 μm in Ra. The maximum etched depth was 100–150 μm in (110). In the figure, in addition to the optical images, etched roughness values are indicated by regions graded in a color bar [21].

The following are concluded from the roughness map:

- The etched roughness strongly depends on the crystallographic orientation.
- The (100) orientation showed the smoothest surface among all orientations.
- A smooth surface appears in a region including (100), (211), and (311).
- A very rough surface appears in a region including (320) and (210).

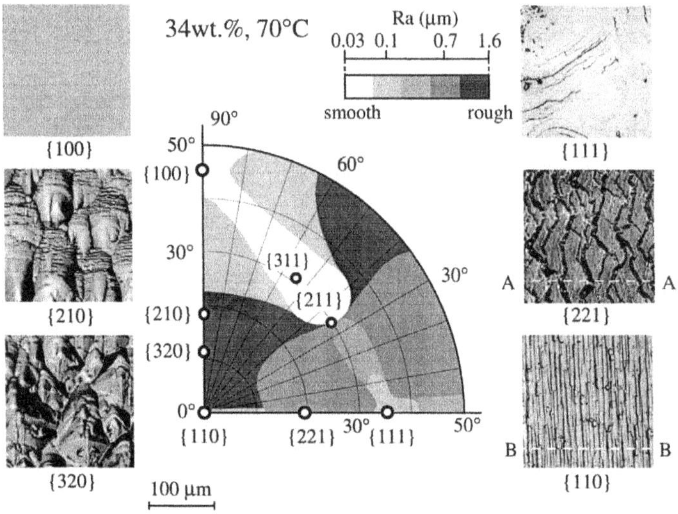

Figure 4.15 Surface roughness and textures related to crystallographic orientation etched in 34 wt% KOH at 70°C. Reprinted from Ref. [21], Copyright (1999), with permission from Elsevier.

They further analyzed the profile of the etched surface for (110) and (221) using a mechanical profiler: Form Talysurf (product of Talor Hobson). Figure 4.16 shows the results. The etched surface textures are composed of facets of certain crystallographic orientations. In the case of the etched (110) silicon surface, it is fully covered with (551) facets. It looks like a cross section of a triangular rooftop whose surface is of (551), as schematically shown in the figure. They later reported additionally that the facet was in a range between (441) and (551) [26].

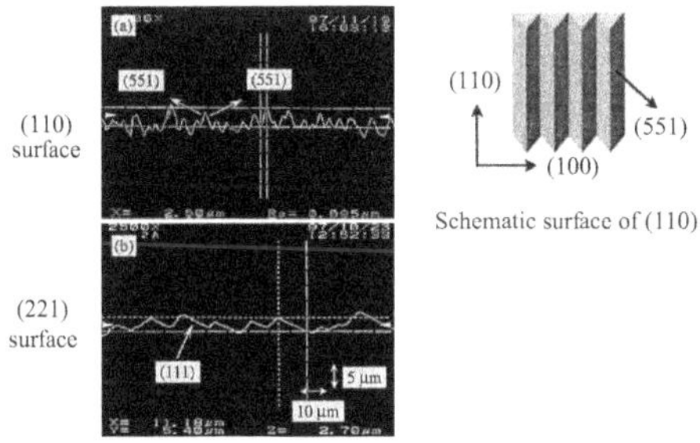

Figure 4.16 Etched surface roughness of a (110) and a (221) surface corresponding to lines B–B and A–A in Fig. 4.15, respectively. At the top right is a schematic morphology of the etched (110). Reprinted from Ref. [21], Copyright (1999), with permission from Elsevier.

In the case of an etched (221) surface, the texture is composed of steps and terraces, as observed in the photo in Fig. 4.15. The roughness profile is shown in Fig. 4.16. The step-and-terrace texture is consistently observed for orientations existing between (111) and (221), though the step and terrace sizes are different. The terrace surface is of (111). According to the deviation from (111) toward (221), the tilting angle of the terrace increased correspondingly to an increase of the off-angle from the (111) orientation.

From the observation of etched surface roughness, we know that the etching of a crystal surface does not take place by a random removal of atoms from an ideally flat orientation but by a local removal of atoms from some specific site of microscopic shapes of roughness.

The example of site-specific etching was investigated for Si(111) and its vicinal orientations. Allongue et al. [22] observed a Si(111) vicinal surface during alkaline etching using in situ scanning tunneling microscopy (STM) and found a series of atomic steps moved in lateral directions. This is the way how the (111) surface is etched. Elwenspoek [23] modeled the anisotropic etch rate, as shown in Fig. 4.17. This is a schematic cross section of Si(111). The real Si wafer surface has atomic steps, partly due to the mechanical polishing process, dislocation ends appearing on the surface, thermodynamic equilibrium, and the off-angle from an ideally flat crystal surface. The steps can be easily etched in the solution, while the terrace surface is hardly etched. Thus the steps move toward the left-hand side in Fig. 4.17. We have already seen macroscopic steps and terraces on the etched (221) orientation in Fig. 4.15, where macroscopic steps are of bunched atomic steps and terraces are of slightly tilted (111).

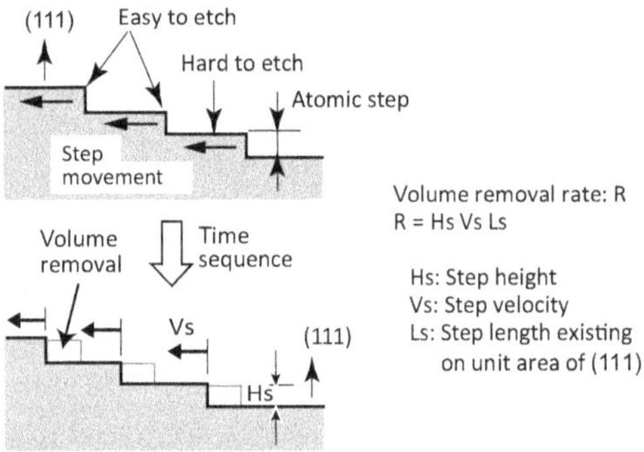

Figure 4.17 Step flow model of Si(111) etching: a cross sectional view of a wafer. Atom removal takes place mostly at atomic step edges, resulting in lateral motion of atomic steps, because smooth terraces of an ideal (111) are hard to etch.

The etch rate of Si(111) is governed by the step-etching rate because there is extremely small chance of atom removal from an ideally flat terrace surface. The volume removal rate R in this model is a product of H_S (step height), V_S (step velocity), and L_S (step length) existing in a unit area of the Si(111) surface, as shown in the figure. It is easily understood that the etch rate proportionally increases by an increase of the off-angle of a wafer surface deviating from ideal Si(111). This is due to an increase in the number of steps in the cross section in Fig. 4.17, resulting in the linear increase of the step length L_S.

The averaged etch rate of each orientation is dominated by the step type and step density. It is also strongly influenced by the stability of Si atoms at step edges, depending on the types of etching solution. It should be noted that the etch rate of some orientation should not be discussed by the stability of atoms on an ideally flat crystal surface. In other words, counting the number of dangling bonds of the surface atom on an ideally flat orientation is useless to compare the etch rates of different orientations. Further, it is suggested when any ions or molecules in the etching solution interact with some site of morphologies, that is, atomic steps, the macroscopic etch rate can easily vary and, thus, etch rate anisotropy as well as surface morphology can vary.

In the following sections, it is discussed that the anisotropy in etch rates dramatically changes when alcohol (or surfactant) is added into the etchant. Such changes in etch rate anisotropy can be explained by selective interactions between some components in the etching solution and a type of atomic steps appearing on crystal orientations.

4.1.8 Comparison of KOH and TMAH Etching Characteristics

KOH and TMAH solutions show different anisotropies in etching. Though details of TMAH-based etching are described in Chapter 5, differences in anisotropic etching characteristics between the two solutions are introduced in this section. Macroscopic differences are first introduced based on the spherical specimen etch results [15]. On the other hand, atomistic step-etching behaviors of Si(111) and its vicinal orientations are presented showing a difference between the two solutions [16]. Then macroscopic differences in anisotropy between the two solutions will be explained by the difference in atomistic etching behaviors.

Macroscopic etch rate anisotropy was characterized for both KOH and TMAH solutions by using hemispherical Si crystal specimens mentioned in Section 4.1.3. The etch rate contour maps for two solutions are compared in Fig. 4.18. One can see common features of the etch rate with two etching solutions:

- Si(111) shows extremely small etch rates among other orientations. The contour patterns are in threefold symmetry around (111), reflecting the crystal symmetry.
- Si(100) shows a local minimum in etch rate. The contour patterns are in fourfold symmetry around the <001> axis, also reflecting the crystal symmetry.
- Si(110) is located in high-etch-rate regions.

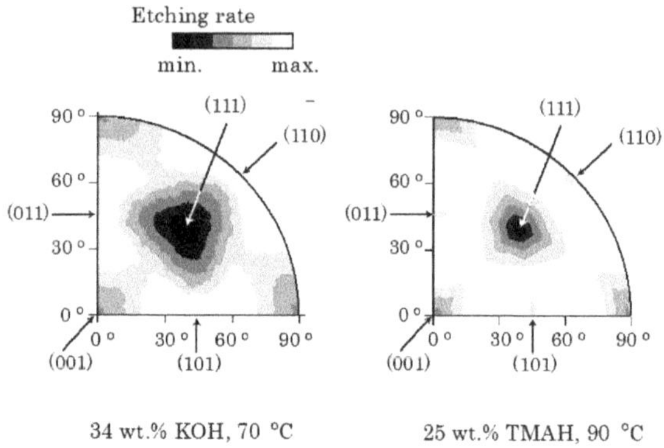

Figure 4.18 Etch rate contour maps of KOH and TMAH, showing differences in etch rate anisotropy between the two solutions. Reproduced from Ref. [24] with permission.

On the other hand, a clear difference between the two solutions is found on the contour map. The contour patterns around (111) are triangular shaped for both solutions, reflecting the threefold symmetry of the crystal. However, the triangles are different in their rotational orientation. The triangle tips are pointing toward (110), (101), (011) with KOH, while they are pointing toward (001), (010), (100) with TMAH. This means KOH and TMAH differently react with silicon crystal orientations.

The difference in anisotropy appears as a difference in etch pit profiles on Si(111). Figure 4.19 shows optical microscopy images of etch pits—one is grown in KOH and the other in a TMAH solution. Pit diameters are about 100 μm, while the depth at the center is 1 μm [24]. Although both are somehow triangular shaped, the triangle tips are pointing in different directions, and consequently the triangle orientations are in mirror symmetry. This is explained in relation to the difference in contour map patterns around (111). The etch pits in Fig. 4.19 were grown originating at an end of dislocation, which perpendicularly penetrate the (111) surface. A schematic cross section of the etch pit is shown in Fig. 4.20. Silicon atoms are removed initially from a dislocation end on the (111), and etching advances to the depth direction. At the same time, atomic steps appear on (111). The steps are quickly etched and move in lateral directions on the (111) surface, while the top surface is stable because it is an ideally flat (111). In other words, etching advances in the lateral direction rather than in the depth direction. After series of atomic steps move in lateral directions, shallow etch pits are formed on Si(111), as we observed in Fig. 4.19. Atomic steps are continuously generated from the center of the pit, as shown in Fig. 4.20, until the etch pit reaches the bottom end of the dislocation. They are quickly etched and propagate outward, looking like a wave propagation on the water surface. Note that the etch pit in Fig. 4.20 is not in symmetry between left and right because the (111) surface is not in fourfold symmetry but in threefold. The shallow pit wall on the right in Fig. 4.20 is composed of steps having two dangling bonds, while the wall on the left is composed of single-dangling-bonded steps.

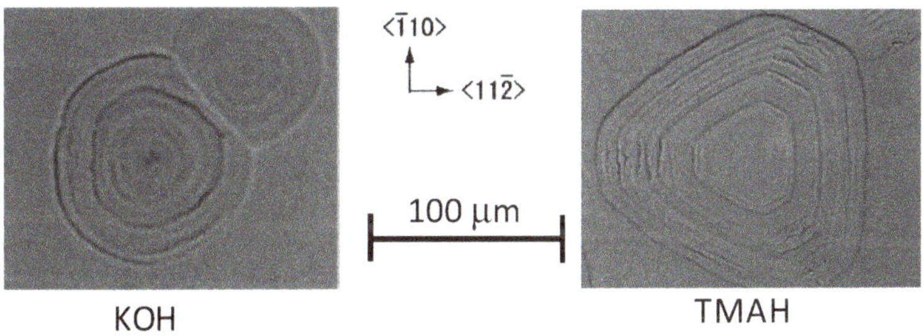

Figure 4.19 Top views of etch pits grown on Si(111). Diameters and depths are 100 μm and 1 μm, respectively. Top profiles are both triangular but in a mirror relationship between KOH and TMAH. Reproduced from Ref. [24] with permission.

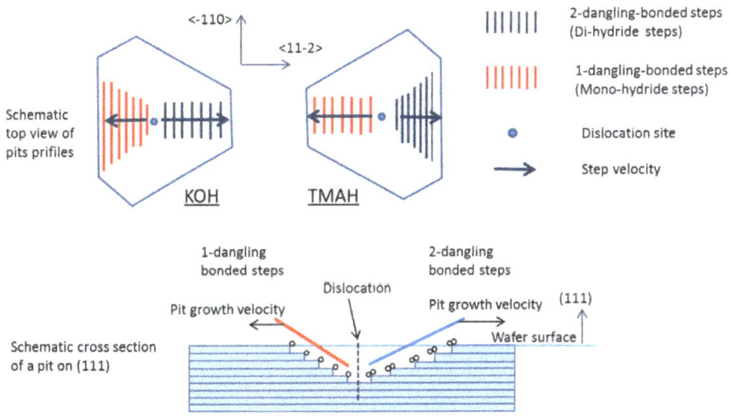

Figure 4.20 Schematic explanation of differently oriented etch pit growth on (111) with KOH and TMAH solutions.

When we look at the top view of a pit profile of KOH in Fig. 4.19, the right-hand side pit profile is advancing faster than the left. This is due to the fact that the two-dangling-bonded steps are etched faster than single-dangling-bonded ones, as shown in Fig. 4.20. It seems reasonable, considering the stability of the step atoms. On the other hand, an etch pit grown in TMAH is different. The right-hand side of the pit grows slower than the left-hand side in Fig. 4.19. The pit profile of TMAH is clearly hexagonal. Three sides of the hexagon composed of two-dangling-bonded steps are apparently etched slower than single-dangling-bonded steps.

The difference in etch rate contour maps between KOH and TMAH around (111) in Fig. 4.18 clearly agree with the difference of the etch pit growth profile. The etch rate of vicinal (111) is equivalent to the pit sidewall velocities shown as straight lines in Fig. 4.20 in the bottom.

The phenomena above are independently supported by etch rates directly measured in the vicinity of (111). Sato [25] measured the vicinal (111) etch rates using a group of fan-shaped groove patterns on a Si(110) wafer, as shown in Fig. 4.21. Total 41 groove mask patterns in the bottom right are made on the Si(110) surface, and deep grooves are wet-etched, as schematically shown in the top right in the same figure. The center groove wall is aligned close to (111), while the other grooves have an incremental angular deviation by 0.1° from (111). Etch rates are measured from an increment of the groove width covering the vicinal (111) in a range of ±2°. Relative etch rates are plotted as a function of off-angles from (111) in the left in the figure. It is clear the etch rates linearly increase by an increase in the off-angle from (111), meaning a linear increase in the number of atomic steps on the vicinal (111). An asymmetric increase in the etching rate is visible by comparing the left and right sides of the graph. A positive deviation angle in the right half of the graph means an increase of two-dangling-bonded steps, and a negative deviation means an increase of single-dangling-bonded steps. It is apparent that KOH and TMAH show reversed asymmetry when comparing the gradients between the left and right sides of the graph.

There arises a question of why two-dangling-bonded steps are etched slower than single-dangling-bonded steps with TMAH. This still is an open question. Some component in the TMAH solution may have blocked the etching reaction of two-dangling-bonded steps selectively. A potential blocker is the TMA^+ ion, which is quite large in size compared to the K^+ ion. It could be an effective barrier against the approach of OH^-. This explanation still is a speculation to be proved.

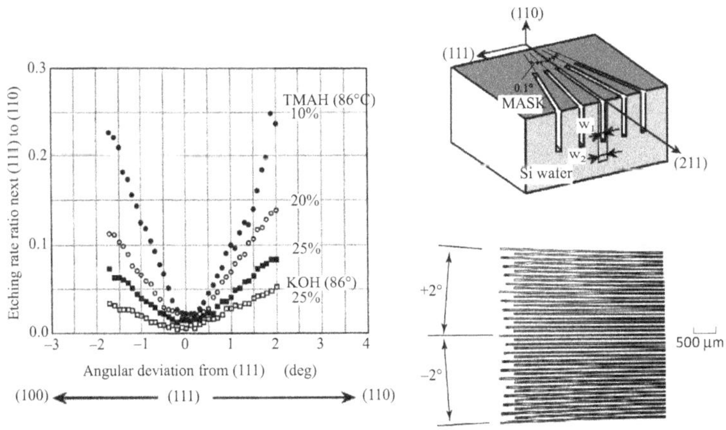

Figure 4.21 Asymmetric increase in etching rate by angular deviation from (111) orientation. KOH and TMAH show reversed asymmetry with each other. Etch rates are measured by an increment of groove width deeply etched on the Si(110) wafer in the top right. Fan-shaped 41 grooves in the right bottom, each having an angular deviation of 0.1°, are covering (111) vicinal orientations in a range of ±2°. Reprinted from Ref. [25], Copyright (1999), with permission from Elsevier.

Such a macroscopic difference in the anisotropy can easily appear as a difference of the etch profiles [15]. From the etch rate characterization in total orientation by using a spherical Si specimen, one can simulate or even graphically draw etch profiles for each etching solution. For example, in the case of deep groove etching on a Si(110) wafer, groove profiles are graphically drawn for two etching solutions on the basis of the Wulff–Jaccodine method [17], as shown in Fig. 4.22. Information necessary to draw the 2D profile is just a planar etch rate distribution in the plane of the groove cross section. The angular etch rate distribution in the analytical plane is compared with KOH and TMAH solutions in the figure. A difference in the etch rate distributions is observed as larger local minima in (311) for the KOH solution. This is why the bottom of a groove on (110) etched with 40% KOH is wedge shaped, while that with TMAH is flat shaped, shown in the bottom.

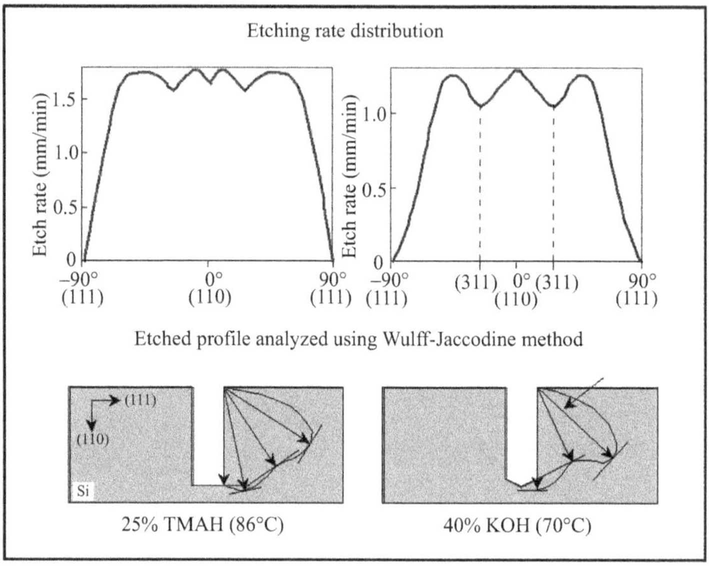

Figure 4.22 Different etch profiles due to a difference in etch rate anisotropy between TMAH and KOH solutions. Deep groove profiles on a Si(110) wafer were graphically analyzed (bottom) on the basis of the Wulff–Jaccodine method [17] for the two solutions. Etch rate distributions as a function of rotational angle are compared at the top. Reprinted from Ref. [25], Copyright (1999), with permission from Elsevier.

4.2 Effect of Alcohols and Surfactant Additives*

4.2.1 Etching in KOH Solutions Saturated with Alcohols

4.2.1.1 Properties of KOH solutions saturated with alcohols

Etching of monocrystalline Si(*hkl*) substrates in KOH solutions is anisotropic in character and the surfaces obtained as a result of etching in low-concentrated (below 5 M) solutions are usually rough and not much useful in semiconductor device technology. However, the etched substrates with (111) orientation, characterized with the lowest etch rate, usually reveal relatively smooth surfaces. In practice, ca. 10 M KOH solutions are used, and roughness of etched substrates with (100)-oriented and vicinal surfaces, such as (113),

*Section 4.2 is written by Prof. Irena Zubel, Wroclaw University of Technology, Poland.

is remarkably improved in relation to the substrates etched in low-concentrated KOH [13, 21, 26, 27]. Relative etch rates $R(hkl)/R(100)$ (in a form of a plot) and morphologies (SEM [21] images) of Si(hkl) substrates belonging to [110] crystal zone are shown in Figs. 4.23 and 4.24. In practice, due to their fourfold symmetry, (100)-oriented Si substrates are the most popular but also (110) and (111) substrates are used in particular applications.

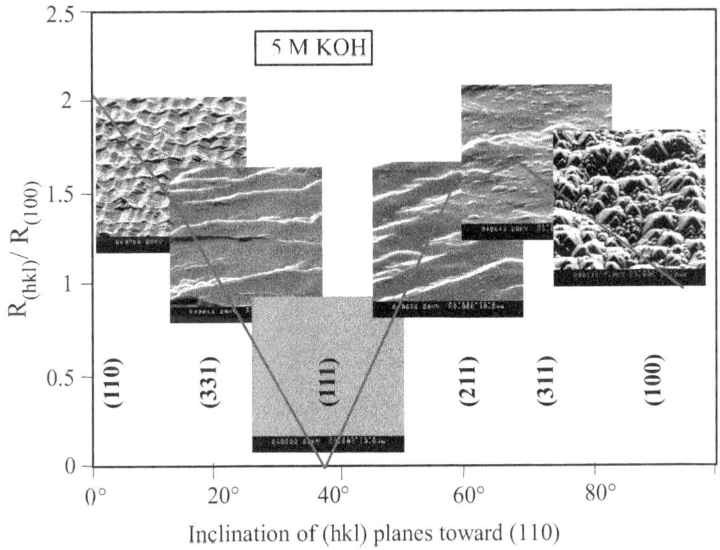

Figure 4.23 SEM images and etch rates of Si substrates with different crystallographic orientations etched in 5 M KOH.

Not without significance are both the roughness and etch rates of the planes with other crystallographic orientations as they form sidewalls and corners of convex and concave structures etched in substrates with the basic crystallographic orientations; therefore, there is a continuous search for additives to KOH solutions enabling reduction of roughness of etched Si(hkl) surfaces [28–31]. A factor that appeared very successful in modifying the properties of KOH solutions was alcohols. The historically first additive was isopropyl alcohol (IPA) [32]. It was shown that the alcohol added to a KOH solution causes reduction of the etch rate of (110)-oriented substrates and results in improvement of their roughness. The ratio of $R(110)/R(100)$ depends on KOH concentration [33]. The obtained results show that the largest difference between etch rates of $R(100)$ and $R(110)$ occurs at low KOH concentrations (Fig. 4.25) and the ratio of $R(110)/R(100)$, known as **anisotropy ratio**—often this ratio is defined reversely as $R(100)/R(110)$ or defined in relation to $R(111)$.

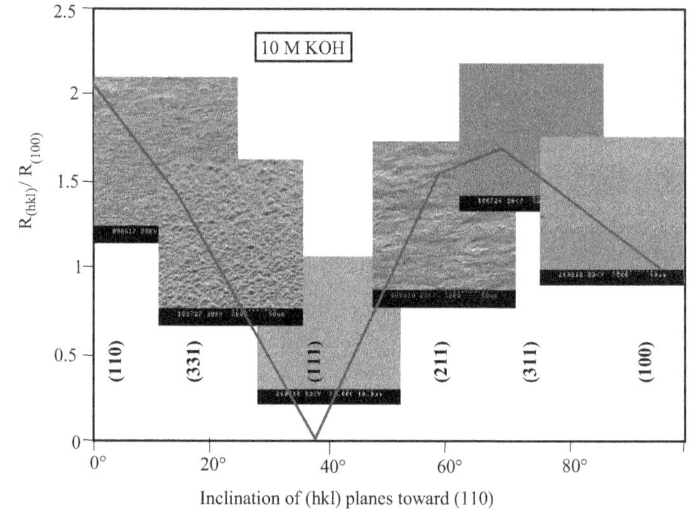

Figure 4.24 SEM images and etch rates of Si substrates with different crystallographic orientations etched in 10 M KOH.

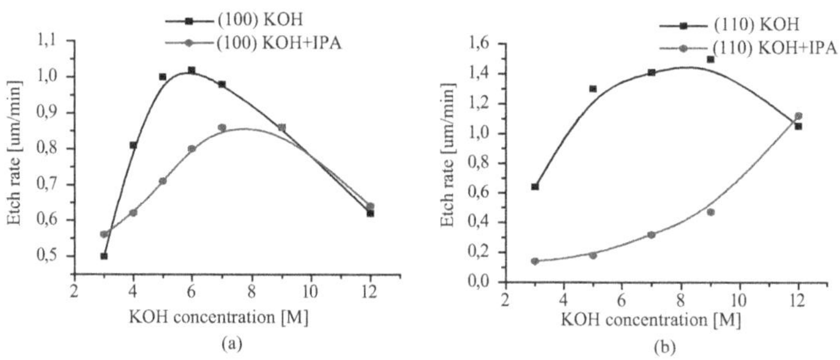

Figure 4.25 Etch rates of (a) (100) and (b) (110) silicon substrates in KOH and KOH + IPA solutions. Reprinted from Ref. [51], Copyright (2011), with permission from Elsevier.

Investigations on etching Si substrates with different (*hkl*) orientations in low-concentrated (3–5 M) KOH solutions modified with IPA were presented in numerous research reports [27, 34–36]. KOH solutions oversaturated with isopropanol (i.e., the ones in which the alcohol concentration exceeds its solubility in the KOH solution) were used. An alcohol film maintained on the solution surface due to the oversaturation prevented the solution from evaporation (boiling point of the alcohol is 82°C). It allowed keeping a stable composition of the solution. The results of etching in a KOH solution saturated with IPA are shown in Fig. 4.26.

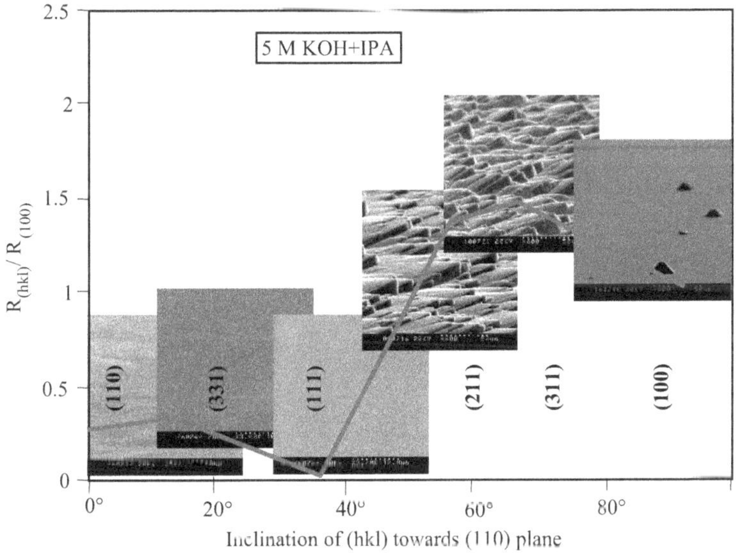

Figure 4.26 SEM images and etch rates of Si substrates with different crystallographic orientations etched in 5 M KOH + IPA.

It can be seen that IPA improves not only the roughness of the (110) plane but also the vicinal planes (331) and (221), simultaneously reducing their etch rates. The alcohol only slightly reduces etch rates of the planes such as (211) and (311) but remarkably deteriorates the morphology of the surfaces which are covered by numerous hillocks. Much smoother planes from the {211} and {311} families can be obtained by etching in pure 10 M KOH. In the following part of this chapter concerning etching in alcohol-containing solutions, these planes will not be taken into account. It is worth noting that the roughness of the (100) plane etched in alcohol-containing solutions is improved in relation to the plane etched in pure KOH, in spite of single hillocks which appear there [27, 37].

As can be seen in Fig. 4.27, alcohol efficiently interacts with the planes that are considered as possessing step monohydride (SM) or step monohydride + terrace monohydride (SM + TM) type of bonds, such as (110) and (331), generally {*hh*1}, causing an improvement of their roughness and reduction of the etch rate. The planes with the bonds of the step dihydride (SD) + TM type reveal a tendency to be covered by hillocks. Arrangements of atoms on the planes with various orientations and classification of their surface bonds have been frequently discussed in the literature, especially the literature concerning the models of etching [38–42].

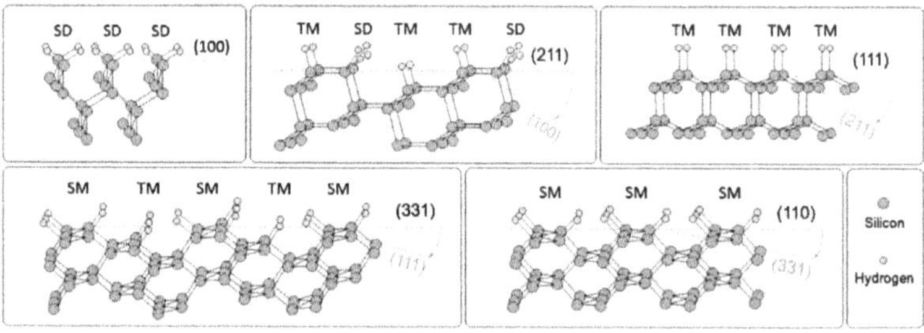

Figure 4.27 Representation of step dihydride (SD), terrace monohydride (TM), and step monohydride (SM) surface bonds at (100), (211), (111), (331), and (110) planes. From Ref. [43]. @ IOP Publishing. Reproduced with permission.

It is interesting to analyze how the properties of alcohol-containing solutions affect the etching of spatial silicon structures on Si substrates with basic crystallographic orientations, for example, (100). In Fig. 4.28a a fragment of a stereographic projection on the (001) plane with marked Si(*hkl*) planes, including the plane ((110), (331), (221) (111) (211), (311), (100)) from the [110] crystal zone is shown. The morphologies of these planes are shown in Figs. 4.23, 4.24, and 4.26. These planes form the sidewalls of structures etched through the masks arranged in different directions on the (100) substrate (Figs. 4.28b and 4.28c). In the figures, besides the mask patterns used for the etching, the layout of the (111) and (100) planes, defined from the stereographic projection, is shown. It images the fourfold symmetry of the (100) substrate [33, 36, 44]. A simplified mechanism of evolving the planes with a specified crystallographic orientation on the sidewalls of etched structures, either concave (holes) or convex (mesas), depending on etch rates of the planes, was presented in Refs. [33, 45], and then the issue has been many times addressed in the literature [46, 47]. In Figs. 4.28d, 4.28e, and 4.28f, concave and convex structures etched through the mask patterns shown in Figs. 4.28b and 4.28c are displayed. The planes developing in <110> and <310> crystallographic directions have been marked. The crystallographic orientation of the sidewalls developing in these directions depends on the etch rate ratio $R\{110\}/R\{100\}$ or $R\{331\}/R\{311\}$, that is, etching anisotropy in the particular solution. When $R\{110\} < R\{100\}$ and $R\{331\} < R\{311\}$, then the sidewalls {110} and {331}, inclined to the (100) substrate at a low angle (ca. 45°) develop. When $R\{110\} > R\{100\}$ and $R\{331\} > R\{311\}$, then the sidewalls {100} and {311}, inclined to the (100) substrate at a large angle (ca. 70°) develop (see stereographic projection) [36, 44]. As can be seen, the shapes of the etched structures depend on substrate crystallographic orientation (arrangement of Si(*hkl*) planes on the stereographic projection of the substrate), arrangement of the mask on the substrate, and composition of the etching solution. The angle of inclination of sidewalls to the substrate and their roughness are important from the point of view of operation of semiconductor devices.

Low etch rates of {110}, {331}, and {221} planes are of significance due to arrangements of these planes in a Si crystal (see stereographic projection). The planes are situated symmetrically along the <100> crystallographic directions corresponding to the corners of typical convex structures bounded by {111} planes (Fig. 4.29a). It is highly desired that these planes have a possibly low etch rate and roughness. This would result in a reduction of the convex corner undercut and low roughness of etched surfaces, as shown in Fig. 4.29 [48].

The improvement of roughness of some planes etched in KOH solutions containing alcohol allows using them as micromirrors reflecting a light beam at an angle of 90° in relation to the substrate in MEMS structures—the mirror is made of the (110) plane inclined at an angle of 45° to the (100) substrate (Fig. 4.30) [49, 50].

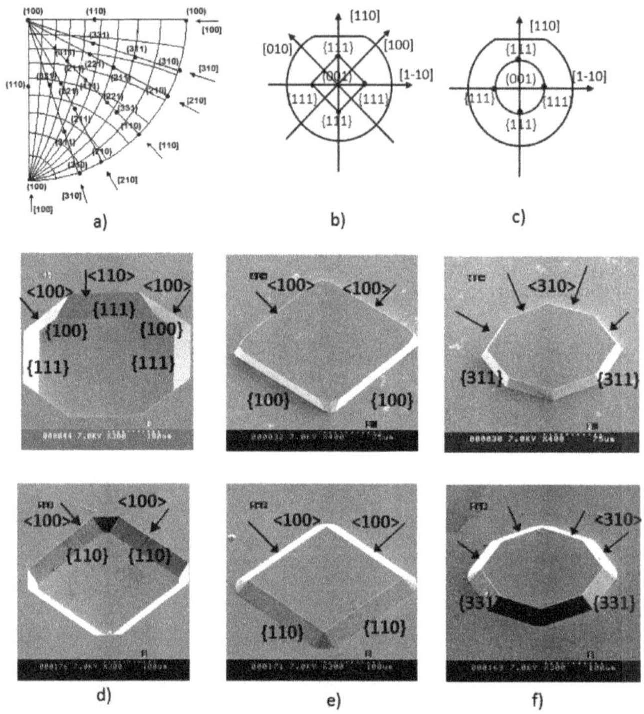

Figure 4.28 Comparison of structures etched in KOH and KOH + IPA. (a–c) Stereographic projection and mask patterns applied in a (001)-oriented substrate, (d) concave structures, and (e, f) convex structures etched in KOH (upper row) and KOH + IPA (lower row). Reprinted from Ref. [44], Copyright (2001), with permission from Elsevier.

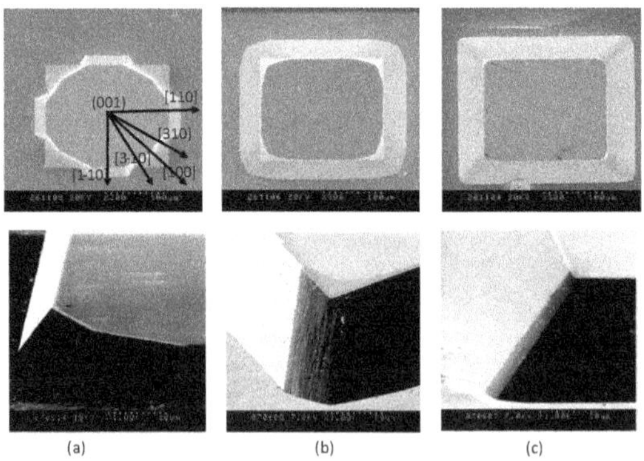

Figure 4.29 Convex corners etched in KOH solutions containing alcohols: (a) pure KOH, (b) KOH + isopropanol, and (c) KOH + *tert*-butanol. Reprinted from Ref. [48], Copyright (2002), with permission from Elsevier.

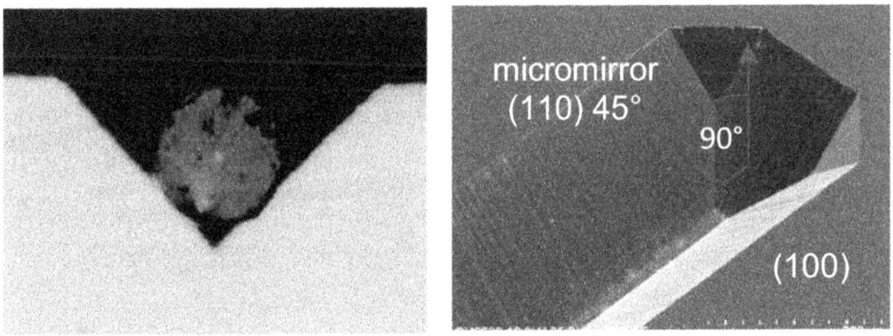

(a) (b)

Figure 4.30 Micromirrors {110} inclined at the angle of 45° to the substrate. (a) V-groove with a waveguide inserted at the bottom and (b) V-groove bound with {110} sidewalls etched in KOH + IPA [50] (light beam direction is indicated) [50].

4.2.1.2 Results of etching in KOH solutions containing alcohols with one hydroxyl group

Analysis of etching in KOH solutions saturated with IPA shows both advantages as well as disadvantages of these solutions:

- The (100) plane is much smoother in comparison to the plane etched in pure (low-concentrated) KOH, but frequently uncontrolled hillocks appear on the surface.
- The planes of $(hh1)$ type, with SM bonds, have etch rates lower than $R(100)$ $(R(hh1)/R(100) < 1)$; therefore, we can observe so-called **reversal anisotropy** in relation to the etching in pure KOH (Figs. 4.23 and 4.26), and on sidewalls of etched structures and convex corners the planes (hkl), inclined at low angles to the substrate, appear. The planes $(hh1)$ are, however, not ideally smooth; especially the (110) plane preserves its characteristic striped structure. An improvement of the surface roughness and reduction of the etch rate would be advisable in terms of the convex corner undercut.
- Limited IPA solubility in KOH causes the etching process to be conducted in low-concentrated solutions to obtain the largest difference between etch rates of $R(100)$ and $R(110)$. To improve the roughness and anisotropy it would be advantageous to use more concentrated KOH solutions and alcohol additives of better solubility.

- A significant limitation is the low boiling point of IPA. It does not allow to raise the temperature of the process so as to increase the etch rate, which is especially important in the case of etching of deep MEMS structures. Additives with a higher boiling point should be then used.

The IPA analyzed above is soluble in water without limits, but its solubility in a KOH solution depends on the hydroxyl concentration. The solubility decreases rapidly with hydroxyl concentration and it also depends slightly on temperature. In Fig. 4.31, the so-called **solubility limit** of isopropanol versus KOH concentration is shown. Above the solubility limit, the alcohol forms supersaturated solutions. In 3 M KOH, the limit of alcohol concentration is below 2 M and in 5 M KOH it is below 1 M. In solutions with higher KOH concentration, the concentration of alcohol is so low that the anisotropy, typical of KOH + IPA solutions (Fig. 4.25), disappears: $R(110) \cong R(100)$. Alcohols with molecules lager than propanols (such as butanols and pentanols) are much less soluble in water (Fig. 4.32), so their solubility limit in KOH solutions is also much lower than that of isopropanol. In fact, alcohols are not very good soluble additives to KOH solutions. So, rather low-concentrated KOH solutions should be used with these additives. Some selected properties of alcohols are shown in Table 4.2.

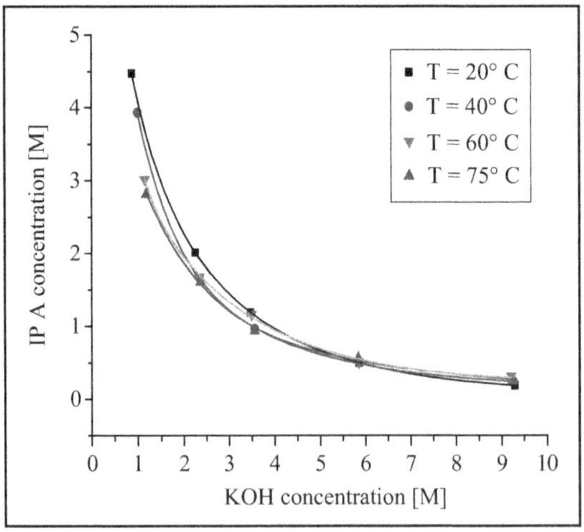

Figure 4.31 IPA concentration at saturation level versus KOH concentration at different temperatures. Reprinted from Ref. [51], Copyright (2011), with permission from Elsevier.

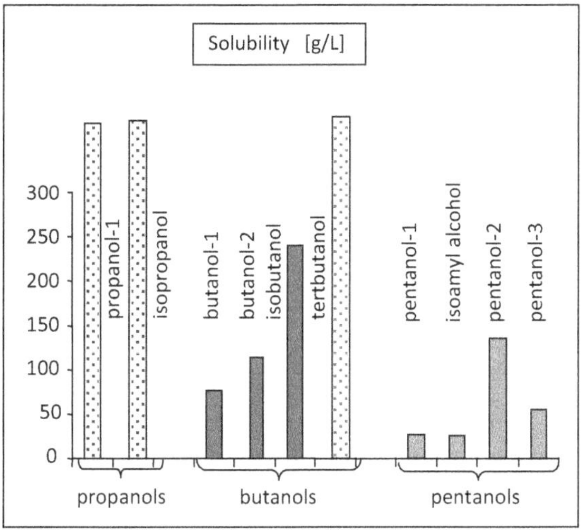

Figure 4.32 Alcohol solubility in water (g/L) (the bar above 300 g/L represents unlimited solubility).

Various groups of alcohols (propanols, butanols, and pentanols) differ among each other by the number of carbon atoms in the molecule. Within each group, the molecules differ with the location of –OH group (order of alcohol). Alcohols belonging to the group of butanols and pentanols have bigger and heavier molecules than isopropanol (their molar mass is larger, and so is larger their boiling point). It is worth noting that in every group the highest boiling point has the alcohol with the –OH group situated at the first carbon atom (1-propanol, 1-butanol, 1-pentanol). Such an increase in the boiling point is beneficial for the course of the etching process.

With an increase in molecular weight, solubility of alcohols in water decreases (Fig. 4.32). The same concerns hydroxide solutions but in a higher degree. The saturated concentration of heavier alcohols in KOH solutions is lower than that of isopropanol. A fall in the solubility limit with an increase in the hydroxide concentration is shown on the example of a convex corner etched in *tert*-butanol-saturated KOH solutions of various concentrations (Fig. 4.33). The smallest underetching is observed in the solution with the lowest KOH concentration. An increase in the KOH concentration is associated with a fall in the alcohol concentration and a reduction of anisotropy, and as a result, with a more spectacular convex corner undercut (compare Figs. 4.25, 4.31, and 4.33). It should be noted that it is connected with roughness improvement of the (100) surface.

Table 4.2 Selected properties of alcohols

<table>
<tr><th colspan="12">Schematic structures of molecules
Chemical formula and molar mass</th></tr>
<tr><td colspan="12">
C_3H_8O M = 60.1 g/mol</td></tr>
<tr><td colspan="6">1-Propanol
T_B = 97 °C
γ = 23.78 mN/m</td><td colspan="6">Isopropanol
T_B = 81.83 °C
γ = 21.7 mN/m</td></tr>
<tr><td colspan="12">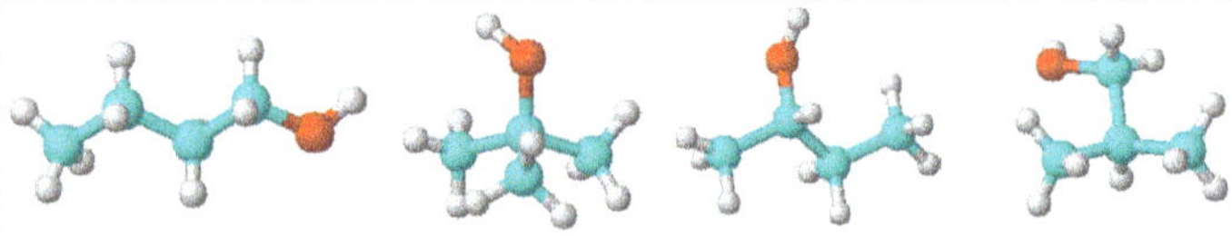
$C_4H_{10}O$ M = 74.12 g/mol</td></tr>
<tr><td colspan="3">1-Butanol
T_B = 117.6 °C
γ = 24.66 mN/m</td><td colspan="3">tert-Butanol
T_B = 83 °C
γ = 19.56 mN/m</td><td colspan="3">2-Butanol
T_B = 99.5 °C
γ = 23.47 mN/m</td><td colspan="3">Isobutanol
T_B = 108 °C
γ = 22.8 mN/m</td></tr>
<tr><td colspan="12">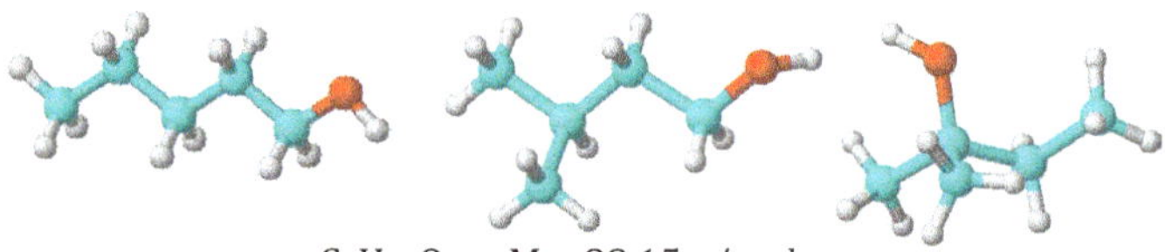
$C_5H_{12}O$ M = 88.15 g/mol</td></tr>
<tr><td colspan="4">1-Pentanol
T_B = 138 °C
γ = 25.5 mN/m</td><td colspan="4">Isopentanol
T_B = 130 °C
γ = 23.8 mN/m</td><td colspan="4">tert-Pentanol
T_B = 102 °C
γ = 23.5 mN/m</td></tr>
<tr><td colspan="12">T_B [°C] – Boiling point
Surface tension - γ[mN/m]
Molar mass - M [g/mol]</td></tr>
</table>

Figure 4.33 Comparison of SEM images of convex corners obtained in a Si(100) substrate as a result of etching in nonagitated 3–9 M KOH solutions saturated with *tert*-butanol and pure 3 M KOH solution. From Ref. [43]. @ IOP Publishing. Reproduced with permission.

In low-concentrated (3 M) KOH solutions saturated with selected butanols and pentanols, etching of Si substrates with different crystallographic orientations has been carried out. Example results of the etching in the solution containing *tert*-butanol are shown in Fig. 4.34. In all tested solutions, the results were similar to the results obtained for a KOH solution saturated with isopropanol (Fig. 4.26):

- Reduction of etch rates and roughness improvement of the same crystal planes—(110), (331), and (221)—have been achieved,
- The same crystal planes—(311) and (211)—have been covered by hillocks.

The obtained results show that the etching process in all KOH solutions saturated with alcohols follows the same rules; however, differences in roughness of (100) surfaces etched in different solutions have been observed: the lowest surface roughness has been achieved for the (100) plane etched in KOH + isopropanol. For improving the roughness of the (100) plane the solution saturated with isopropanol is supreme. In all KOH solutions saturated with butanols and pentanols, the (100) plane has been covered by hillocks. As it was already mentioned, (311) and (211) planes etched in alcohol-containing solutions are useless due to the bulky hillocks on their surfaces and therefore will be not considered now. To evaluate the properties of KOH solutions saturated with various alcohols it is useful to compare etch rates and roughness of the planes with SM bonds—(110, (331), (221)—with that of the (100) plane, which usually is a substrate. Etch rates of (110), (331), and (221) are comparable (Fig. 4.34), whereas the roughness assessed using an AFM is the lowest for the {331} plane (Ra ≅ 5 nm). The

{110} planes in all cases exhibit a striped microstructure (Ra ≅ 50 nm, Fig. 4.35). It would be beneficial to obtain smooth and slowly etching SM planes, such as {110}, {331}, and {221}. Due to high usefulness of the (110) plane (in such applications as micromirrors 45° on (100), {111} planes perpendicular to the (110) substrate) this plane has been chosen among {110}, {331}, and {221} planes for further analysis. Not ideal roughness of this plane was the reason of searching for new solutions with alcohol additives.

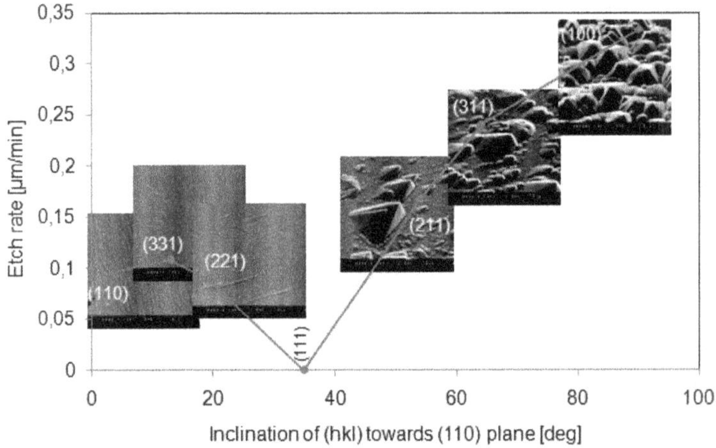

Figure 4.34 SEM images and etch rates of selected Si(*hkl*) substrates etched in an agitated 3 M KOH solution saturated with *tert*-butanol versus the angle of inclination of the (*hkl*) plane to the (110) one. From Ref. [43]. @ IOP Publishing. Reproduced with permission. All rights reserved.

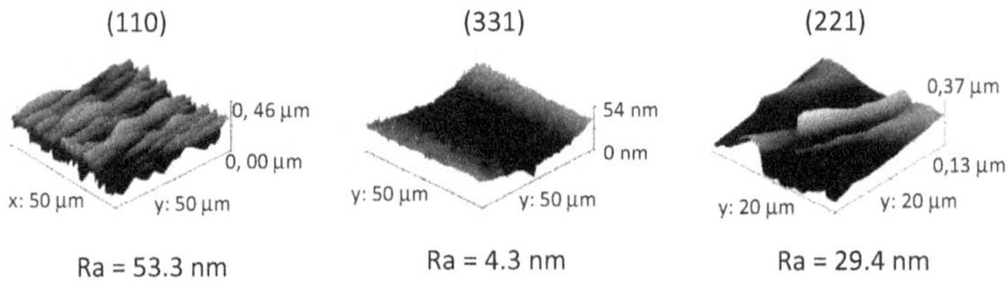

Figure 4.35 AFM images of (110), (331), and (221) planes etched in an agitated 3 M KOH solution saturated with *tert*-butanol. From Ref. [43]. @ IOP Publishing. Reproduced with permission. All rights reserved.

To compare the properties of solutions saturated with alcohols, in the following part of this chapter the studies concerning the properties of two selected planes will be presented:

- Planes (100) (used as substrates in MEMS structures)
- Planes (110) (used in many MEMS and micro-opto-electro-mechanical systems [MOEMS] applications)

Etch rates of (100) and (110) planes etched in 3 M KOH solutions saturated with different alcohols, and anisotropy ratios ($R(100)/R(110)$) have been estimated and compared (Figs. 4.36 and 4.37). The roughness of etched surfaces was assessed using SEM images of concave structures etched in (100) substrates, bound with (110) sidewalls (Fig. 4.38) [52, 53]. SEM images of (100) and (110) substrates etched in selected solutions representing particular groups of alcohols (propanols, butanols, pentanols) are shown in Fig. 4.39. It would be desirable to obtain smooth (110) surfaces with the roughness of 10 nm, comparable with {331} planes, characterized by a possibly low etch rate as well as {100} planes without any hillocks.

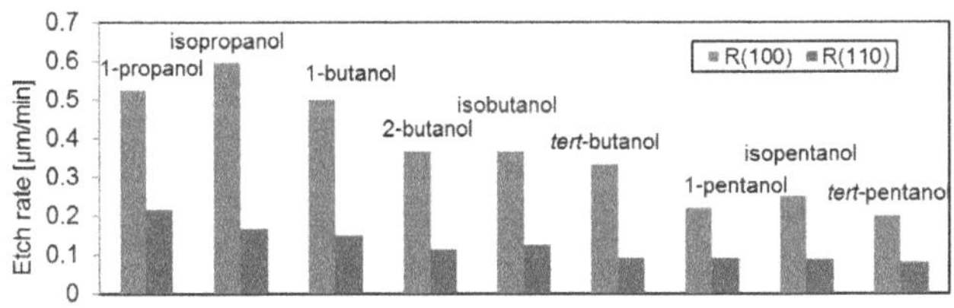

Figure 4.36 Etch rates of (100) and (110) planes etched in 3 M KOH solutions saturated with different alcohols. Reprinted from Ref. [55], Copyright (2016), with permission from Elsevier.

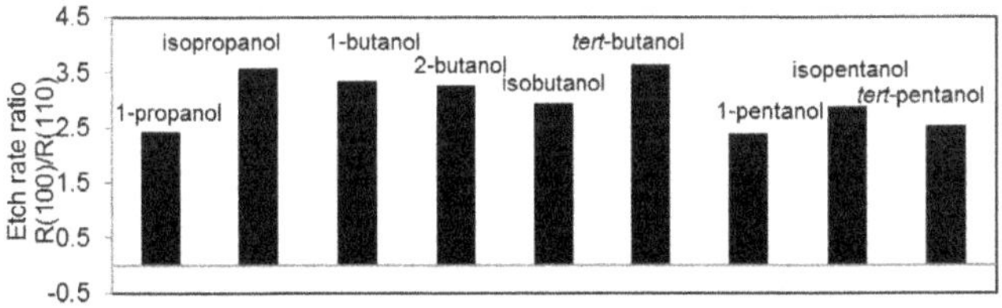

Figure 4.37 Etch rate ratios $R(100)/R(110)$ in 3 M KOH solutions saturated with different alcohols. Reprinted from Ref. [55], Copyright (2016), with permission from Elsevier.

It can be seen that the etch rate of the (100) plane decreases with the growth of the molecular weight of applied alcohol—almost twofold difference between etch rates of (100) substrates etched in propanol and pentanol containing solutions has been observed. It is not beneficial in terms of etching MEMS structures. However, it does not influence very much the anisotropy ratio $R(100)/R(110)$. This ratio in all studied solutions is higher than 2, which results in {110} planes developing on the sidewalls of the etched structures (minimal value of $R(100)/R(110) = 1.4$ [33]). Furthermore, the roughness of the (100) surface deteriorates with the growth of the molecular weight of alcohol molecules. This may be connected with reduced solubility of heavier alcohols and enhanced adsorption of their molecules on Si(100) and some other reasons. Obviously, the best results in terms of etch rate and roughness of the Si(100) surface are obtained in a KOH solution saturated with isopropanol. The roughness of {110} surfaces, forming both substrate and sidewall of etched structures (Figs. 4.38, sidewalls, and Fig. 4.39) does not depend on the weight of alcohol molecules.

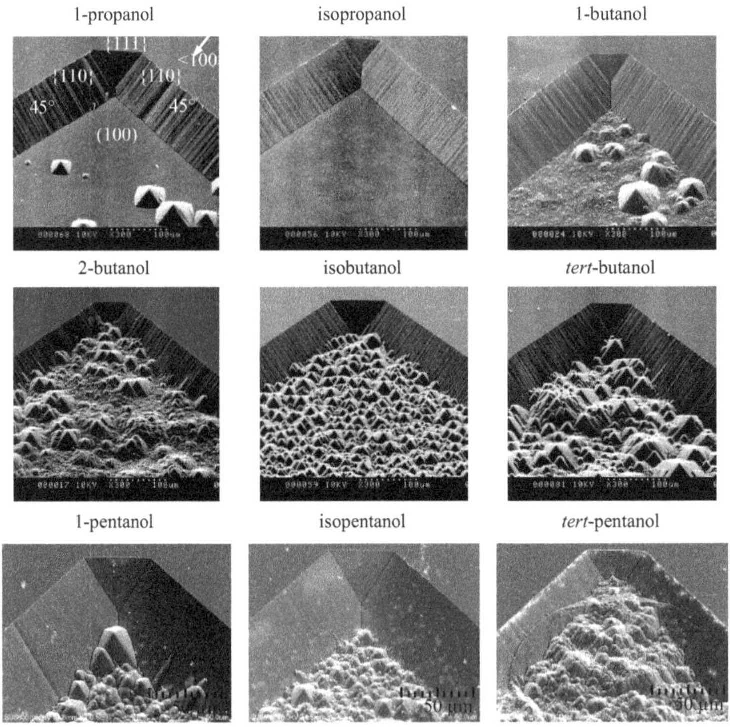

Figure 4.38 SEM images of structures with {110} sidewalls fabricated in (100) substrates in 3 M KOH solutions saturated with different alcohols [52, 53].

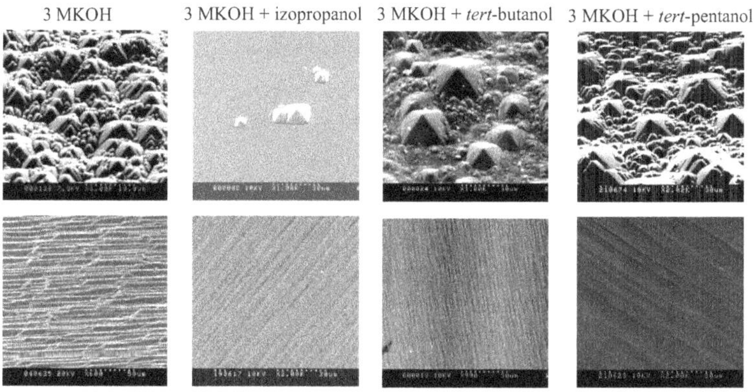

Figure 4.39 SEM images of (100) (upper row) and (110) (lower row) substrates fabricated in 3 M KOH solutions saturated with different alcohols.

The final conclusion from these considerations is that using of heavier alcohols (butanols and pentanols) as additives to etching solution does not improve the roughness of the (100) plane in comparison to pure KOH and even worsens it in comparison to KOH containing isopropanol solutions. The etch rate of the (100) plane has not increased, the roughness of {110} planes has not improved, and also the anisotropy ratio $R(100)/R(110)$ has changed only slightly. Thus, there is no reason for using anything other than isopropanol additives to KOH solutions in etching processes.

4.2.1.3 Results of etching in solutions containing alcohols with two hydroxyl groups (diols)

Alcohols with more than one hydroxyl group contain two or more –OH groups in a molecule. The names of the alcohols stem from the number of these groups, for example, diols contain two –OH groups (the simplest diol is ethylene glycol: $C_2H_4(OH)_2$) and triols contain three –OH groups (the simplest triol is glycerol: $C_3H_5(OH)_3$). The alcohols have relatively small molecules and low boiling points; therefore they are rather useless from the point of view of anisotropic etching.

For the studies, pentanediols and hexanediols with a boiling point above 200°C have been chosen. These diols contain five or six carbon atoms, and to every carbon atom an –OH group may be attached. There is a wide range of such molecules, so for their selection the following criteria have been used:

- Both –OH groups are situated at the one end of the molecule, at the first and second carbon atoms (1,2-pentanediol and 1,2-hexanediol). Such molecules have the highest polarity.
- Both –OH groups are situated at opposite ends of the molecule (1,5-pentanediol, 1,6-hexanediol).

Two-hydroxide groups cause that diols have much higher solubility in water than one-hydroxyl alcohols and the two properties solubility and boiling point cause that they are used for modification of KOH solutions. In Table 4.3 basic properties of selected diols have been collected.

Table 4.3 Basic properties of selected diols

Name and chemical formula	Schematic structures of molecules	Molar mass (g/mol)	Boiling point (°C)	Surface tension (mN/m)
1,2-Pentanediol $C_5H_{12}O_2$	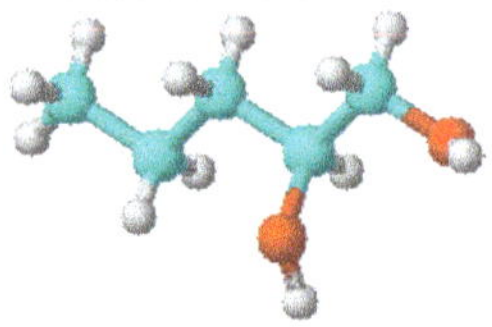	104.15	206	29.16
1,5-Pentanediol $C_5H_{12}O_2$	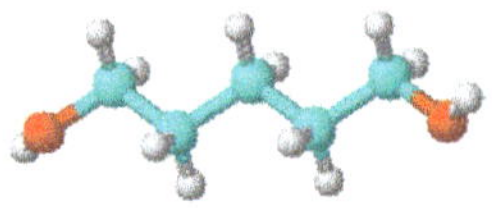	104.15	242	43.3
1,2-Hexanediol $C_6H_{14}O_2$	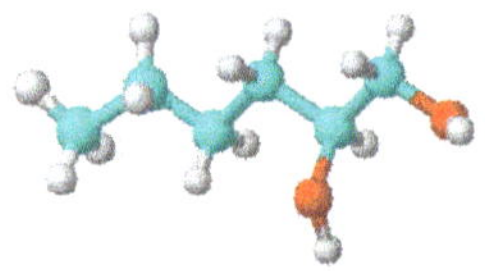	118.17	223	36.2
1,6-Hexanediol $C_6H_{14}O_2$	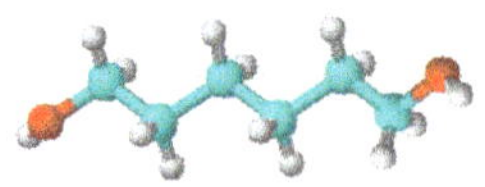	118.17	250	37.97

Diols with a concentration of 2 M have been added to a 3 M KOH solution. The amount of the diols was larger than the corresponding IPA concentration necessary for saturation of the 3 M KOH solution. A saturated solution was obtained only in the case of 1,2-hexanediol. The other solutions did not achieve the saturation state. It means that the solubility limit of those alcohols in a 3 M KOH solution was

higher than that of isopropanol. Due to the high boiling point of the diols it can be assumed that in spite of a lack of saturation (lack of an alcohol layer on the solution surface), the composition of the solution was stable during the whole process (75°C, 120 min).
Etch rates of (100) and (110) planes and the anisotropy ratio $R(100)/R(110)$ have been estimated. The results are shown in Figs. 4.40. and 4.41. Etch rates in the solutions containing 1,5-pentanediol and 1,6-hexanediol are comparable with the etch rates in KOH solutions saturated with IPA (ca. 0.5 μm/min). Etch rates in solutions containing 1,2-pentanediol and 1,2-hexanediol are slightly lower (comparable to the solutions saturated with pentanols, that is, 0.3 μm/min).

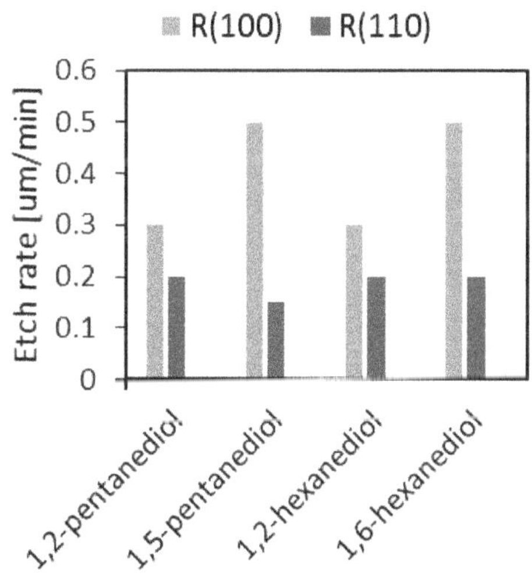

Figure 4.40 Etch rates of (100) and (110) planes in 3 M KOH solutions with different diols.

It is clearly seen that much lower differences in etch rates of (100) and (110) planes occur in the solutions with additives having –OH groups at one end of their molecule (in the 1,2 position) in comparison with solutions containing 1,5-pentanediol and 1,6-hexanediol. The anisotropy ratio $R(100)/R(110)$ for solutions containing 1,5-pentanediol and 1,6-hexanediol is similar to solutions containing alcohols with one hydroxyl group (close to 3).

Such arrangement of –OH groups in a molecule seems to be more advantageous in terms of etch rates and anisotropy. SEM images of (100) and (110) substrates etched in 3 M KOH solutions with different diols are shown in Fig. 4.42.

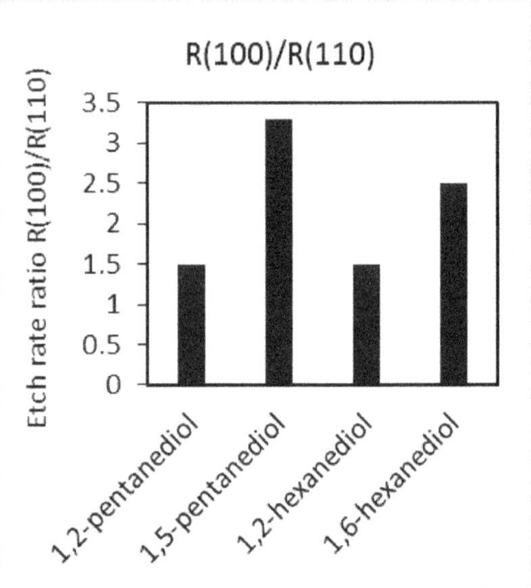

Figure 4.41 Etch rate ratios *R*(100)/*R*(110) in 3 M KOH solutions with different diols.

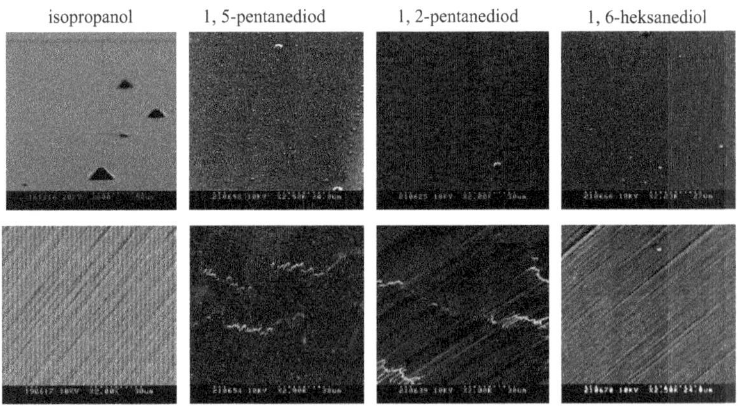

Figure 4.42 SEM images of (100) (upper row) and (110) (lower row) substrates fabricated in 3 M KOH solutions with different diols (concentration of the diols was 2 M).

The roughness of Si(100) substrates etched in diol-containing solutions is much lower than that obtained as a result of etching in solutions saturated with butanols and pentanols, in spite of the fact that the sizes of the molecules of the used surface-active compounds (butanols, pentanols, and diols) are comparable. Roughness of the etched surfaces is comparable with the roughness obtained in KOH solutions containing isopropanol, though the sizes of IPA and diol molecules differ from each other significantly. It seems that the concentration of these tensioactive compounds has a much greater effect on the roughness of etched surfaces than the type of the compound (in the case of IPA and diols the concentration was relatively high, at the level of 2 M). The roughness of (110) substrates was significantly improved in comparison to pure 3 M KOH, but the results did not differ much from the results obtained in the case of alcohols with one hydroxyl group. However, an improvement in the roughness would be highly desired.

It can be concluded that the results of etching (100) and (110) substrates in a solution with selected diols (1,5-pentanediol, 1,6-hexanediol) are comparable with the results obtained in KOH solutions saturated with IPA. The advantages of diols are their high boiling point and good solubility. It makes it possible to perform optimization of solution composition and process parameters. Results concerning this issue will be presented in the following section.

4.2.2 Etching in KOH Solutions with the Alcohol Concentration Below Saturation

4.2.2.1 Effect of alcohol concentration on the course of etching

Etching in solutions saturated with alcohols is beneficial in terms of practical applications. This has been discussed in the previous section in detail. However, an objective investigation of such solutions is difficult as they consist of two layers which are dependent on each other, that is, a layer of KOH solution saturated with alcohol and an alcohol layer saturated with KOH floating over it. Therefore, it is impossible to assess the influence of alcohol concentration on the solution properties. The studies presented in this section concern the analysis of solutions and etching processes in hydroxide solutions with an alcohol concentration below saturation.

Alcohols used for modification of etching solutions are tensioactive in character and, added to water and hydroxide solutions, cause lowering of their surface tension. Alcohols and surfactants consist of a hydrophobic hydrocarbon chain and a hydrophilic group. Polar water molecules remove the hydrophobic chains from the bulk of the solution and shift them to the surface, which causes the excessive (in relation to the bulk of the solution) surfactant molecules to gather on the solution surface with their hydrophobic part directed outward and the hydrophilic part immersed in the solution. The surface tension of organic compounds is usually lower than that of water; therefore lowering of surface tension of the solution occurs. The relation between surface tension γ and concentration of tensioactive compound, $\gamma = f(c)$, is a basic relation describing the properties of solutions containing tensioactive compounds. A surfactant layer gathered on the surface is often considered as a surface excess or the density of the adsorption layer, Γ (mole/cm^2). The surface excess is a difference between the molar concentration of a substance on the air–liquid interface and its concentration in the bulk solution. It can be expressed by Gibbs' equation:

$$\Gamma = -\frac{c}{RT} \cdot \frac{d\gamma}{dc}$$

where Γ is the surface excess in mole/cm^2, c is the solution concentration, R is the gas constant, T is the temperature, and γ is the surface tension of the solution.

The mentioned phenomena occur on the solution surface after addition of a tensioactive substance. A simplified model of a Si substrate immersed in a solution containing a tensioactive compound is shown in Fig. 4.43a. The phenomena described above take place in the region denoted as A in the figure, that is, on the solution surface. Results of measurement of surface tension in hydroxide solutions versus the KOH concentration for different concentrations of IPA are shown in Fig. 4.43b. In the studies, a tensiometer EasyDyne by Kruss was used; the measurements were carried out with the use of a platinum ring by the du Nouy method. For comparison, the plot of the limit concentration of IPA above which the solution becomes saturated, depending on the hydroxide concentration, has been attached (Fig. 4.43c). It is worth noting that the minimum of surface tension for 3, 5, and 8 M KOH solutions is very close to the limit solubility of isopropanol. Measurements of surface tension can therefore, to some extent, replace the laborious measurements of the limit solubility. In Fig. 4.43d, the density of the adsorption layer (surface excess) Γ is presented. The faster drop of surface tension (higher KOH concentration), the thicker the adsorption layer, and the lower concentration necessary for solution saturation.

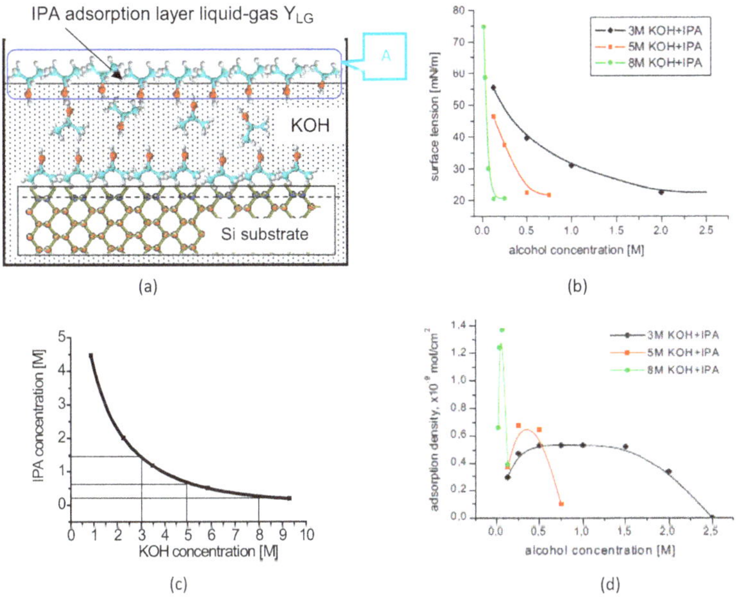

Figure 4.43 (a) Model of a Si substrate immersed in a KOH solution containing isopropanol and effects observed at the solution surface (region A) when the alcohol concentration increases. (b) Drop of surface tension, (c) drop of the solubility limit, and (d) adsorption density on the liquid–gas surface (surface excess).

Much more difficult to describe are the phenomena on the solid–liquid interface, especially in a KOH solution where a dynamic etching process takes place (the region denoted by B, Fig. 4.44a). The tensioactive compound, which forms the adsorption layer on the solution surface, is in direct contact with the Si substrate. It is also confirmed by the decrease in the contact angle due to alcohol. Pure monocrystalline Si is not wettable by water; a water drop does not spread on the Si surface and makes with the substrate a large contact angle. Tensioactive compounds facilitate solution spreading on the substrate. Alcohol molecules directed with their hydrophobic part of carbon chain toward the hydrophobic Si substrate interact with it with dispersion forces and assist in its wetting. It is very important for the process of etching. The contact angle is evaluated from the force balance (Fig. 4.44c):

$$\gamma_{SG} = \gamma_{SL} + \gamma_{LG}\cos\theta$$

where γ_{SG} is the solid–gas surface tension, γ_{SL} is the solid–liquid surface tension, and γ_{LG} is the liquid–gas surface tension.

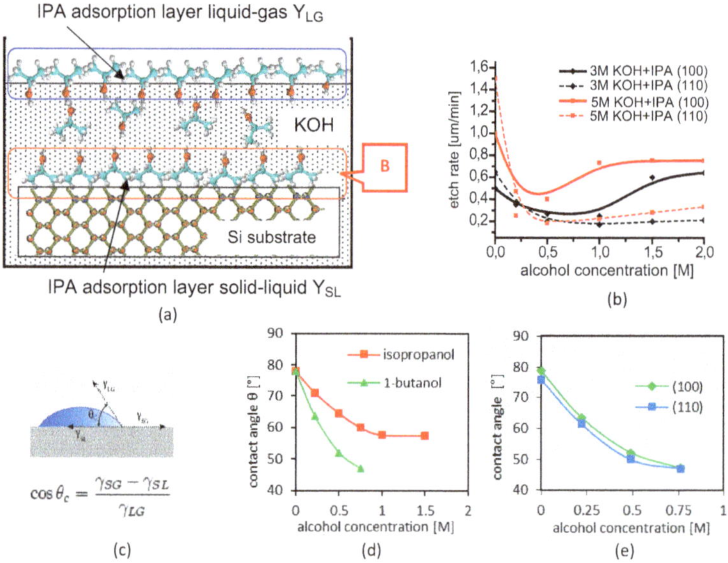

Figure 4.44 (a) Model of a Si substrate immersed in an etching solution containing isopropyl alcohol and effects observed on the Si surface (region B) upon the growth of alcohol concentration. (b) Nonlinear changes in etch rates of Si(100) and Si(110), (c) definition of contact angle, and (d, e) reduction of contact angle.

Measurements of contact angles of solutions containing alcohols on the Si surface give some information about the phenomena on the Si–solution interface. It has been stated that the contact angle decreases (wetability increases) with an increase in alcohol concentration and depends on the length of the carbon chain (Fig. 4.44d). However, the dependence of contact angle on Si substrate orientation (bond types on the surface) has not been evidenced in the experiments, which would confirm the anisotropy of adsorption of alcohol molecules and thus the etching anisotropy (Fig. 4.44e).

Nonetheless, it is most important to establish the relation between the etch rate and morphology of Si substrates and the alcohol concentration on the Si–solution boundary surface (region B, Fig. 4.44a). On this basis and on the basis of previously mentioned measurements of surface tensions and contact angles, the role of alcohol in the etching process should be established. However, it must be remembered that the etching process is conducted at high temperatures, whereas the measurements of surface tensions and contact angles versus alcohol concentration are usually performed at room temperature.

Analysis of etching results at different alcohol concentrations (in nonsaturated solutions), especially for alcohols with low boiling points (like isopropanol), creates some difficulties. As alcohols with molecules larger than IPA have a higher boiling point, studies have been mainly based on butanols and diols. In the studies, the results concerning isopropanol were also taken into account as they did not differ from the other ones. In Fig. 4.44b exemplary plots of etch rates of Si(100) and Si(110) etched in 3 M and 5 M KOH solutions containing IPA versus alcohol concentration have been shown. The alcohol concentration was gradually increased even above the saturation. Analysis of these results will be presented in the following part of this chapter.

From the comparison of measurement results of surface tension and contact angle versus concentration of tensioactive compounds in water solutions and KOH solutions with different hydroxyl concentrations (Figs. 4.45 and 4.46), the following conclusions can be drawn:

- In KOH solutions the decrease in surface tension with the concentration of alcohol is much more rapid than in water solutions (Fig. 4.45a), which is connected with a decrease in alcohols solubility in KOH (Fig. 4.43b,c).
- In KOH solutions, the rate of decrease of surface tension depends on the KOH concentration and on the kind of alcohol (Fig. 4.45a,b).

- The rate of decrease of surface tension and contact angle with the concentration of alcohol in a water solution depends on the kind of alcohol (the length of the carbon chain of that alcohol) (Fig. 4.46a,b).
- Alcohols with a straight carbon chain (propanol-1, butanol-1) cause a more rapid decrease in surface tension and contact angle than alcohols with branched carbon chains (isopropanol, *tert*-butanol) (Fig. 4.46a,b).

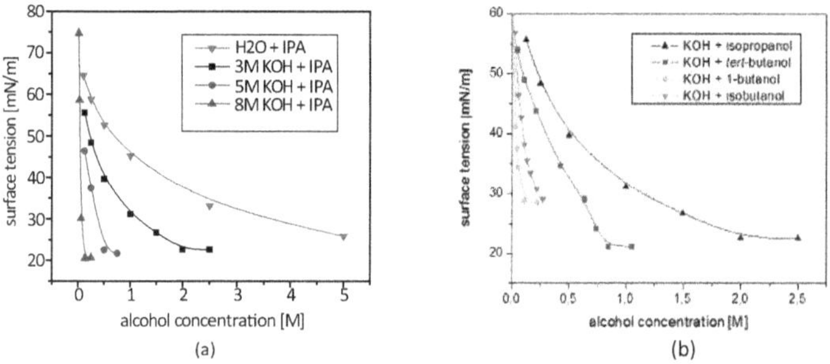

Figure 4.45 Reduction of surface tension versus alcohol concentration in water and KOH solutions with different concentrations: (a) IPA + H_2O, IPA + KOH [51] (reprinted from Ref. [51], copyright (2011), with permission from Elsevier) and (b) 3M KOH + different alcohols [54].

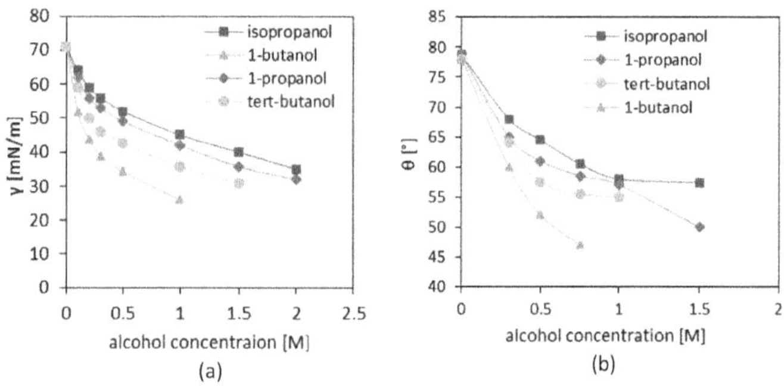

Figure 4.46 Comparison of surface tension (a) and contact angles (b) of water solutions containing different alcohols. Reprinted from Ref. [55], Copyright (2016), with permission from Elsevier.

Very interesting conclusions follow from a comparison of the measurements of surface tension and contact angle (Figs. 4.46a and 4.46b). Results of measurements of surface tension at the air–

solution interface and contact angle at the Si–solution interface point out explicitly the presence of the tensioactive compound on both the interfacial surfaces. Alcohol molecules must be present there because they cause some effects (reduction of γ and θ). This suggests that adsorption of these molecules takes place on both interfacial surfaces, and additionally, as follows from the plots, alcohols with longer and straighter alkyl chains contribute to stronger adsorption, which manifests itself in a more rapid decrease in γ and θ. The consideration points out on the usefulness of adsorption studies in silicon anisotropic etching.

The etching in 3 M KOH solutions containing alcohols of different concentrations has been analyzed. The experiments were conducted at a temperature of 75°C for t = 60 min under mechanical stirring using the following alcohols:

- Isopropanol
- 1-Butanol
- Isobutanol
- *tert*-Butanol
- *tert*-Pentanol

For the solutions, the etch rate and surface tension versus alcohol concentration have been assessed. In addition, SEM images of etched (100) and (110) substrates have been compared. Detailed results were reported in Refs. [51, 53, 54]. Below, only results for isopropanol (Fig. 4.47) and selected butanol (Fig. 4.48) and pentanol (Fig. 4.49) are presented. The results obtained for other alcohols were similar.

The lowest roughness of Si(110) surfaces etched in solutions with different KOH concentrations (3 M and 5 M) was obtained for an alcohol concentration corresponding to the minimum etch rate. For a 3 M KOH solution this concentration is 1 M (Fig. 4.47). At higher alcohol concentrations, the surface gains a remarkably striped structure, though its etch rate does not rise significantly. A remarkable improvement of the roughness of the (100) surface was observed for the alcohol concentration exceeding the concentration corresponding to the minimum etching rate, where the significant increase in the (100) etch rate started. In this range, alcohol concentration tends to saturation (ca. 2 M). This finds its reflection on the plot of surface tension versus IPA concentration since the surface tension of a 3 M KOH solution containing 2 M of alcohol decreases to the level of pure alcohol (corresponding to solution saturation). Detailed relations between these phenomena will be discussed in a further part of this section.

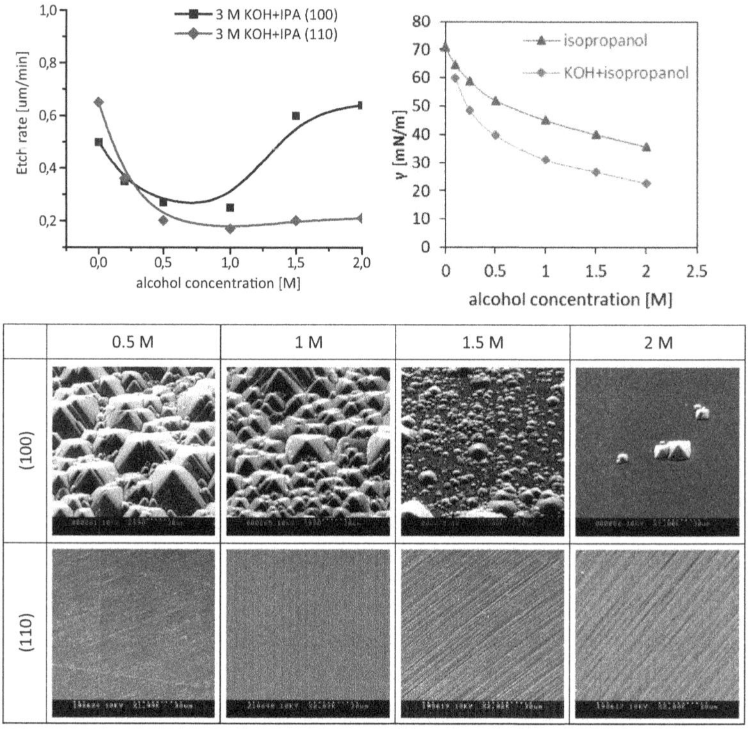

Figure 4.47 Etch rates, surface tensions (H_2O + alcohol and 3 M KOH + alcohol), and SEM images of Si(100) and Si(110) substrates etched in 3 M KOH solutions at different isopropanol concentrations. Reprinted from Ref. [51], Copyright (2011), with permission from Elsevier.

In spite of the fact that the quality of (100) surfaces etched in butanol-containing solutions was not satisfactory in either case, and the etched surfaces were covered by hillocks, the phenomena associated with the etching process were similar to those observed in the KOH + IPA solution. A remarkable minimum in etch rates of both etched planes occurred, however, at a significantly lower concentration of butanol than isopropanol (0.1 M) due to the fact that the rate of reduction of surface tension was higher in the case of butanol. The roughness improvement of the (110) surface was also observed at the minimum etch rate (Fig. 4.48). The model of the phenomena described for a KOH solution containing isopropanol can be thus extended to solutions containing other alcohols.

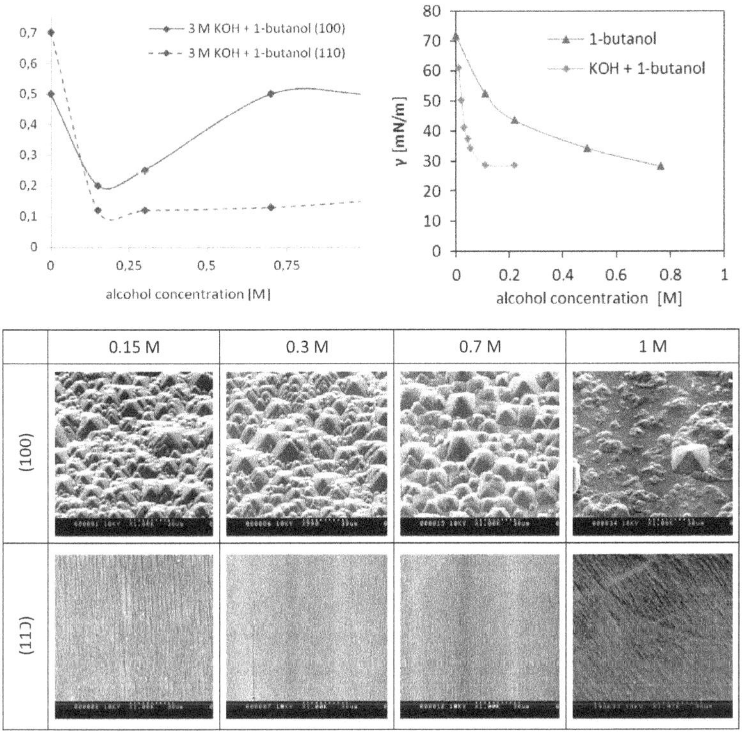

Figure 4.48 Etch rates, surface tensions (H_2O + alcohol and 3M KOH + alcohol), and SEM images of Si(100) and Si(110) substrates etched in 3 M KOH solutions for different concentrations of 1-butanol [54].

In the case of pentanols, the minimum etch rate and the lowest roughness of Si(110) occur at an even lower alcohol concentration (below 0.1 M) and surface tension in these solutions rapidly decreases with the alcohol concentration. Also etch rates of both the planes decrease in comparison to solutions with propanols and butanols (Fig. 4.49).

Very interesting are studies of the effect of concentration of alcohols with more than one hydroxyl groups (diols) on the properties of etching solutions, as the high boiling point of these alcohols prevents from changes in solution composition during the etching process. The diols mentioned in Section 4.1 have been used as additions to 3 M KOH solutions. Surface tensions of the solutions were measured, etch rates of (100) and (110) planes were estimated, and the morphologies of etched surfaces were assessed. The etching was carried out at 75°C, similar to the alcohols with one hydroxyl group. During the etching, mechanical stirring was employed. In Fig. 4.50 results of measurements of the surface tension of water solutions and 3 M KOH solutions both containing diols are presented, and in Fig. 4.51 there are plots of the etch rate versus the diol concentration.

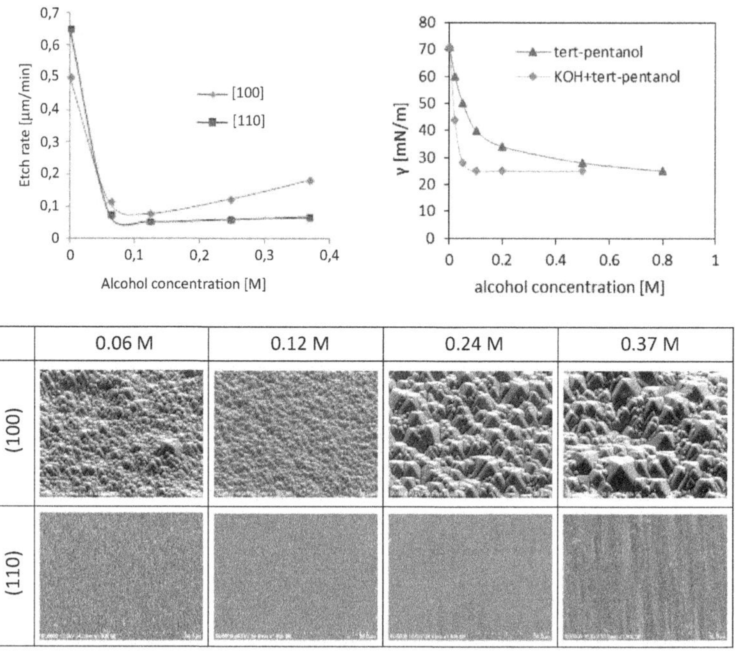

Figure 4.49 Etch rates, surface tensions (H_2O + alcohol and 3M KOH + alcohol), and SEM images of Si(100) and Si(110) substrates etched in 3 M KOH solutions for different concentrations of *tert*-pentanol [53].

The performed studies led us to the following conclusions:

- The surface tension of water solutions of the mentioned diols is higher than that of propanols and butanols (Tables 4.2 and 4.3) and decreases with the growth of concentration, similarly to alcohols (Fig. 4.50a) (however, as opposed to alcohols, the surface tension of diols approaches a constant value at concentration <0.5 M).

- The surface tension of KOH solutions containing diols decreases with the growth of alcohol concentration much more rapidly than that of KOH solutions containing propanol and butanol (Fig. 4.50b,c and Fig. 4.48) (at concentration <0.5 M it approaches a constant value).
- For all measured KOH + diol solutions (Fig. 4.50d) surface tension decreases rapidly with the increase of diol concentration. The faster the surface tension decreases, the stronger the surfactant is. Thus, diols are more efficient tensioactive compounds than alcohols with one hydroxyl group.

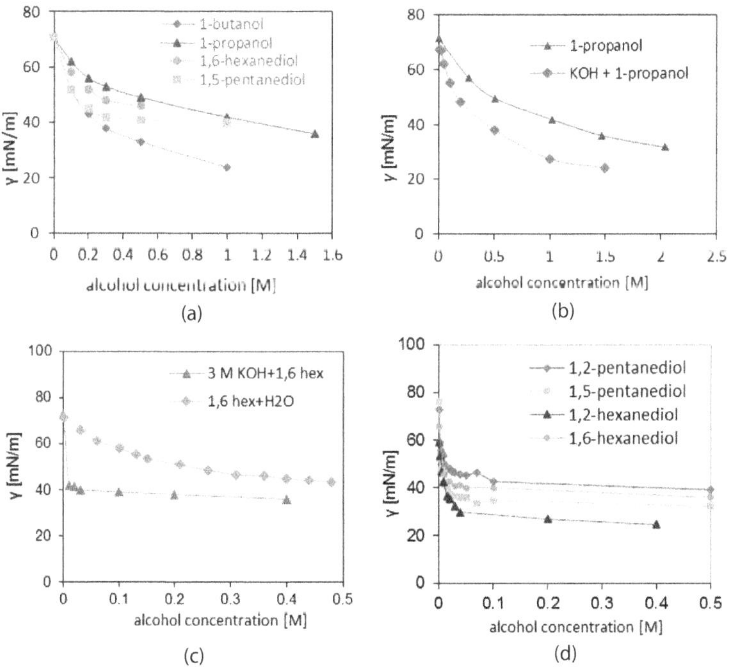

Figure 4.50 Surface tension of water solutions and 3 M KOH solutions both containing surface-active agents: (a) alcohols and diols, (b) alcohol and 3 M KOH + alcohol, (c) diol and 3 M KOH + diol, and (d) 3 M KOH + diols [56].

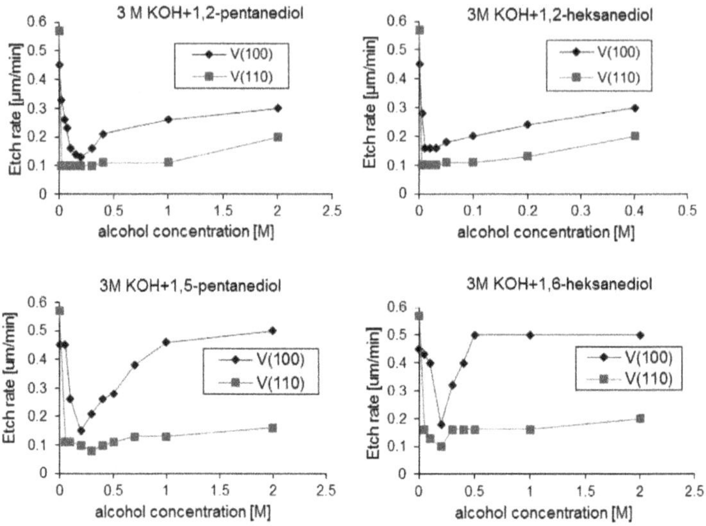

Figure 4.51 Etch rates of Si(100) and Si(110) planes etched in a 3 M KOH solution containing diol additions [56].

A comparison of SEM images of surfaces etched in solutions containing diols shows that they do not differ much among each other. As an example, a set of images obtained for the surfaces etched in a solution containing 1,6-hexanediol is presented (Fig. 4.52). The characteristic elements concerning $R_{\min}$ (smooth (110) and textured (100) surfaces) have been common for all diols. For this reason, in Fig. 4.53, only images of surfaces obtained at the concentrations at which $R_{\min}$ for each solution was achieved have been shown. They include (110) surfaces with the lowest roughness and corresponding textured (100) surfaces.

From the analysis of etch rates versus diol concentrations it follows that:

- The minimum of etch rate is better recognizable than for alcohols with one hydroxyl group.
- For 1,5 pentanediol and 1,6-hexanediol the changes in etch rates around $R_{\min}$ are very intense and the ratio $V(100)/V(110) > 2$.
- With an increase in diol concentration a significant improvement in Si(100) roughness is observed.

- Similar to alcohols with one hydroxyl group, the connection between surface roughness of Si(110) and the minimum of etch rate is observed.

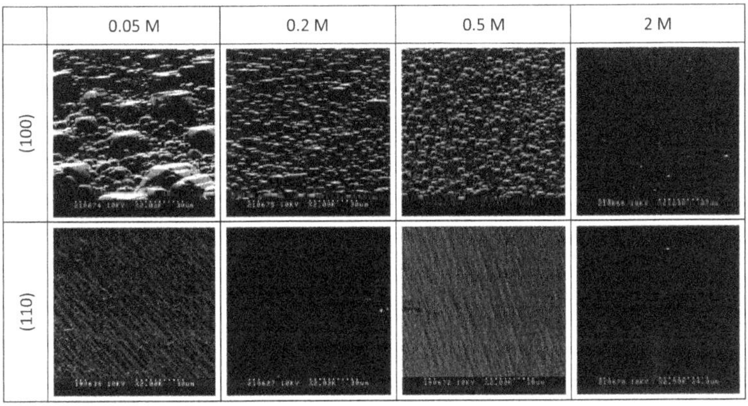

Figure 4.52 Morphologies (SEM images) of (100) and (110) substrates etched in a 3 M KOH solution containing 1,6-hexanediol versus diol concentration.

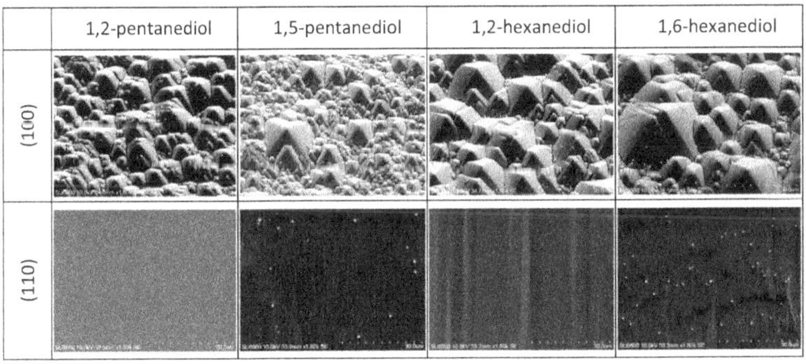

Figure 4.53 SEM images of (100) and (110) substrates etched in 3 M KOH solutions containing diol at a concentration corresponding to R_{min}.

The presented results concerning different alcohols are very consistent, though not quite obvious. From the experiments it follows that hillocks on the (100) surface disappear at a high alcohol concentration in a solution containing alcohol with small molecules (such as IPA) or diols exhibiting high solubility in the solution. However, in solutions containing alcohol of poor solubility, the hillocks are stable even at high alcohol concentrations, since the alcohol addition does not lead to increasing its concentration but to oversaturation of the solution. The minimum of the etch rate, which obviously occurs in the case of all the alcohols, is correlated with the maximum drop of surface tension (i.e., surface excess Γ, or maximum density of adsorption layer). This suggests that the minimum of the etch rate of the (110) surface is connected with the maximum adsorption of alcohol molecules, which makes that at the minimum etch rate, the roughness of the (110) surface is the lowest. But why is the (100) surface at the highest alcohol adsorption (the lowest etch rate) covered by hillocks? Why must the alcohol concentration be increased so as to obtain a good-quality (100) surface? An attempt to answer these questions on the basis of the literature considerations concerning interaction between water and alcohol will be discussed in Section 4.2.2.3.

Solutions with the composition corresponding to R_{min} can be used for obtaining smooth {110} surfaces, but due to the low etch rate and large roughness of the (100) surface, the solutions are not appropriate for deep etching of MEMS structures. However, the solutions, especially diols with a high boiling point, can be successfully used for texturing of Si(100) substrates for solar cell applications. Using of surface-active compounds for texturing of Si surfaces will be discussed in Section 4.2.2.2.

4.2.2.2 Using unsaturated solutions for texturing Si(100) substrates

Texturing of Si(100) substrates for solar cell applications is a particular case of using alcohols for etching in unsaturated solutions. Solutions used for texturing described in the literature are typically solutions with low concentrations of hydroxide and alcohol [57–59]. Usually, these are 1–3 M KOH solutions containing some wt% of alcohol from the propanols or butanols group. The concentration of the alcohols expressed in moles ranges from 0.1 to 1 M. It is the range of alcohol concentrations where the etch rate of the (100) plane approaches its minimum value and, independently of the kind of used alcohol, the Si(100) surface is effectively textured and covered by pyramids.

If a low concentration (below saturation) of alcohols is used there is a danger of their rapid evaporation (boiling point of IPA is 82°C) as the etching process is conducted at high temperature (70°C–80°C). Replacing alcohols by diols is an ideal solution to this problem, thanks to the high boiling point of diols, which highly exceeds the temperature of the process, which guarantees a stable composition of the etching solution during the whole process (Table 4.3).

Examples of substrates textured in KOH solutions containing alcohols with one or more hydroxyl groups are shown in Figs. 4.54–4.57.

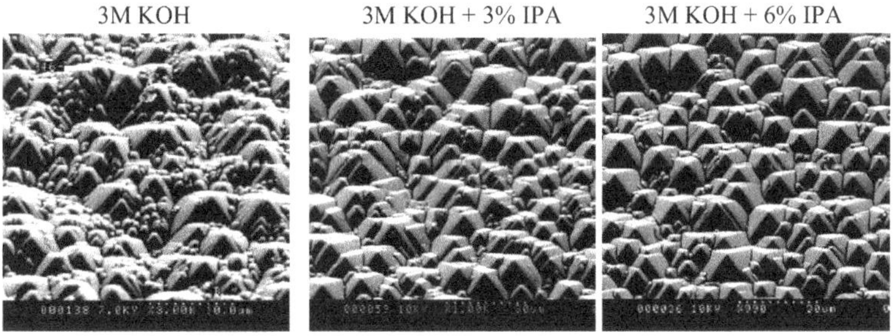

Figure 4.54 SEM images of (100) substrates textured in KOH solutions containing isopropyl alcohol (etch time 60 min, temp. 75°C, mechanical stirring).

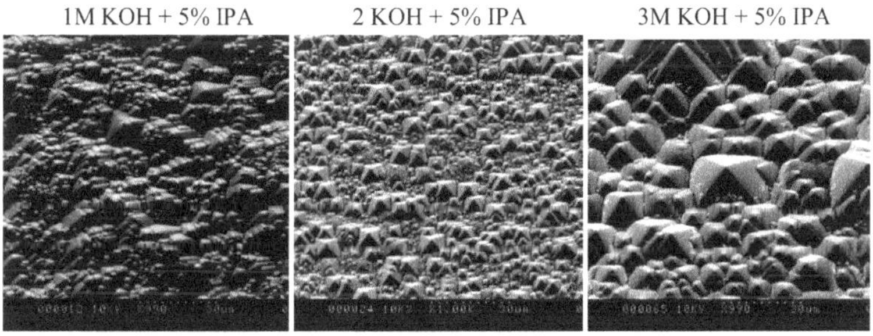

Figure 4.55 SEM images of (100) substrates textured in KOH solutions containing isopropyl alcohol (etch time 15–30 min, temp. 75°C, without stirring).

The presented results of etching in KOH solutions containing alcohols with one hydroxyl group come from different processes which were not optimized. In Figs. 4.54–4.56, general trends following from various process parameters are shown. There are many laboratories in the world which perform the process, and each of them gives its own optimal conditions (or keeps them in secret). In our laboratory, we have optimized texturing of (100) substrates in KOH solutions containing diols. As a basis, we used the results of etching at the minimum etch rate, described in the preceding section. On the basis of a preliminary selection among the four studied solutions, a KOH solution containing 1,2-pentanediol (Fig. 4.52) was chosen. The best texturing results were achieved with 1 M KOH + 1,2-pentamediol.

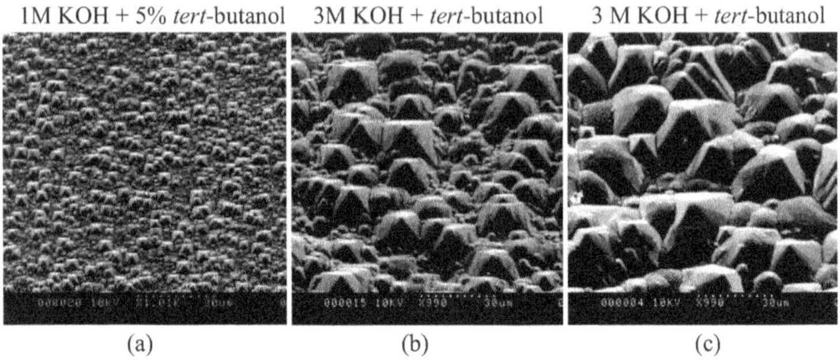

Figure 4.56 SEM images of (100) substrates textured in KOH + *tert*-butanol solutions: (a) nonsaturated, without stirring, (b) saturated, without stirring, and (c) saturated, with stirring (etch time 15–30 min, temp. 75°C). Reprinted from Ref. [60], Copyright (2012), with permission from Elsevier.

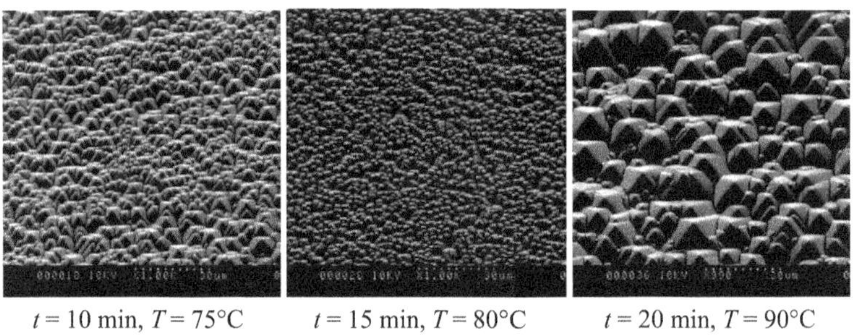

Figure 4.57 SEM images of (100) substrates textured in 1M KOH + 2% 1,2-pentanediol. Reprinted from Ref. [60], Copyright (2012), with permission from Elsevier.

The results of texturization have been analyzed in terms of:

- Diol concentration (1%, 2%, 4%)
- Process duration (10 min, 15 min, 20 min)
- Process temperature (70°C, 80°C, 90°C)

Results of SEM observations of some substrates selected after etching are shown in Fig. 4.57. Optimum conditions were chosen on the basis of spectral measurements of the coefficient of light reflectance, which in the studied wavelength range approached the value of 10%. Deposition of an antireflection coating (Si_3N_4) significantly improved this coefficient. Detailed information can be found in Ref. [60]. In Figs. 4.58 and 4.59 the reflectance versus process parameters is shown.

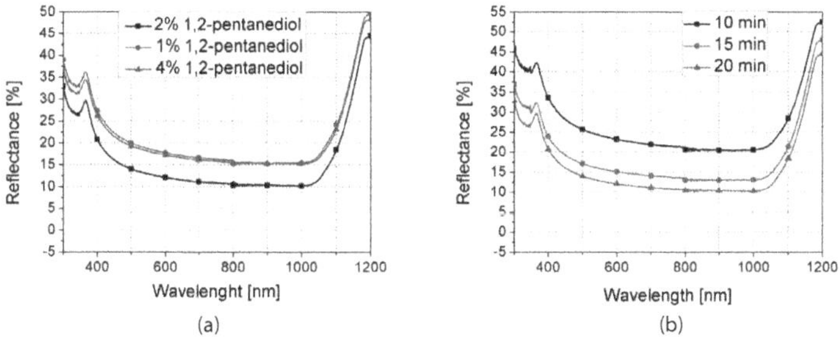

Figure 4.58 Coefficient of light reflectance of Si(100) substrates etched in 1 M KOH containing 1,2-pentanediol, depending on (a) concentration of 1,2-pentanediol and (b) process duration. Reprinted from Ref. [60], Copyright (2012), with permission from Elsevier.

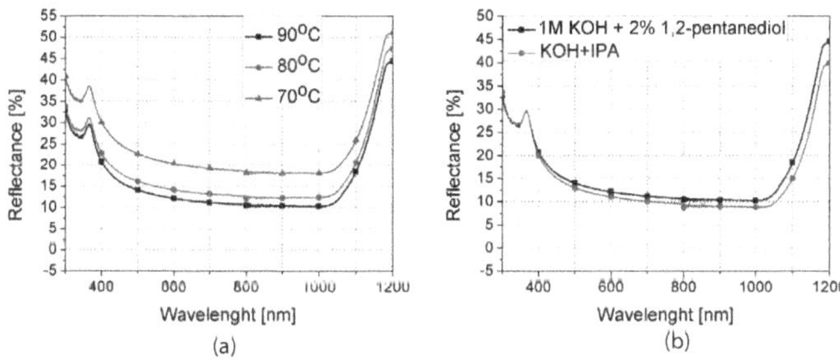

Figure 4.59 Coefficient of light reflectance of Si(100) substrates etched in 1 M KOH + 2% 1,2-pentanediol, depending on (a) process temperature and (b) comparison of obtained results with commercial ones etched in a KOH + IPA solution (Fraunhofer Institute for Solar Energy System). Reprinted from Ref. [60], Copyright (2012), with permission from Elsevier.

It can be seen that the texturing with the use of 1,2-pentanediol provides parameters which are not worse than the parameters obtained with the use of a commercial solution containing IPA. As an optimal solution, 1 M KOH + 2% 1,2-pentanediol has been selected. The process duration was 20 min at a temperature of 80°C–90°C. Thanks to using diol, the time can be shortened by increasing the process temperature. Composition of the solution is stable in contrast to alcohols containing solutions; it does not require frequent control and supplementing the tensioactive compound. Employing diols instead of propanols is thus beneficial from the technological, economical, and ecological points of view.

Pyramids covering the surface can not only serve for substrate texturing in order to reduce the light reflectance coefficient, the pyramids are also are island structures resembling the structures obtained in KOH solutions saturated with alcohols, as shown in Fig. 4.28 [37]. By appropriate choosing of the island sizes and distances between them as well as the process duration, it is possible to obtain an ideal structure covered by pyramids which can be used in numerous applications (Fig. 4.60).

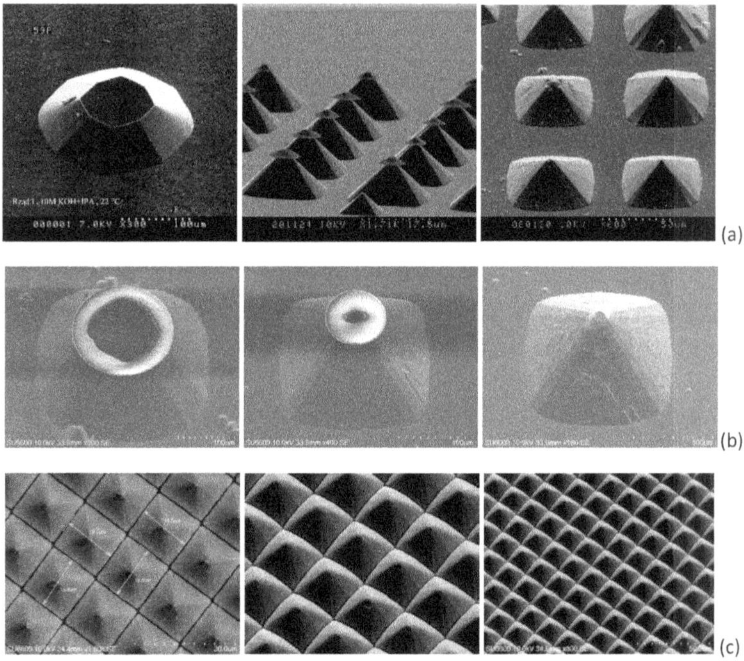

Figure 4.60 Islands etched through a circular mask in a 3 M KOH solution saturated with IPA: (a, b) successive etching stages and (c) surface covered with pyramids.

4.2.2.3 Analysis of the etching mechanism in KOH solutions containing alcohols

The phenomena at the interface solution (adsorption layer)–silicon substrate play an essential role in the process of silicon etching and are responsible for the relationship between etch rate and alcohol concentration. The run of the plots displaying the minimum may be surprising, taking into account the fact that the factor causing the changes is the alcohol, which in fact does not participate in the etching reaction.

In the literature we have not found any description of the phenomena occurring in a KOH–alcohol solution, while a lot of reports (also the newest ones) are related to the processes taking place in alcohol–water solutions [61–65]. The phenomena occurring in such solutions are very complex and, in spite of the unlimited solubility of light alcohols in water, depend heavily on the concentration of dissolved alcohol. The authors of those papers usually consider three ranges of alcohol concentration in which different phenomena connected with alcohol–water interactions occur. In terms of silicon etching, the range with a very high alcohol concentration is not very much interesting since such concentrations are impossible to achieve in KOH–alcohol solutions due to limited solubility of alcohols in KOH. Phenomena occurring at low alcohol concentrations are ascribed to strong hydrogen bonds in water, which cause the high surface tension of water (ca. 72 mN/m). The continuous hydrogen bond network impedes introduction of any molecules to the bulk as this could disturb the internal order. Alcohol molecules, introduced in a small amount to water, are pushed to the surface, which reduces the surface tension of the solution, whereas the remaining ones are surrounded by water molecules, forming around them cages, inside which an alcohol molecule is trapped. The process is known as **hydrophobic hydration**. It is continued until a dense monolayer of the alcohol is formed on the surface. The presence of this layer is involved with the maximum concentration of alcohol on the surface and is related to maximum of the Γ function, determined for the water–alcohol solution (Fig. 4.61, I). Successive alcohol molecules added to the solution cannot be situated on the surface; therefore they pull together, forming large aggregates with the alcohol molecules gathered in the bulk solution (so-called **micellar aggregates**). A reconstruction of the internal water structure takes place (hydrogen bonds are broken, percolation of water disappears), which consists of forming cages composed of water molecules around the multimolecular alcohol aggregates and releasing a part of water which has been previously used for trapping the single alcohol molecules. The released molecules may trap the following alcohol molecules situated at the surface (Fig. 4.61, II). This leads to vanishing of the surface adsorption layer. Therefore, the surface tension of the water–alcohol solution decreases with raising alcohol concentration and hydrogen bonds are weakened. The reason causing gathering of alcohol on the solution surface disappears. The authors of the publications connect this with a moment at which the surface excess function Γ starts decreasing and claim that at higher alcohol concentrations the surface adsorption layer completely disappears [62] (Fig. 4.61, III).

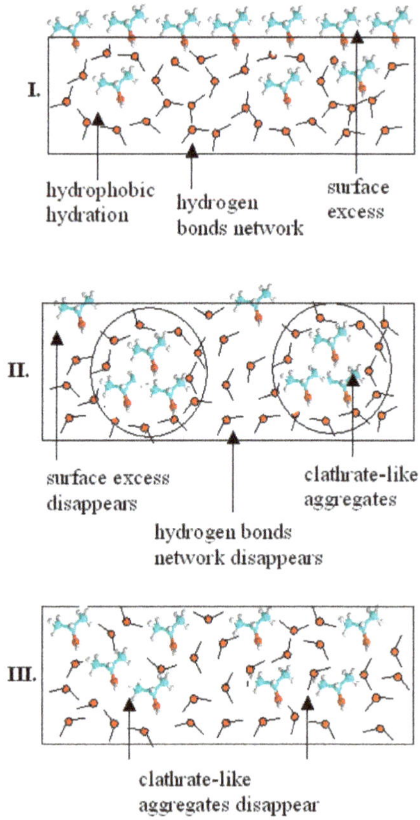

Figure 4.61 Model of phenomena occurring during increasing of alcohol concentration in water: (I) formation of adsorption layer and hydrophobic hydration, (II) disappearance of the adsorption layer and formation of clathrate-like aggregates, and (III) breaking of the water network structure. Reprinted from Ref. [51], Copyright (2011), with permission from Elsevier.

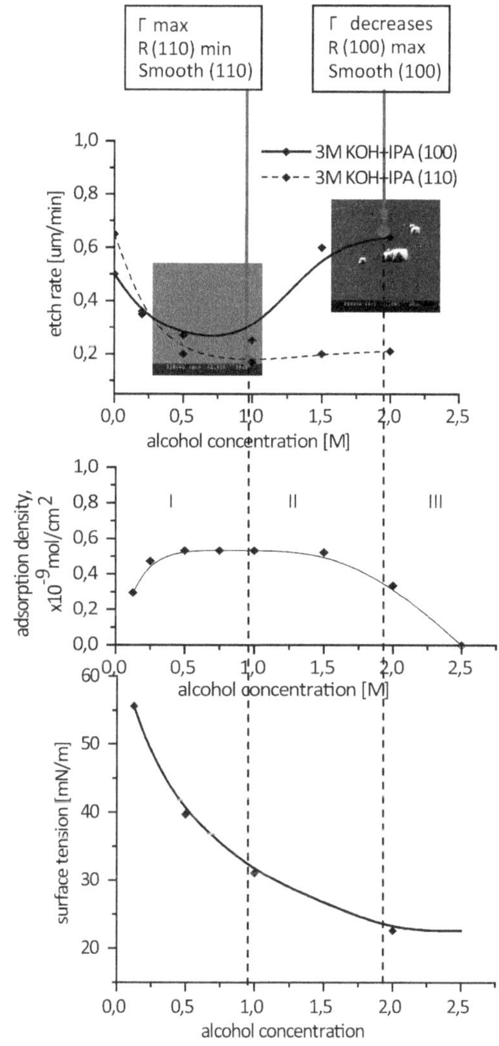

Figure 4.62 Correlation between etch rates of Si(110) and Si(100) in a 3 M KOH + IPA solution, density of adsorption layer and surface tension (ranges I, II, and III correspond to those in Fig. 4.61). Reprinted from Ref. [51], Copyright (2011), with permission from Elsevier.

On the basis of these considerations, supported by the results published in the literature, some phenomena observed during silicon etching in the solutions with systematically increasing alcohol concentration can be explained (Fig. 4.62). An alcohol concentration corresponding to the minimum etch rate depends on the KOH concentration in the solution. The higher the KOH concentration, the lower the alcohol concentration necessary to achieve the minimum etch rate (Fig. 4.46). It is also seen that the minimum etch rate and the maximum value of the Γ function occur at the same alcohol concentration for a particular concentration of KOH in the solution. Thus, there is a connection between the minimum etch rate and the maximum density of the adsorption layer, corresponding to a maximum of Γ.

The alcohol monolayer formed on the surface efficiently blocks the access of etching agents to the surface, which results in decreasing of etch rates of both the planes. The initial rapid drop of the etch rate may also be connected with networking of the water structure (hydrophobic hydration) around alcohol molecules, causing the water concentration in the solution to decrease dramatically, and the water molecules cannot participate actively in the etching process. At alcohol concentrations corresponding to etch rates above R_{min} (maximum Γ function), the adsorption layer disappears and part of the water is released, which results in an increase of etch rates, in particular $R(100)$. Why then does the etch rate of the (110) plane not increase? Some explanation can be found in the works [66–68] by K. Sato and M. Gosalvez et al. on the adsorption of large surfactant molecules (e.g., Triton) on (100)- and (110)-oriented Si substrates. According to the authors, the (110) surface, being more hydrophobic, shows a stronger tendency to lasting adsorption of this surfactant than the (100) one. The results of etching in KOH solutions containing alcohols suggest that a similar phenomenon takes place also in the case of other molecules of tensioactive compounds, smaller than Triton. The hydrophobic part of an alcohol molecule may adsorb more easily on the (110) surface than on the (100) one, analogously to Triton. Unfortunately, we are unable to confirm this phenomenon for alcohols by measurement of the contact angle on those substrates (in Fig. 4.44e the difference between the contact angle of (100)- and (110)-oriented silicon substrates is almost negligible) but these measurements showed a small difference between the contact angle of (100)- and (110)-oriented silicon substrates for the case of Triton (Section 4.2.3). It is therefore probable that alcohol molecules remain hydrophobic on the (110) surface with dispersion forces, even at large IPA concentrations, when the hydrogen bond network disappears, and contribute to the still low etch rate of the (110) plane. The phenomena have also implications for the roughness of etched (100) and (110) substrates (Fig. 4.46). The (110) surfaces with the lowest roughness have been obtained in all studied solutions (independent of KOH concentration) at the alcohol concentration corresponding to the lowest etch rate, that is, at the highest alcohol concentration, Γ on the surface. At lower alcohol concentrations, the surface is remarkably worse; at higher ones, when the nanolayer resulting from high surface tension of water disappears, the roughness deteriorates only slightly. On the other hand, a significant improvement in the roughness of the (100) surface is observed when the alcohol concentration exceeds the value corresponding to the minimum etch rate. It is associated with a remarkable increase in the etch rate of the (100) plane. It corresponds to the alcohol concentration range where the solution starts to saturate. Micellar aggregates composed of alcohol molecules that are then formed may be the reason of developing micromasks on the etched surface. Further growth of the alcohol concentration results in vanishing of the hydrogen bond network in water, breaking of the aggregates, and improving of the Si(100) surface roughness. The increase in alcohol concentration, especially in the case of alcohols with larger molecules, does not necessarily result in vanishing of hillocks. Probably the concentration of those alcohols, even in the solution close to saturation, is too low due to poor solubility of those alcohols (e.g., butanol) to change the water structure as in the case of lighter alcohols.

The presented model explains some phenomena observed during the etching of Si in KOH solutions containing alcohol below the saturation level. It connects the etch rate and roughness of etched surfaces with the properties of etching solutions containing alcohols, which have been directly measured in the laboratory (Y, Γ, θ) and with the properties described in the literature (hydrophobic hydration, aggregation, hydrogen bond network). The presented model does not explain definitely the anisotropy of silicon etching in KOH solutions containing alcohols. It is also not quite consistent with the description of the phenomena occurring during the etching in solutions of TMAH, TMAH containing typical surfactants (like Triton), or TMAH containing surfactants along with alcohols [69]. To enhance the studies on silicon etching in KOH solutions modified by tensioactive compounds, the results of etching in KOH solutions containing Triton will be presented in the next section.

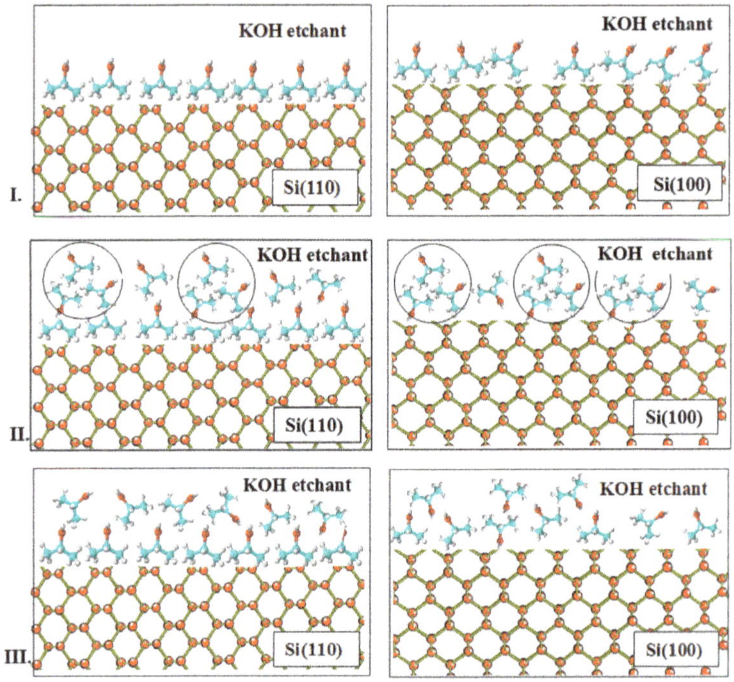

Figure 4.63 A model of phenomena occurring on Si(110) and Si(100) surfaces immersed in etchants with different alcohol concentrations (stages I, II, and III are analogous to those in Fig. 4.61). (I) Low concentration of alcohol—the alcohol layer is adsorbed with its hydrophobic part on silicon surfaces. (II) Higher alcohol concentration (adsorption layer disappears)—the more hydrophobic (110) surface adsorbs the alcohol molecules from the solution; the (100) surface is attacked by micellar aggregates formed in the solution. (III) Further increase in alcohol concentration—the (110) surface covered with adsorption layer; the (100) surface exposed to solution. Reprinted from Ref. [51], Copyright (2011), with permission from Elsevier.

4.2.3 Etching in KOH Solutions Containing Surfactants

As additions modifying silicon etching in organic solutions, such as TMAH, typical surfactants are used. The most popular is Triton X-100, first reported as an etching addition in Ref. [70]. It is a nonionic surfactant with the formula $C_{14}H_{21}(OC_2H_4)_nOH$, consisting of a hydrophobic part and a polyoxyethylene group with n = 9–10 (Fig. 4.64.) The Triton molecule is large in comparison to formerly discussed alcohols. Its molar mass is 646 g/mol, so it is about six times heavier than pentanediol. The boiling point of Triton is 270°C (similar to diols) and its surface tension is 33 mN/m (for a 1000 ppm concentration).

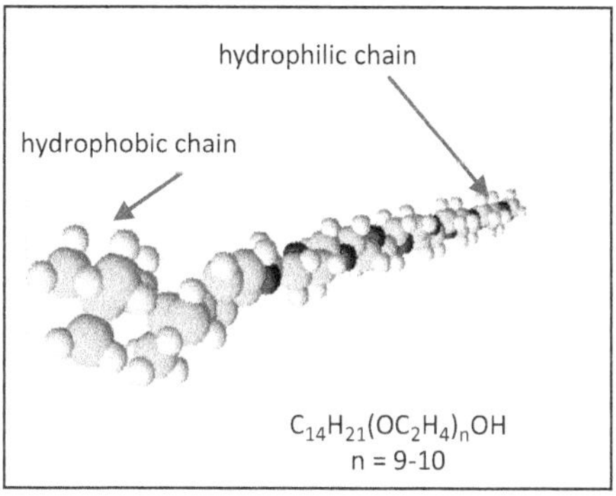

Figure 4.64 A model of the Triton X-100 molecule.

Addition of the Triton X-100 surfactant to a water solution of potassium hydroxide results in lowering its surface tension (Fig. 4.65). The solubility of Triton in KOH is lower than in water, so the surface tension of a KOH solution containing Triton decreases more rapidly with Triton concentration than the surface tension of water + Triton solutions (the behavior resembles that of alcohols). A remarkable reduction of a critical molecular concentration (CMC), from ca. 150 ppm to ca. 30 ppm, connected with the presence of KOH in the solution is observed. Due to the large size of the molecule, its structure and low concentration in the solution (at a level of parts per million), the effect of Triton on the etching process is different than that of alcohols.

It has been stated that the Triton X-100 surfactant results in wetting improvement of silicon surfaces. Similar to alcohols, the contact angle decreases with raising the Triton concentration and slightly changes in time (Fig. 4.66). The difference in contact angles of (100) and (110) substrates is more remarkable than in the case of alcohols, but it does not exceed 5°. So, the measurements of contact angle do not confirm unambiguously the anisotropic character of Triton adsorption on these substrates. However, in all cases the contact angle of the (110) surface is lower than that of the (100) one, both for water and for water solutions of tensioactive compounds (alcohols and surfactants).

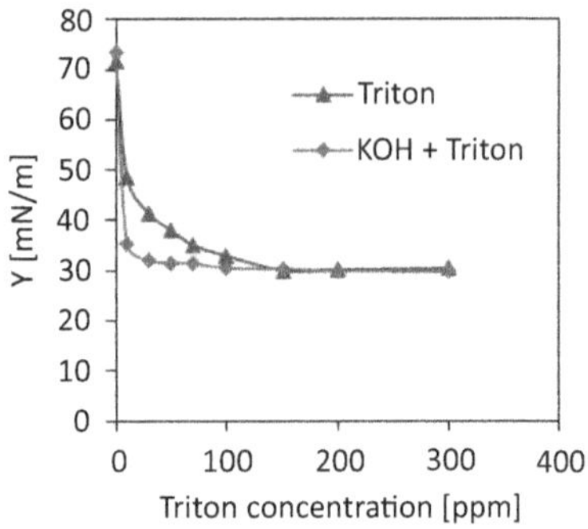

Figure 4.65 Surface tension of water and 2 M KOH solutions containing Triton versus concentration of Triton X-100 surfactant.

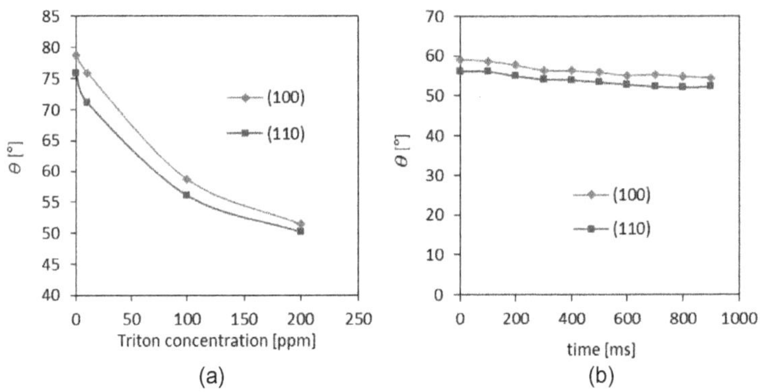

Figure 4.66 Contact angles of a water solution of the Triton X-100 surfactant on Si(100) and Si(110) substrates (a) versus surfactant concentration and (b) versus time (for a Triton concentration of 100 ppm, dynamic measurements) [71].

A decrease in contact angle with the time of residence of a surfactant drop on the silicon surface can be observed (Fig. 4.67). It means that the surfactant molecules interact with the silicon surface and the higher the concentration of Triton, the better the wetting of the surface by the solution. (When the surfactant concentration increases, the contact angle decreases).

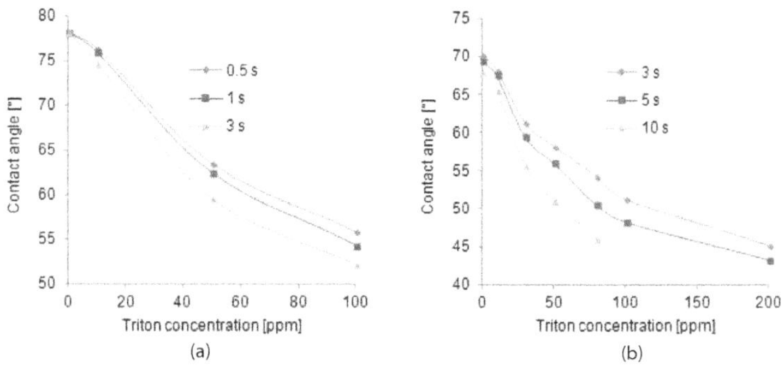

Figure 4.67 Contact angle of (a) aqueous solution of Triton and (b) 2 M KOH aqueous solution containing Triton on the Si(100) surface versus surfactant concentration for different periods of time [71].

The Triton X-100 surfactant has not been earlier used as an addition to the KOH solution in the anisotropic etching process. Preliminary studies of a 3 M KOH solution containing Triton addition have been performed using a similar procedure as in the case of alcohols.

It has been stated that addition of the Triton X-100 surfactant to a 3 M KOH solution in the amount of 10 ppm significantly reduces etch rates of (100) and (110) planes (Table 4.4). The reduction is significantly larger for the (110) plane, which results in the value of the anisotropy ratio $R(100)/R(110) > 1$. This causes a change of the shapes of spatial structures etched in Si(100). The convex corner undercut is reduced (Fig. 4.68, upper row), and in <100> and <310> crystallographic directions, the walls {110} and {331}, inclined to the substrate at low angles, develop (Fig. 4.68, lower row). Such walls have been also observed after etching in KOH solutions with alcohol additions (Fig. 4.28). A significant improvement in (100) and (110) surface roughness has been observed after addition of a small amount of the surfactant (10 ppm) as well (Fig. 4.69). However, on the (100) surface some solitary hillocks develop; therefore in the following part of this chapter, the studies aiming at roughness reduction of this surface by, for instance, increasing surfactant concentration, will be presented. The roughness of {110} surfaces etched in this solution seems to be satisfactory, for example, the roughness of {110} planes etched in a KOH + Triton solution, developing as the sidewalls of structures etched in the (100) substrate, is lower than the roughness of corresponding walls etched in KOH + IPA (Fig. 4.70). It is a great advantage of Triton-containing solutions.

Figure 4.68 SEM images of convex corners and islands etched in Si(100) in a 3 M KOH solution and a 3 M KOH solution + 10 ppm Triton X-100 surfactant. © [2013] IEEE. Reprinted, with permission, from Ref. [72].

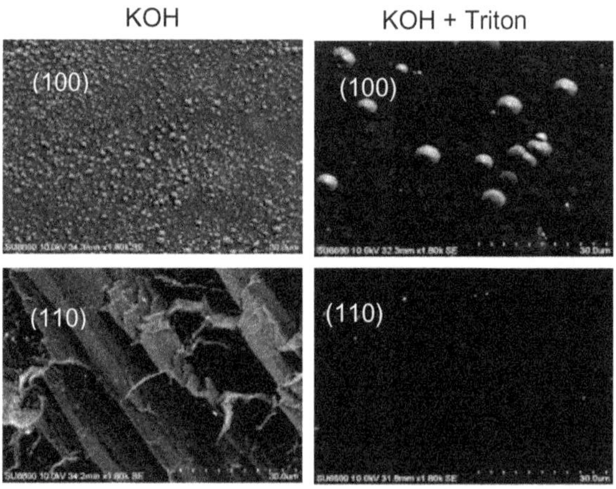

Figure 4.69 SEM images of Si surfaces (110) and (100) etched in a 3 M KOH solution without additions and a 3 M KOH solution + 10 ppm Triton X-100 surfactant. © [2013] IEEE. Reprinted, with permission, from Ref. [72].

Figure 4.70 SEM images of Si(100) surfaces etched in a 3 M KOH solution (a) saturated with IPA and (b) containing 10 ppm Triton X-100 surfactant. © [2013] IEEE. Reprinted, with permission, from Ref. [72].

The obtained results show that modification of KOH solutions by Triton gives positive results. A disadvantage is the low etch rate of the (100) plane, though the ratio $R(100)/R(110)$ is appropriate.

Table 4.4 Etch rates of Si(100) and Si(110) in 3 M KOH and 3 M KOH containing 10 ppm of Triton X-100

	$R(100)$ (μm/min)	$R(110)$ (μm/min)	$R(100)/R(110)$
3 M KOH	0.5	0.65	0.77
3 M KOH + Triton X-100	0.1	0.06	1.61

Increasing KOH concentration above 3 M, though resulting in enhancing etch rates of both (100) and (110) planes (Fig. 4.71), significantly deteriorates the roughness of etched surfaces. On the (100) planes the number of hillocks dramatically increases, and on the {110} sidewalls a striped structure develops (Fig. 4.72). For 4 M and 5 M KOH the measurement results are loaded with a considerable error, which is marked in the plots.

Surfaces with the lowest roughness were obtained at low KOH concentrations (2–3 M), and in such solutions the concentration of Triton affected the roughness and etch rate of the etched surfaces only very slightly. The plots of etch rates of (100) and (110) planes and anisotropy ratio $R(100)/R(110)$ versus Triton concentration are shown in Fig. 4.73.

It is seen that the addition of 1 ppm of the surfactant to a 2 M KOH solution results in considerable reduction of etch rates of (100) and

(110) planes and ensures an anisotropy ratio $R(100)/R(110) = 4$. A further increase in the Triton concentration does not affect much the etch rates of both the planes and the anisotropy ratio, which in the whole studied range of Triton concentration is ca. 2.

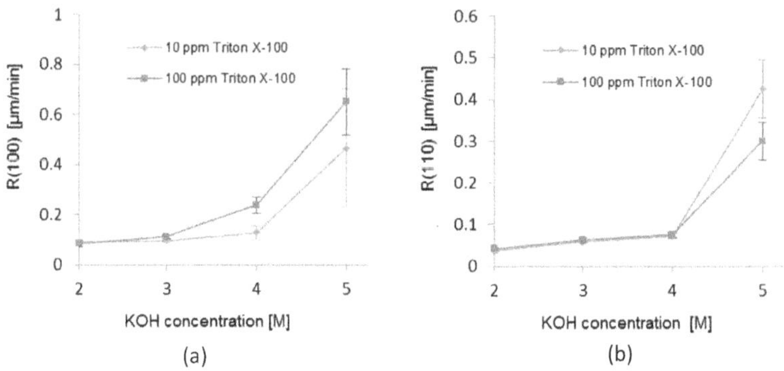

Figure 4.71 Etch rates of (a) (100) and (b) (110) planes etched in KOH solutions containing Triton X-100 surfactant [72].

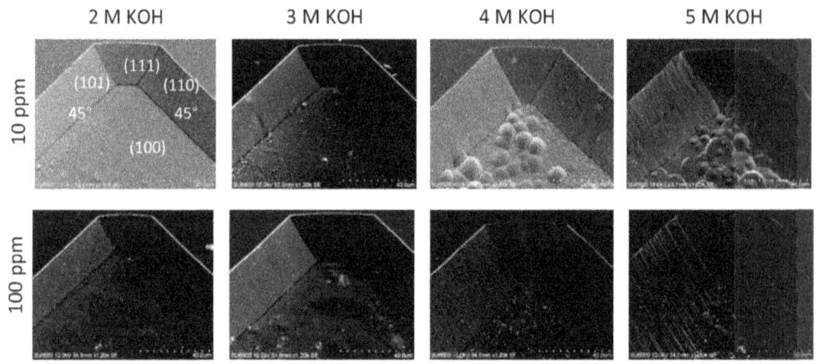

Figure 4.72 SEM images of fragments of concave structures bounded by {110} planes inclined at an angle of 45° to the (100) substrate, etched in KOH solutions (2–5 M) containing 10 and 100 ppm Triton X-100 [72].

Addition of 1 ppm of the surfactant to a 2 M KOH solution has not ensured appropriate roughness of both (100) and (110) surfaces (Fig. 4.74). However, for a Triton concentration above 10 ppm (in a range of 10–200 ppm) both the surfaces are smooth, independent of the Triton concentration. An objective measure of surface roughness is its roughness measured with an AFM (Fig. 4.75).

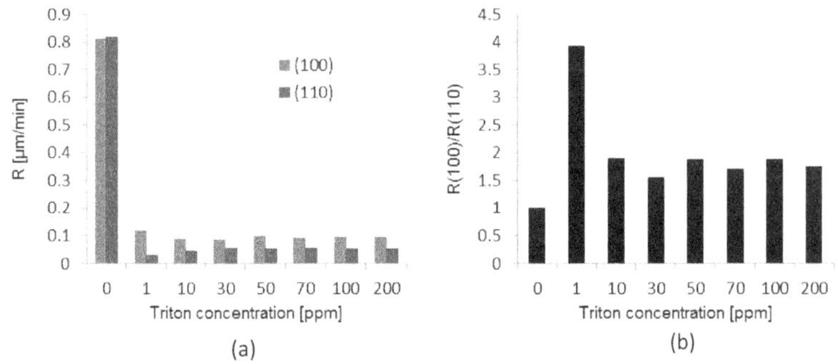

Figure 4.73 (a) Etch rates of (100) and (110) planes and (b) anisotropy ratio $R(100)/R(110)$ in a 2 M KOH solution containing the Triton X-100 surfactant in the concentration range of 0–200 ppm [71].

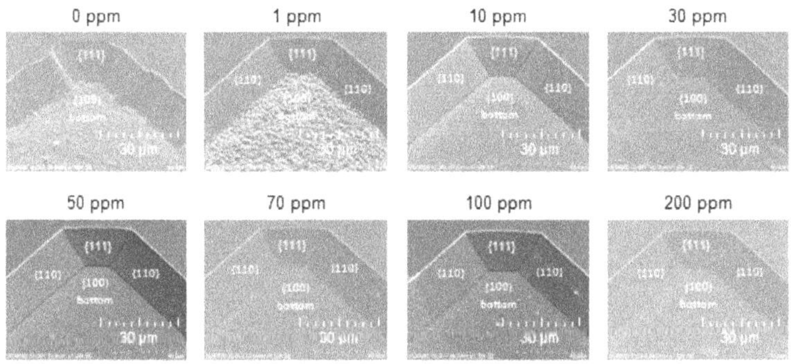

Figure 4.74 SEM images of fragments of concave structures bounded by {110} planes etched in (100) substrates in a 2 M KOH solution containing 10–200 ppm Triton X-100 [71].

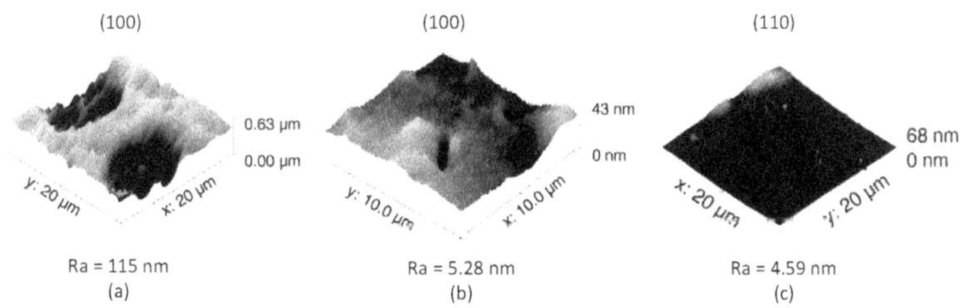

Figure 4.75 AFM images of Si(100 and Si(110) surfaces etched in (a) 2 M KOH solution + 1 ppm Triton X-100, (b) 2 M KOH solution + 50 ppm Triton X-100, and (c) 2 M KOH solution + 100 ppm Triton X-100 [71].

The average roughness of a surface etched in 2 M KOH + 1 ppm Triton is Ra ≅ 100 nm, and this result is not satisfactory, which is also confirmed by the SEM images. In a 2 M KOH solution containing Triton additions at the level of 10–200 ppm, the roughness of (100) and (110) surfaces is of a few nanometers. SEM images confirm that the surfaces are smooth and can be successfully used for MEMS applications. The structures were etched at a temperature of 75°C. It was also stated that stirring of the etching solution resulted in a roughness reduction of the etched surface, while an increase of temperature (up to 90°C) had a negative effect on the roughness of etched surfaces, though etch rates of both the planes increased.

Results obtained in KOH + Triton solutions resemble the results of etching in KOH solutions containing alcohols at the concentrations below the saturation level (corresponding to R_{min}). In both cases, after addition of the tensioactive compound, a rapid drop in etch rates of both (100) and (110) planes occurs and the best quality (the lowest roughness) of the (110) plane is observed. An advantage of KOH solutions containing Triton is low roughness of both (100) and (110) planes, while neither alcohols with one hydroxide group nor diols reveal such characteristics. The Si(100) surface with low roughness can be obtained only at high alcohol concentrations, but the quality of the (110) surface deteriorates remarkably at such conditions (Figs. 4.35 and 4.75). The low etch rate of the (100) plane can be compensated by an increase of etching process temperature. It is obvious that for potassium hydroxide solutions, various modifying additions can be applied, including not only alcohols with small molecules but also other tensioactive compounds with more carbon atoms in the molecule (from five, like pentanediol, to several tenths, like Triton). There are many possibilities which need further investigation.

References

1. Seidel, H., Csepregi, L., Heuberger, A., and Baumgartel, H. (1990a). Anisotropic etching of crystalline silicon in alkaline solutions I. Orientation dependence and behavior of passivation layers, *J. Electrochem. Soc.*, **137**(11), pp. 3612–3626.

2. Seidel, H., Csepregi, L., Heuberger, A., and Baumgartel, H. (1990b). Anisotropic etching of crystalline silicon in alkaline solutions II. Influence of dopants, *J. Electrochem. Soc.*, **137**(11), pp. 3626–3632.
3. Cornely, R. H., and Marcus, R. B. (1991). Formation of silicon micromirrors by anisotropic etching of silicon, *Sens. Actuators A*, **29**, pp. 241–250.
4. Mehregany, M., and Senturia, S. D. (1988). Anisotropic etching of silicon in hydrazine, *Sens. Actuators*, **13**, pp. 375–390.
5. Baum, T., and Schiffrin, D. J. (1997). Kinetic isotopic effects in the anisotropic etching of p-Si(100) in alkaline solutions, *J. Electroanal. Chem.*, **436**, pp. 239–244.
6. Tabata, O., Asahi, R., Funabashi, H., and Sugiyama, S. (1992). Anisotropic etching of silicon in TMAH solutions, *Sens. Actuators A*, **34**, pp. 51–57.
7. Palik, E. D., Bermudez, V. M., and Glembocki, O. J. (1985). Ellipsometric study of orientation-dependent etching of silicon in aqueous KOH, *J. Electrochem. Soc.*, **132**(4), pp. 871–884.
8. Wind, R. A., and Hines, M. A. (2000). Macroscopic etch anisotropies and microscopic reaction mechanisms: a micromachined structure for the rapid assay of etchant anisotropy, *Surf. Sci.*, **460**, pp. 21–38.
9. Ueda, T., Kohsaka, F., Ito, T., and Yamazaki, D. (1986). Theory to predict etching shapes in quartz crystal and its application to design devices, *Trans. Soc. Instrum. Control Eng.*, **23**(12), pp. 1233–1238 (in Japanese).
10. Weirauch, D. F. (1975). Correlation of the anisotropic etching of single-crystal silicon and spheres and wafers, *J. Appl. Phys.*, **46**(4), pp. 1478–1483.
11. Kendall, D. L., De Guel, G. R., and Torres-Jacome, A. (1982). The Wagon Wheel method applied around around the [011] zone of silicon, *Electrochem. Soc. Extend. Abstr.*, **82**(1), pp. 209–211.
12. Sato, K., Koide, A., and Tanaka, S. (1993). Anisotropic etching simulation system as a design tool for fabrication 3D microstructures in single crystal silicon, *Proc. 1st IFToMM Int. Symp. Micromechanism*, Tokyo, Japan, pp. 155–160.
13. Sato, K., Shikida, M., Matsushima, Y., Yamashiro, T., Asaumi, K., Iriye, Y., and Yamamoto, M. (1998). Characterization of orientation dependent etching characteristics of single crystal silicon: effects of KOH concentration, *Sens. Actuators A*, **64**(1), pp. 87–93.
14. Sato, K., Asaumi, K., Kobayashi, G., Iriye, Y., and Shikida, M. (2000). Development of an orientation-dependent anisotropic etching

simulation system MICROCAD, *Electron Commun. Jpn. Part 2*, **83**(4), pp. 13–22.

15. Sikida, M., Sato, K., Tokoro, K., and Uchikawa, D. (2000). Differences in anisotropic etching properties of KOH and TMAH solutions, *Sens. Actuators*, **80**, pp. 179–188.
16. Pal, P., Ashok, A., Haldar, S., Xing, Y., and Sato, K. (2015). Anisotropic etching in low-concentration KOH: effects of surfactant concentration, *Micro Nano Lett.*, **10**(4), pp. 224–228.
17. Jaccodine, R. J. (1962). Use of modified free energy theorem to predict equilibrium growing and etching shapes, *J. Appl. Phys.*, **33**(8), pp. 2643–2647.
18. Koide, A., Sato, K., Tanaka, S., and Kato, S. (1995). Orientation-dependent etch rate of single crystal silicon related to etching temperature, *J. Jpn. Soc. Precis. Eng.*, **61**(4), pp. 547–551 (in Japanese).
19. Tanaka, H., Yamashita, S., Abe, Y., Shikida, M., and Sato, K. (2004). Fast etching of silicon with a smooth surface in high temperature ranges near the boiling point of KOH solution, *Sens. Actuators A*, **114**, pp. 516–520.
20. Kloeck, B., Collins, S. D., de Rooji, N. F., and Smith, R. L. (1989). Study of electro-chemical etch-stop for high precision thickness control of silicon membranes, *IEEE Trans. Electron Devices*, **36**(4), pp. 663–669.
21. Sato, K., Shikida, M., Yamashiro, T., Tsunekawa, M., and Ito, S. (1999). Roughening of single-crystal silicon surface etched by KOH watersolution, *Sens. Actuators A*, **73**, pp. 122–130.
22. Allongue, P., Costa-Kieling, V., and Gericher, H. (1993). Etching of silicon in NaOH solutions, *J. Electrochem. Soc.*, **140**(4), pp. 1009–1026.
23. Elwenspoek, M. (1993). On the mechanism of anisotropic etching of silicon, *J. Electrochem. Soc.*, **140**(7), pp. 2075–2080.
24. Sato, K., Masuda, T., and Shikida, M. (2003). Difference in activated atomic steps on (111) silicon surface during KOH and TMAH etching, *Sens. Mater.*, **15**(2), pp. 93–99.
25. Sato, K., Shikida, M., Yamashiro, T., Asaumi, K., Iriye, Y., and Yamamoto, M. (1999). Anisotropic etching rats of single-crystal silicon for TMAH water solutions as a function of crystallographic orientation, *Sens. Actuators A*, **73**, pp. 131–137.
26. Shikida, M., Tokoro, K., Uchikawa, D., and Sato, K. (2000). Surface morphology of anisotropically etched single-crystal silicon, *J. Micromech. Microeng.*, **10**, pp. 522–527.

27. Zubel, I., and Kramkowska, M. (2004). Etch rates and morphology of silicon (hkl) surfaces etched in KOH and KOH saturated with isopropanol solutions, *Sens. Actuators A*,**115**, pp. 549–556.

28. Divan, R., Moldovan, N., and Camon, H. (1999). Roughning and smoothing dynamics during KOH silicon etching, *Sens. Actuators A*, **74**, pp. 18–23.

29. Yang, C.-R., Chen, P.-Y., Chiou, Y.-C., and Lee, R.-T. (2005). Effects of mechanical agitation and surfactant additive on silicon anisotropic etching in alkaline KOH solution, *Sens. Actuators A*,**119**, pp. 263–270.

30. Yang, C.-R., Chen, P.-Y., Yang, C.-H., Chiou, Y.-C., and Lee, R.-T. (2005). Effects of various ion-typed surfactants on silicon anisotropic etching properties in KOH and TMAH solutions, *Sens. Actuators A*, **119**, pp. 271–281.

31. Yang, C.-R., Yang, C.-H., and Chen, P.-Y. (2005). Study on anisotropic silicon etching characteristics in various surfactant-added tetramethyl ammonium hydroxide water solutions, *J. Micromech. Microeng.*, **15**, pp. 2028–2037.

32. Price, J. B. (1973). In *Semiconductor Silicon*, Huff, H. R., and Brugess, R. R., eds., *The Journal of Electrochemical Society Softbond, Proceeding Series* (Princeton, New York), pp. 339–351.

33. Zubel, I., and Barycka, I. (1998). Silicon anisotropic etching in alkaline solutions I. The geometric description of figures developed under etching Si (100) in various solutions, *Sens. Actuators A*, **70**, pp. 250–259.

34. Barycka, I., and Zubel, I. (1995). Silicon anisotropic etching in KOH-isopropanol etchant, *Sens. Actuators A*, **48**, pp. 229–238.

35. Zubel, I. (1998). Silicon anisotropic etching in alkaline solutions II: on the influence of anisotropy on the smoothness of etched surfaces, *Sens. Actuators A*, **70**, pp. 260–268.

36. Zubel, I. (2000). Silicon anisotropic etching in alkaline solutions III: on the possibility of spatial structures forming in the course of Si(100) anisotropic etching in KOH and KOH+IPA solutions, *Sens. Actuators A*, **84**, pp. 116–125.

37. Zubel, I., and Kramkowska, M. (2008). Development of etch hillocks on different Si(hkl) planes in silicon anisotropic etching, *Surf. Sci.*, **602**, pp. 1712–1721.

38. Gosálvez, M., Zubel, I., and Viinikka, E. (2010). In *Handbook of Silicon Based MEMS Materials and Technologies*, Lindroos, V., ed. (Elsevier, New York), pp. 375–407.

39. Gosalvez, M. A., and Sato, K. (2008). Analytical solution of the continuous cellular automaton for anisotropic etching, *J. Microelectromech. Syst.*, **17**, pp. 410–431.

40. Gosálvez, M. A., Sato, K., Foster, A. S., Nieminen, R. M., and Tanaka, H. (2007). An atomistic introduction to anisotropic etching, *J. Micromech. Microeng.*, **17**, pp. S1–S26.

41. Gosálvez, M. A., Xing, Y., Sato, K., and Nieminen, R. M. (2009). Discrete and continuous cellular automata for the simulation of propagating surfaces, *Sens. Actuators A*, **155**, pp. 98–112.

42. Zubel, I. (2003). The model of etching of (hkl) planes in monocrystalline silicon, *J. Electrochem. Soc.*, **150**, p. C391.

43. Rola, K. P., and Zubel, I. (2011). Investigation of Si(hkl) surfaces etched in KOH solutions saturated with tertiary-butyl alcohol, *J. Micromech. Microeng.*, **21**, p. 115026.

44. Zubel, I. (2001). The influence of atomic configuration of (hkl) planes on adsorption processes associated with anisotropic etching of silicon, *Sens. Actuators A*, **94**, pp. 76–86.

45. Zubel, I., Barycka, I., Kotowska, K., and Kramkowska, M. (2001). Silicon anisotropic etching in alkaline solutions IV, *Sens. Actuators A*, **87**, pp. 163–171.

46. Pal, P., Haldar, S., Singh, S. S., Ashok, A., Yan, X., and Sato, K. (2014). A detailed investigation and explanation of the appearance of different undercut profiles in KOH and TMAH, *J. Micromech. Microeng.*, **24**, p. 095026.

47. Chien, W. T. C., Chang, C. O., Lo, Y. C., Li, Z. W., and Chou, C. S. (2005). On the Miller-indices determination of Si{100} convex corner undercut planes, *J. Micromech. Microeng.*, **15**, pp. 833–842.

48. Zubel, I., and Kramkowska, M. (2002). The effect of alcohol additives on etching characteristics in KOH solutions, *Sens. Actuators A*, **101**, pp. 255–261.

49. Bäcklund, Y. (1997). Micromechanics in optical microsystems: with focus on telecom systems, *J. Micromech. Microeng.*, **7**, pp. 93–98.

50. Rola, K. P., Ptasiński, K., Zakrzewski, A., and Zubel, I. (2013). Silicon 45° micromirrors fabricated by etching in alkaline solutions with organic additives, *Microsyst. Technol.*, **20**, pp. 221–226.

51. Zubel, I., Rola, K., and Kramkowska, M. (2011). The effect of isopropyl alcohol concentration on the etching process of Si-substrates in KOH solutions, *Sens. Actuators A*, **171**, pp. 436–445.

52. Rola, K. P., and Zubel, I. (2011). In *2011 International Students and Young Scientists Workshop "Photonics and Microsystems,"* Cottbus, pp. 110–114.

53. Rola, K., and Zubel, I. (2014). In *Proceedings of 2013 International Students and Young Scientists Workshop "Photonics and Microsystems,"* St. Marienthal, Germany, 12–14 July, 2013 (Oficyna Wydawnicza Politechniki Wrocławskiej, Wrocław), pp. 89–93.

54. Rola, K. P., and Zubel, I. (2012). Impact of alcohol additives concentration on etch rate and surface morphology of (100) and (110) Si substrates etched in KOH solutions, *Microsyst. Technol.*, **19**, pp. 635–643.

55. Rola, K. P., and Zubel, I. (2016). Effect of molecular structure of alcohols on wet anisotropic etching of silicon, *Sens. Actuators A*, **242**, pp. 18–26.

56. Zubel, I., Rola, K., and Zalewska, J. (2013). In *Proceedings of SPIE: The International Society for Optical Engineering*, Szczepanski, P., Kisiel, R., and Romaniuk, R. S., eds., Vol. 8902, p. 89022I.

57. Fulong, Y., Yongfeng, G., Yingchun, L., Yongda, Y., Honggang, F., Kai, C., and Xichun, L. (2004). Micro-fabrication of crystalline silicon by controlled alkali etching, *J. Mater. Proc. Technol.*, **149**, pp. 567–572 (2004).

58. Wu, C. J., Wei, P.-J., and Lin, J. F. (2009). Reflectivity of an etched silicon surface with pyramids: II. Experimental results from different etching conditions, *J. Micromech. Microeng.*, **19**, p. 125019.

59. Forniés, E., Zaldo, C., and Albella, J. M. (2005). Control of random texture of monocrystalline silicon cells by angle-resolved optical reflectance, *Sol. Energy Mater. Sol. Cells*, **87**, pp. 583–593.

60. Zubel, I., Granek, F., Rola, K., and Banaszczyk, K. (2012). Texturization of Si(100) substrates using tensioactive compounds, *Appl. Surf. Sci.*, **258**, pp. 9067–9072.

61. Yano, Y. F. (2002). Surface structure of aqueous 2-butoxyethanol mixtures studied by x-ray reflection, *J. Chem. Phys.*, **116**, pp. 8093–8096.

62. Yano, Y. F. (2005). Correlation between surface and bulk structures of alcohol-water mixtures. *J. Colloid Interface Sci.*, **284**, pp. 255–259.

63. Koga, Y., Nishikawa, K., and Westh, P. (2004). "Icebergs" or no "icebergs" in aqueous alcohols? Composition-dependent mixing schemes, *J. Phys. Chem. A*, **108**, pp. 3873–3877.

64. Koga, Y., Siu, W. W. Y., and Wong, T. Y. H. (1990). Excess partial molar free energies and entropies in aqueous tert-butyl alcohol solutions at 25°C, *J. Phys. Chem.*, **94**, pp. 7700–7706.

65. Wong, T. Y. H., Wong-Moon, K. C., Beach, L. J., Chuang, Y.-F., and Koga, Y. (1992). Excess partial molar free energies, enthalpies, and entropies in 2-butanone-H2O mixtures: solute-solute interactions, *J. Phys. Chem.*, **96**, pp. 10025–10029.

66. Gosálvez, M. A., Tang, B., Pal, P., Sato, K., Kimura, Y., and Ishibashi, K. (2009). Orientation- and concentration-dependent surfactant adsorption on silicon in aqueous alkaline solutions: explaining the changes in the etch rate, roughness and undercutting for MEMS applications, *J. Micromech. Microeng.*, **19**, p. 125011.

67. Tang, B., , Pal, P., Gosalvez, M. A., Shikida, M., Sato, K., Amakawa, H., and Itoh, S. (2009). Ellipsometry study of the adsorbed surfactant thickness on Si{110} and Si{100} and the effect of pre-adsorbed surfactant layer on etching characteristics in TMAH, *Sens. Actuators A*, **156**, pp. 334–341.

68. Cheng, D., Gosálvez, M. A., Hori, T., Sato, K., and Shikida, M. (2006). Improvement in smoothness of anisotropically etched silicon surfaces: effects of surfactant and TMAH concentrations, *Sens. Actuators A*, **125**, pp. 415–421.

69. Zubel, I., Kramkowska, M., and Rola, K. (2012). Silicon anisotropic etching in TMAH solutions containing alcohol and surfactant additives, *Sens. Actuators A*, **178**, pp. 126–135.

70. Resnik, D., Vrtacnik, D., Aljancic, U., Mozek, M., and Amon, S. (2005). The role of Triton surfactant in anisotropic etching of {110} reflective planes on (100) silicon, *J. Micromech. Microeng.*, **15**, pp. 1174–1183.

71. Rola, K. P., and Zubel, I. (2013). In *Proceedings of SPIE: The International Society for Optical Engineering*, Szczepanski, P., Kisiel, R., and Romaniuk, R. S., eds., Vol. 8902, p. 89021Y.

72. Rola, K. P., and Zubel, I. (2013). Triton surfactant as an additive to KOH silicon etchant, *J. Microelectromech. Syst.*, **22**, pp. 1373–1382.

Chapter 5

TMAH-Based Anisotropic Etching

5.1 Introduction

The selection of the etchant depends on various factors such as undercutting at mask edges and corners, sidewall profile, complementary metal-oxide-semiconductor (CMOS) process compatibility, etch rate, selectivity to masking layer and its availability, selectivity to metals, anisotropy (crystal plane selectivity), etched surface morphology, control of etch parameters, safety of reactants and products, cost (including disposal and fixed), etch-stop characteristics, handling, etc. CMOS devices are very sensitive to any kinds of mobile or immobile ions; especially threshold voltage (V_T) is adversely affected by these ions. The integration of microelectromechanical system (MEMS) structures and CMOS circuitry on the same chip is basically used to miniaturize the device and to reduce the parasitic effects. Therefore, in some cases, anisotropic etching is performed after the realization of CMOS circuitry. Hence, in the case of post-CMOS anisotropic etching step, the etchant should not contain any kinds of ions. Potassium hydroxide (KOH) and tetramethylammonium hydroxide (TMAH) are the most widely employed wet anisotropic etchants due to their easy handling and less hazardous and

Silicon Wet Bulk Micromachining for MEMS
Prem Pal and Kazuo Sato

ISBN 978-981-4613-72-9 (Hardcover), 978-1-315-36492-6 (eBook)
www.panstanford.com

superior etching characteristics [1–29]. Both etchants have their own set of advantages and disadvantages. KOH provides high etch selectivity between {100} and {111} (i.e., $R_{\{100\}}/R_{\{111\}}$) and the significant etch rate of Si{100}, but potassium ionic contamination and poor selectivity between Si and SiO_2 are its major drawbacks. The etching characteristics of KOH have been discussed in Chapter 4. TMAH does not involve any ionic contamination and provides high etch selectivity between Si and SiO_2; however, the low etch rate ratio of $R_{\{100\}}/R_{\{111\}}$ in comparison to KOH is a disadvantage. Since TMAH does not involve alkali ions, it is compatible with the CMOS fabrication process. Considering the CMOS compatibility and a thermal oxide as the etch mask, TMAH gained widespread popularity [13–33]. Furthermore, it can be modified for the etching of silicon wafers containing aluminum metal patterns as the aluminum etch rate can be reduced to zero by adding some additives, for instance, ammonium peroxodisulfate, $(NH_4)_2S_2O_8$ [34–39]. To modify the etching characteristics of TMAH, isopropyl alcohol (IPA) [40–42] or a surfactant [43–55] is added to it. However, a surfactant is most widely used as an additive for the modification of TMAH etching characteristics. In ionic and nonionic surfactants, nonionic ones are selected to maintain the post-CMOS process compatibility of the etchant. Various kinds of nonionic surfactants, namely NCW-601A (30 wt% polyoxyethylene alkyl phenyl ether), NCW-1002 (10 wt% polyoxyalkylene alkyl ether), NC-200 (100% polyoxethylene-alkyl-phenyl-ether), poly(ethylene glycol) (PEG), and Triton X-100 (iso-octylphenoxy polyethoxyethanol) are studied, but they exhibit almost same effect on the etching characteristics in terms of reducing the undercutting at mask edges and corners. However, the effect and applications of Triton X-100 ($C_{14}H_{22}O(C_2H_4O)_n$, n = 9–10) are most widely investigated [15–17, 44–49]. In addition, Triton remains in liquid phase at room temperature, making it easy to measure volumetrically and to mix in the solution. It has the following molecular structure:

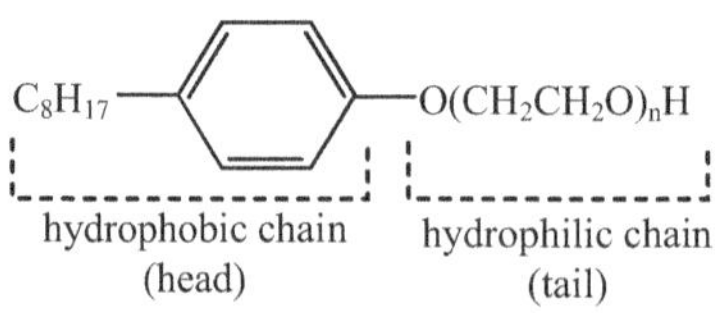

Triton-X-100 (iso-octylphenoxy polyethoxyethanol, n = 9-10)

One end of the surfactant molecule is hydrophilic, while the other is hydrophobic.

This chapter is focused on the etching characteristics of pure and surfactant-added TMAH solutions.

5.2 The Etching Mechanism

Anisotropic etching consists of two alternating reactions, namely (i) oxidation and (ii) etching, with the peculiarity that oxidation is very slow as compared to etching [44, 47].

Oxidation: $(\equiv Si)_3SiH + OH^- + H_2O \rightarrow (\equiv Si)_3SiOH + H_2 + OH^-$ (5.1)

Etching: $(\equiv Si)_3SiOH + 3H_2O \rightarrow 3(\equiv SiH) + Si(OH)_4$ (5.2)

Although these reactions describe anisotropic etching on the Si{111} surface, similar equations are written for Si{110} and Si{100} for a number of alkaline etchants, including NaOH, HF/NH_4F, KOH, and TMAH. Due to the difference in the rates of the two reactions, most OH-terminated silicon atoms resulting from the oxidation reaction are removed soon after they are formed and the surface remains dominantly H terminated. The active species replacing H by OH in the oxidation reaction is the water molecule, while the hydroxyl ion is a catalyzer, which is not consumed but effectively lowers the activation energy for the reaction with water. After becoming OH terminated, the Si-Si back bonds exhibit significant polarity due to the large electronegativity of O and react with the polar water molecules, leading to etching reactions similar to Eq. 5.2. Although the surfaces are dominantly H terminated during etching, small amounts of OH-terminated and other intermediate oxidized-silicon complexes can be detected.

5.3 Etch Rates

The etch rate is a measure of the thickness removed per unit time and is measured along the direction perpendicular to the surface being etched. If the etch depth of the (*hkl*) plane in etching time t is $d_{(hkl)}$, the etch rate (i.e., $R_{(hkl)}$) is expressed as follows:

$$R_{(hkl)} = \frac{d_{(hkl)}}{t}$$

The etch rate is a very useful parameter of an anisotropic etchant as it is used to estimate other parameters such as throughput, anisotropy ratio between two planes (particularly amongst {100}, {110}, and {111} planes), undercutting, etc. The etch rates of principal orientations (i.e., {100}, {110}, and {111}) can easily be determined using {100}, {110}, and {111} silicon wafers, which are commonly available in the market. Most of the etching studies are performed on Si{100} wafers, as this orientation is most extensively used for the fabrication of MEMS and CMOS devices. Figure 5.1 shows the etch rates of n- and p-type Si{100} in a wide range of TMAH concentrations at different temperatures [33]. It can be observed in Fig. 5.1, and is reported in many references, that the etch rate of Si{100} in pure TMAH decreases with increasing concentration [13–17, 29, 31, 33]. The type of silicon (p- or n-type) does not affect the etch rate significantly.

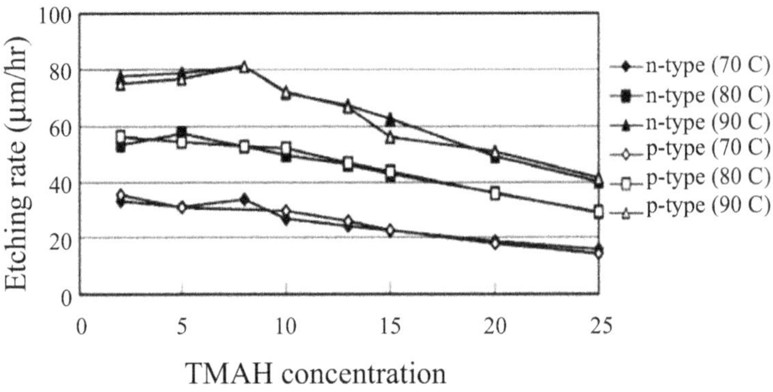

Figure 5.1 Etch rates of p- and n-type Si{100} in various concentration TMAH at different temperatures. Reprinted from Ref. [33], Copyright (2001), with permission from Elsevier.

The etching rates of three principal orientations (i.e., {100}, {110}, and {111}) in 5 wt% and 25 wt% pure and surfactant (NC-200)-added TMAH are shown in Fig. 5.2 [43]. It can be noted in Fig. 5.2a that the etch rates of Si{100} and Si{110} in low-concentration TMAH (5 wt%) is reduced to a very low level when a surfactant around 0.1% v/v is added to the solution. However, in the case of high-concentration TMAH (25 wt%), as presented in Fig. 5.2b, the etch rate of Si{100} is almost unaffected, while the etch rate of Si{110} decreases considerably. It is clearly seen in Fig. 5.2 that the etch rate difference of Si{100} in lower-concentration (5 wt%) TMAH without and with surfactant addition is considerably high, whereas at higher concentration (25 wt%) the difference is very small.

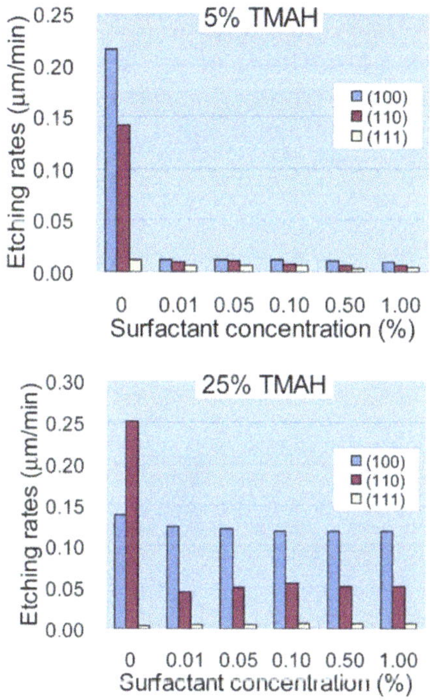

Figure 5.2 Etch rates of Si{100}, Si{110} and Si{111} planes in different concentration of pure and surfactant (NC-200) added TMAH at 60°C. Reprinted from Ref. [43], Copyright (2006), with permission from Elsevier.

The commonly accepted silicon etching process takes place in four steps: (i) diffusion of the reactant molecules to the silicon surface, (ii) adsorption of the reactant molecules on the surface, (iii) chemical reaction between etchant and silicon atoms, and (iv) desorption of the reaction products and dissolution of desorbed products into the bulk etchant. The reduction in the etch rate, as explained in the next section, is caused by the adsorption of the surfactant at the silicon–etchant interface, thus hindering the reaction between OH ions and silicon [31, 43, 47], that is, step 3 is the dominant factor for defining the resultant etch rate in surfactant-added solutions. It can be stated here that the absolute and relative etch rates of different crystallographic planes are a strong function of the etchant concentration and etching temperature. Moreover, the etch rates are significantly affected by the impurities (or additives) in the etchant. The etch rates of Si{100} and Si{110} in 25 wt% TMAH without and with 0.1% v/v Triton X-100 (surfactant) at different temperatures are presented in Fig. 5.3 [47]. To obtain detailed data on the crystal orientation dependence of the etch rate, either a wagon-wheel (WW) structure (a series of narrow and long mask openings rotated by a small angle to each other about a center) or a silicon hemisphere is employed. These methods are briefly described in the following sections.

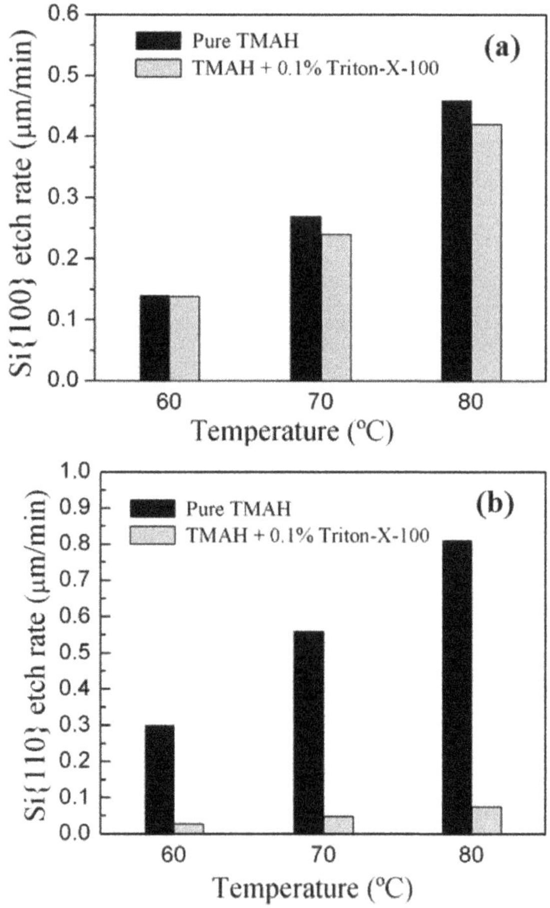

Figure 5.3 Etch rates of (a) Si{100} and (b) Si{110} in pure and surfactant added 25 wt% TMAH at different temperatures. From Ref. [47]. Copyright @ 2009 IEEE.

5.3.1 Etch Rates of Various Crystallographic Planes Using the Wagon-Wheel Structure

A WW structure consists of a series of narrow and long mask openings rotated by a small angle to each other about a center, as shown in Fig. 5.4a. The widest application of the WW pattern is to know the relation between the lateral etch rate and the crystallographic direction on different orientation surfaces (e.g., {100}, {110}). In other words, the orientation-dependent lateral underetch rates on the wafer surface can easily be determined using a WW structure, as presented in Figs. 5.4b and 5.4c for Si{100} and Si{110} wafer surfaces, respectively [3, 56]. The wet-etched shape on the top surface is analyzed using the orientation-dependent lateral underetch rate diagram. This part is discussed in detail in the next chapter.

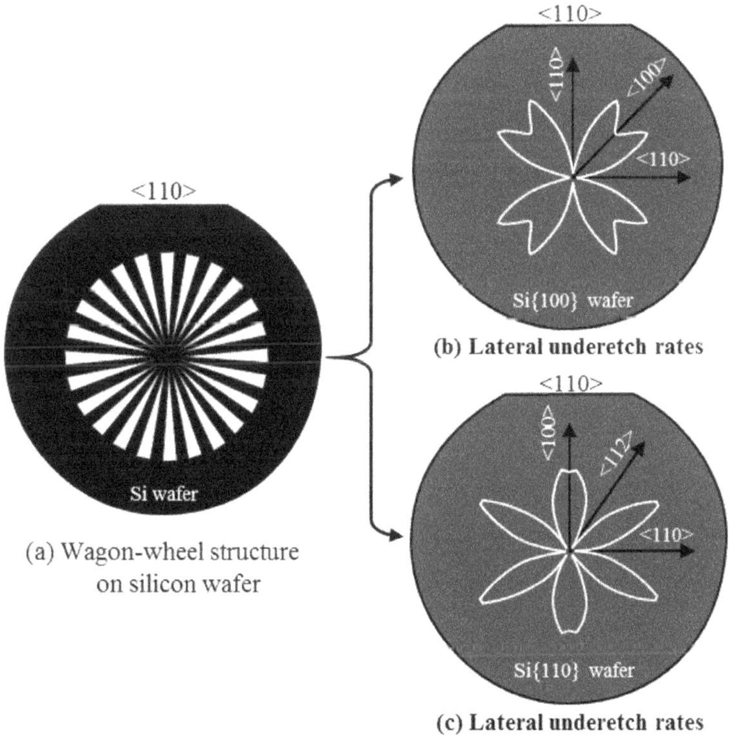

Figure 5.4 The diagram of lateral underetch rates as a function of orientation on silicon wafer using wagon-wheel structure (Etchant: 50% KOH). These patterns provide relative etch rates rather than absolute etch rates.

To extract multiple etch rates using the WW structure, deep reactive ion etching (DRIE) micromachined WW patterns are used [57–61]. The DRIE of the WW structure exposes many crystallographic planes which are perpendicular to the wafer surface. This method is briefly summarized in Fig. 5.5. In this figure, the WW structure is composed of 180 vertically etched, elongated wedges (or spokes) bounded by differently oriented vertical walls. To perform DRIE (Bosch process) of the WW pattern, either a thermal oxide or a photoresist is used as the mask material. The WW pattern is first transferred to a layer of photoresist on the silicon wafer using the standard photolithography process. DRIE is accomplished to etch the unmasked areas of silicon to a depth of about 100 μm. After DRIE, the front mask layer (photoresist or oxide or both) is completely removed chemically. An optical image of a typical DRIE micromachined WW structure is shown in Fig. 5.5b. Scanning electron microscopy (SEM) images present the close-up views of the micromachined WW structure at two different locations.

The micromachined WW structure, as shown in Fig. 5.4b, is etched in a wet anisotropic etchant. Prior to wet etching, the micromachined patterns are chemically cleaned properly to avoid any kind of contamination in the wet etchant and the native oxide is removed by dipping the sample in diluted HF, followed by thorough rinse in de-ionized (DI) water. In the anisotropic etchant, the sides of the wedges etch with a characteristic, face-specific rate. As a result, the wedges become narrower. Due to the variation in etch rates of sidewall planes, wet anisotropic etching leads to the development of a characteristic flower pattern, as shown in Fig. 5.5c. After etching, the flower pattern is quantified by optical microscopy and image analysis.

A simple trigonometric relation using various parameters defined in Fig. 5.5 can be defined to calculate the etch rate (R):

$$R = \frac{L}{t}\tan(\phi/2) \qquad \text{where } L = \frac{W}{\tan(\phi/2)}$$

As illustrated in Fig. 5.5, W, t, ϕ, and L are the distance advanced by the sidewall, the etch time, the angle spanned by the corresponding wedge, and the distance advanced by the wedge tip, respectively. Although the retraction of the spokes along the radial direction from the wheel center (i.e., distance L) can be measured by an optical microscope, it is a tedious task due to a large number of spokes. A software tool (Wagon Wheel Analyzer, WWA) that performs this task automatically for all the spokes appearing on the sample has been developed [60]. The procedure is based on a simple algorithm that analyzes the change in the local color of the image along each radial direction, repeating the operation for all wedges.

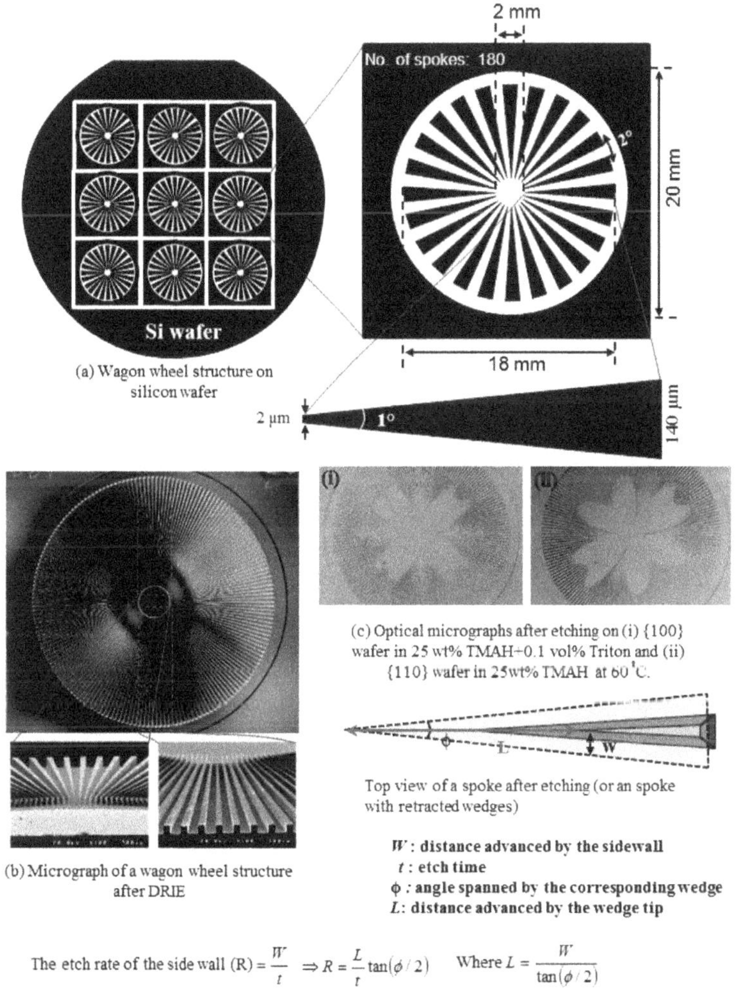

Figure 5.5 Wagon-wheel method for the extraction of multiple etch rates: (a) Mask pattern of WW structures comprising 180 spokes on silicon wafer (note: schematic design contains less number of spokes). (b) Micrograph of WW structure after deep reactive ion etching (DRIE). (c) Optical micrograph of a WW after etching in 25 wt% TMAH. The DRIE etched spoke is bounded by differently oriented vertical walls and the dimensions of the retracted wedges are measured for the calculation of the etch rates. From Ref. [60].

This technique measures the etch rates of the faces that are perpendicular to the wafer surface. Therefore, the orientation of sidewall planes depends on the type of wafer used. Figure 5.5c presents optical micrographs of WW structures etched on (i) a {100} wafer in 25 wt% TMAH+0.1 vol% Triton and (ii) a {110} wafer in 25 wt% TMAH. The figures show the eightfold and fourfold symmetries of the retraction patterns, respectively, thus providing eight and four independent measurements of symmetry-equivalent etch rates for a total of 23 (180/8) and 45 (180/4) vertical sidewalls, respectively. Therefore the choice of wafer orientation is crucial. For instance, the sides of the wedges on {110} are composed of many surfaces in the Si<110> zone, including all of the important close-packed faces of silicon: Si{100}, Si{111}, and Si(110). The limitation in the number of etch rates obtained from the WW has been overcome by a theoretical procedure based on a statistically reliable reconstruction of the complete three-dimensional orientation dependence of the etch rate by carefully interpolating the 45 + 23 etch rates from the Si{110} and Si{100} WWs [62].

5.3.2 Etch Rates of the Whole Range of Crystallographic Planes Using a Convex Hemispherical Specimen

As discussed in the previous section, the WW structure gives the etch rates of a limited number of crystallographic planes. A silicon hemisphere is used for determining the etch rate of all possible crystallographic planes. The profile on a spherical surface is measured using a three-dimensional measuring system (e.g., UPMC550-CARAT from Carl Zeiss) [5, 11, 14, 15, 43]. The surface of the hemisphere is mechanically probed before and after etching, every 2° of latitude ranging from 0° to 70° and every 2° of longitude ranging from 0° to 360°. In this way, etch rates at 6516 (36 × 181) surface orientations are measured. The etch rates at different locations of the hemisphere (or different orientations) are calculated by taking the difference of the values measured before and after the etching step. Figure 5.6 presents the stereographic projection of the etch rates for all surface orientations within the unit sphere for 25 wt% TMAH without and with the addition of Triton [15]. These etch rates are obtained using a hemispherical silicon specimen, as shown in Fig. 5.6a. The range of the planes affected by the surfactant molecules is clearly visible. While the etch rates of the exact and vicinal {100} planes are almost unaffected, the etch rates of the exact and vicinal {110} planes are reduced significantly. Although this method provides complete orientation dependence of the silicon etch rate, it is rather expensive as single-crystal spheres of silicon are not commercially available. Moreover it requires a nonstandard, purpose-modified surface profiler to mechanically probe the spheres.

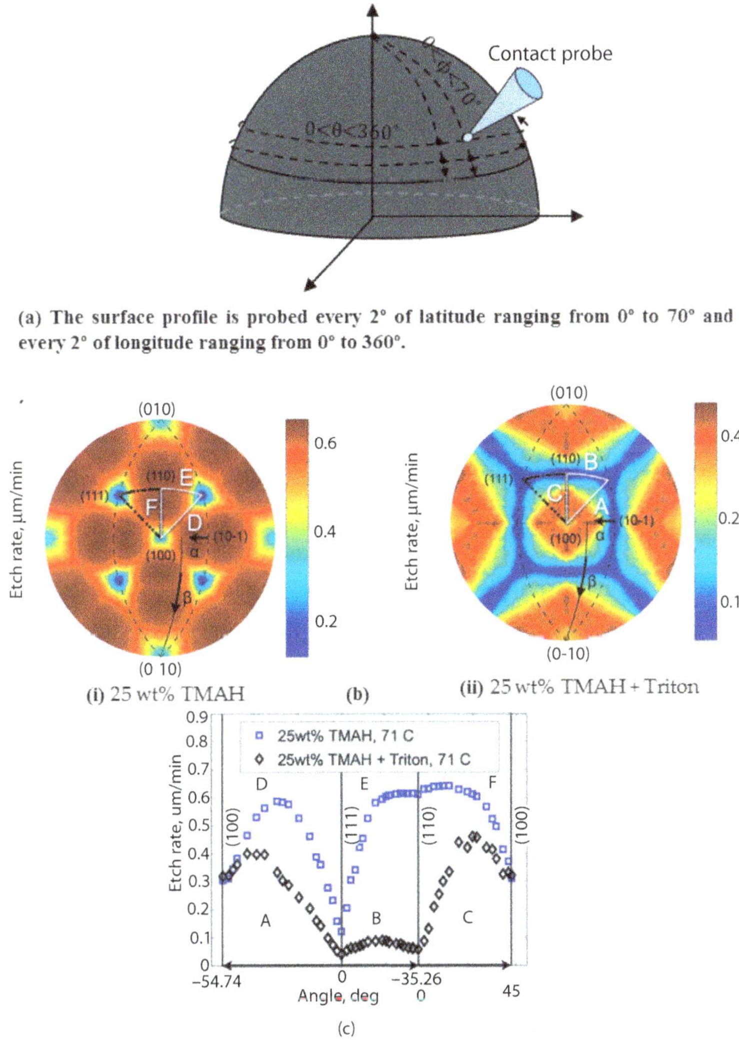

Figure 5.6 Silicon hemisphere for the determination of etch rates of whole range of crystallographic planes: (a) probing of hemisphere; (b) 3D distribution of the etch rates in 25 wt% TMAH (i) without and (ii) with surfactant; (c) 2D plot of the etch rates extracted from hemispherical data. From Ref. [15].

5.4 Surface Morphology

The etched surface morphology of silicon is a major concern, especially in optical MEMS and for high-efficiency solar cells. Figure 5.7 shows the etched surface roughness of three principal orientations etched in pure and different concentrations of the

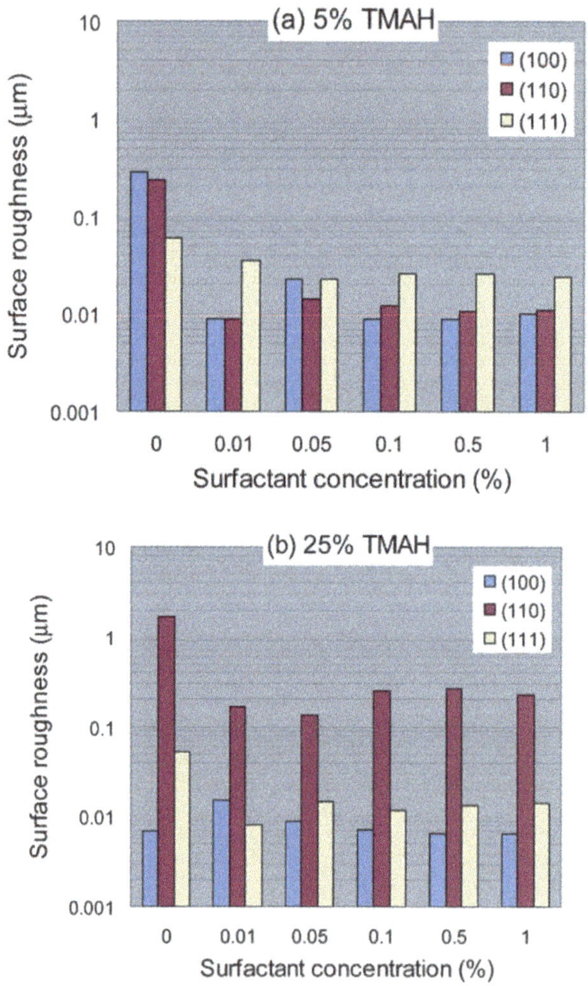

Figure 5.7 Surface roughness of three principal orientations etched in different concentrations of surfactant (NC-200) added (a) 5 wt% and (b) 25 wt% TMAH at 60°C. Reprinted from Ref. [43], Copyright (2006), with permission from Elsevier.

surfactant (NC-200)-added 5 wt% and 25 wt% TMAH at 60°C [43]. Figure 5.8 represents the etched surface roughness of Si{100} and Si{110} etched in different TMAH concentrations with varying amounts of the surfactant (NCW-1002) [63]. The surface roughness of Si{100} and Si{110} orientations after etching in 0.1% v/v Triton and 1% v/v NCW-1002-added different-concentration TMAH are shown in Fig. 5.9 [63]. The etched surface morphologies of Si{100} and Si{110} in pure and surfactant-added TMAH are shown in Figs. 5.10 and 5.11, respectively [31, 43].

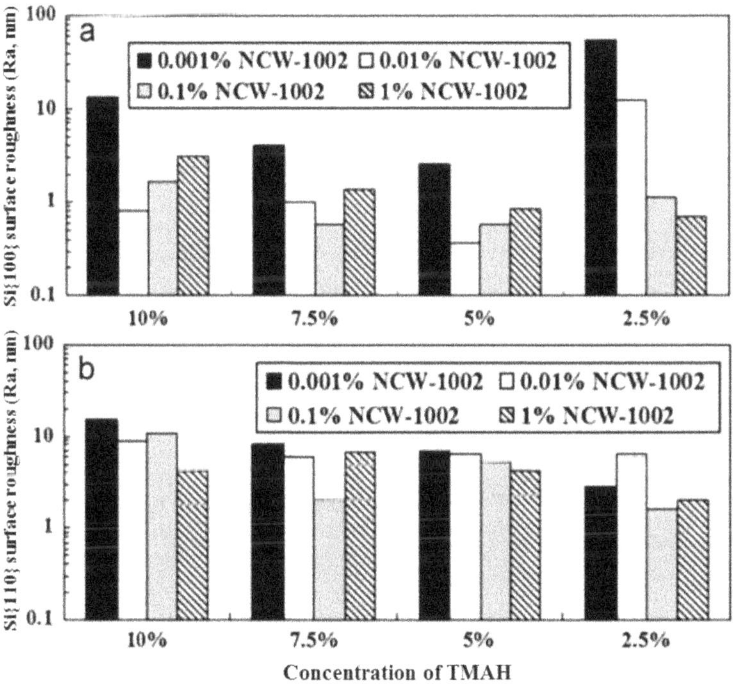

Figure 5.8 Surface roughness of (a) {100}- and (b) {110}-oriented silicon surfaces etched in different concentration TMAH (from 10% to 2.5%) with varying concentration of NCW-1002 (from 0.001% to 1%). Reprinted from Ref. [63], Copyright (2011), with permission from Elsevier.

It can be noted in Fig. 5.7 that the addition of the surfactant reduces the etched surface roughness of all three principal orientations. However, a significant effect is observed in the case of Si{100) for low TMAH concentration and Si{110} for both low and high TMAH concentrations. The required amount of surfactant necessary for the effects is quite small, reaching saturation at 0.1% of volume concentration.

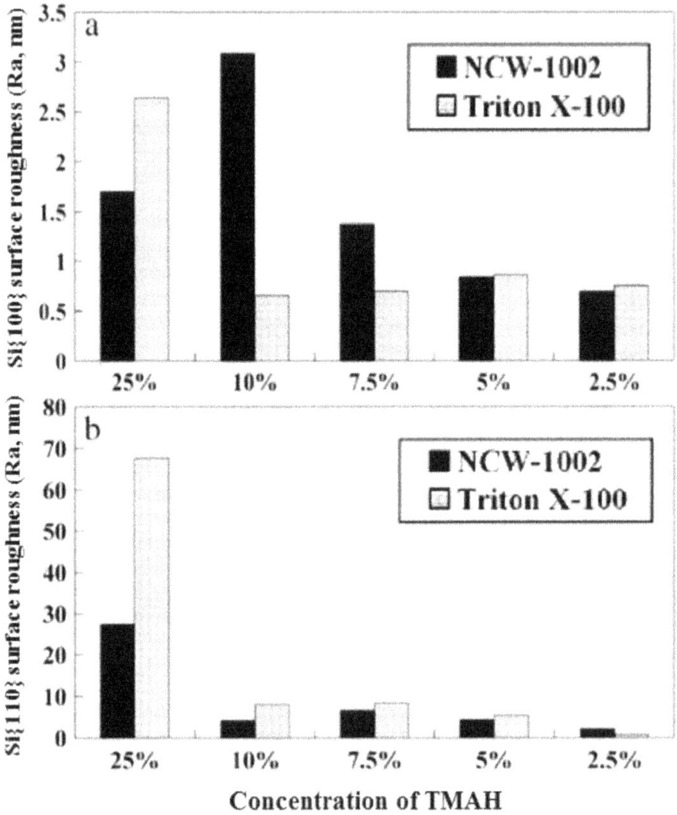

Figure 5.9 Surface roughness of (a) Si{100} and (b) Si{110} surfaces etched in TMAH (25% and 10-2.5%) with NCW-1002 (1%) and Triton (0.1%) surfactants. Reprinted from Ref. [63], Copyright (2011), with permission from Elsevier.

As shown in Fig. 5.10 the etched surface morphology of Si{100} in low-concentration TMAH (or 10 wt%) is improved dramatically on addition of a surfactant, while the etched surface morphology in high-concentration TMAH (or 25 wt%) is almost unaffected by the surfactant (Fig. 5.10c,d). A similar comparison for the Si{110) surface is presented in Fig. 5.11. For the low TMAH concentration, the surfactant changed the surface morphology from zigzags (Fig.

5.11a) to a smooth surface finish (Fig. 5.11b). However, for the high TMAH concentration, the typical linear stripe structures persist (Fig. 5.11c,d), although the roughness is improved to some extent. As a result, it can be concluded that smooth {100} and {110} surfaces can simultaneously be obtained at low TMAH concentrations by adding a surfactant.

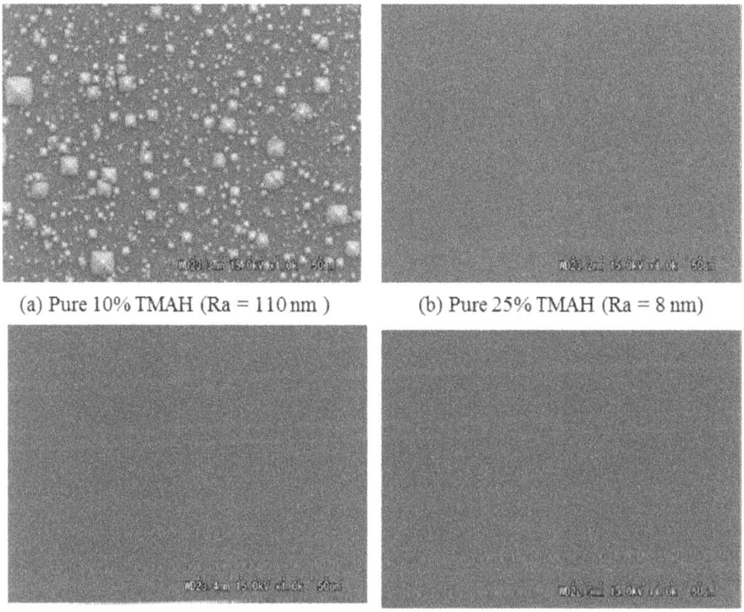

(a) Pure 10% TMAH (Ra = 110 nm) (b) Pure 25% TMAH (Ra = 8 nm) (c) 10% TMAH+surfactant (Ra = 9 nm) (d) 25% TMAH+surfactant (Ra = 9 nm)

Figure 5.10 Etched surface morphologies of Si{100} in pure and surfactant added (a) 10 wt% and (b) 25 wt% TMAH. From Ref. [31]. @ IOP Publishing. Reproduced with permission. All rights reserved.

It must be emphasized here that the etched surface roughness mainly depends on the etchant composition (i.e., type of etchant and its concentration, pure or with additives). Moreover it depends on etching time, etching temperature, mechanical agitation, and the localized impurities in the wafer or on the surface during etching. Therefore it is hard to achieve a reproducible etched surface morphology on a large surface area. Owing to this reason different authors reported different etched surface roughness for the same orientation etched in the same etchant [31, 43, 63].

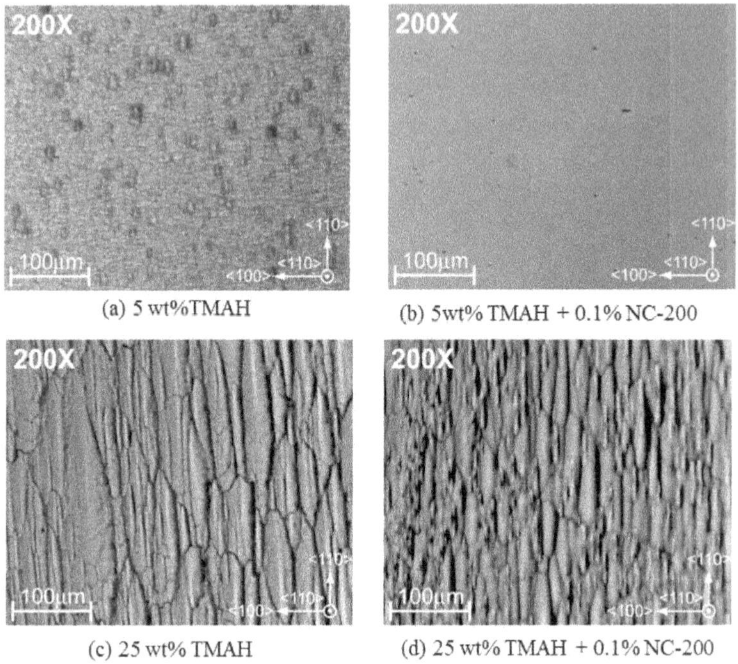

Figure 5.11 Etched surface morphologies of Si{110} in pure and surfactant (NC-200) added 5 wt% and 25 wt% TMAH at 60°C. Reprinted from Ref. [43], Copyright (2006), with permission from Elsevier.

5.5 Undercutting and Etched Profiles

The lateral etching at mask edges and/or corners is termed "undercutting" (or "underetching"). The details of undercutting and underetching at different kinds of edges and corners in wet anisotropic etching are discussed in the next chapter. Since Si{111} planes are the slowest-etch-rate planes, the directions/edges comprising the {111} planes, for instance, the <110> direction on the Si{100} surface, exhibit minimum undercutting, which takes place owing to the finite etch rate of {111} planes. If the sides are slightly misaligned from the <110> direction (or the direction containing the {111} planes), the underetching starts at the edges, resulting in a cavity of bigger dimensions than the requirement, as illustrated in Fig. 5.12. In other words, a stable etched profile in wet anisotropic etching is formed by the slowest-etch-rate planes, that is, {111} planes. Any arbitrary mask opening after prolonged etching in an anisotropic etchant provides a well-shaped structure whose sidewalls are formed by {111} planes, as presented in Fig. 5.12. Thus, to fabricate a cavity of controlled dimensions, the sides of mask edges must be aligned along the directions comprising {111} planes, for instance, for the formation of a square/rectangular cavity on a {100} wafer, the sides of the square/rectangle of mask patterns are aligned along the <110> direction.

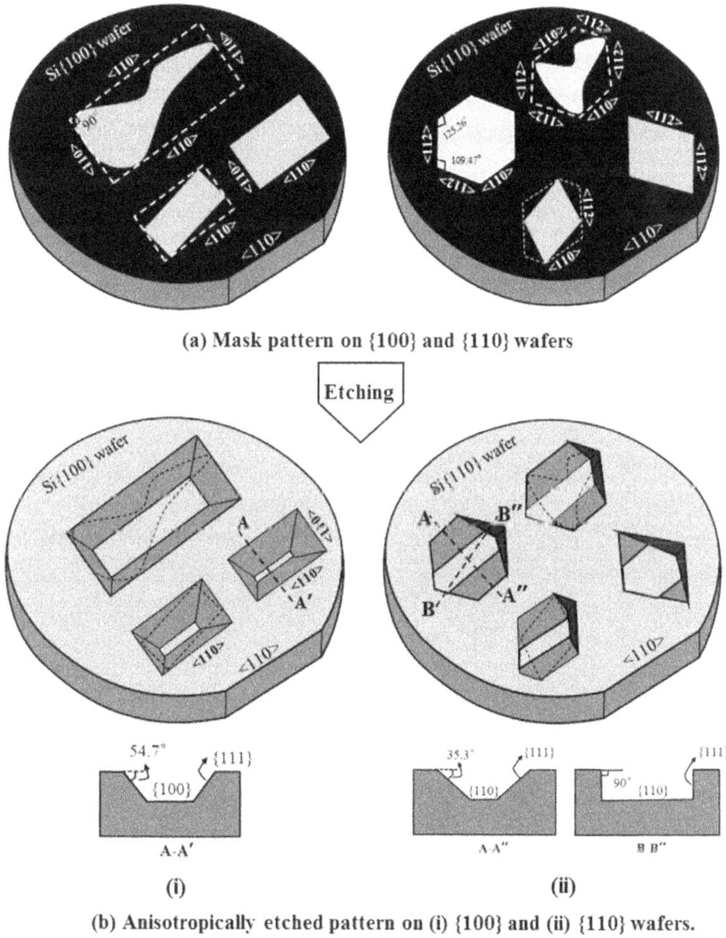

Figure 5.12 Schematic representation of the etched profiles obtained by anisotropic etching of different-shaped mask geometries on Si{100} and Si{110} wafers.

In wet anisotropic etching, the corners formed by the intersection of the directions containing {111} planes are termed "concave" and "convex" corners. In a {100} silicon wafer, these corners are bounded by two mutually perpendicular <110> directions due to its fourfold symmetry. Although both types of corners are formed by the intersection of {111} planes, their etching characteristics are opposite to each other [64–72]. Convex corners in wet anisotropic etching exhibit severe undercutting, while no undercutting takes place at the concave corners. The reason behind this is discussed in the next chapter. Moreover the addition of surfactants/alcohols/impurities and the use of mechanical agitation (e.g., magnetic bead stirring) influence the corner undercut profile. In the case of pure TMAH, undercutting increases with its concentration, while in surfactant-added TMAH, it is almost the same for all ranges of TMAH concentration [31, 53, 54]. Figures 5.13 and 5.14 present the etched profiles of different-shaped mask patterns on Si{100} in low (10 wt%) and high (25 wt%) concentration TMAH without and with a surfactant, respectively [31]. It can be observed in Fig. 5.13 that the addition of a surfactant in 10 wt% TMAH provides very similar etched shapes as those without it. This etching behavior indicates that the addition of a surfactant in 10 wt% TMAH does not change the etch rate anisotropy significantly. As discussed previously, the addition of a surfactant in low-concentration TMAH suppresses the etch rates of almost all planes but improves the etched surface morphology drastically, as can be observed in Fig. 5.10b. In terms of corner undercutting, 25 wt% TMAH provides two options, low undercutting and high undercutting, as can be noticed in Fig. 5.14. A comparison of undercutting at sharp convex corners and rounded concave corners in pure and surfactant-added 25 wt% TMAH is presented in Fig. 5.15. The undercutting in wet bulk micromachining has a favorable effect of removal of underneath material for the formation of suspended microstructures (e.g., cantilever beams) made of materials such as P^+-Si, SiO_2, Si_3N_4, etc. [49, 73, 74]. However, it is unwanted for the fabrication of proof masses for accelerometers, mesa structures, chip isolation grooves, bent V-grooves, serpentine-shaped channels, etc., where severe undercutting distorts the desired shape of the structure [16, 73, 75–77]. Therefore, for the fabrication of such types of structures, the problem of undercutting should be eliminated (or minimized to a significantly low level) by some means. As described earlier, the addition of a surfactant in 25 wt% TMAH does not affect the etched surface morphology and the etch rates of {100} and its vicinal planes. The high and significantly low undercutting in pure and surfactant-added 25 wt% TMAH, respectively, can be explored for engineering applications where the etchant is selected on the basis of its undercutting characteristics. It may be emphasized here that the etch rate of Si{110} and its vicinal planes in surfactant-added TMAH decreases to a very low level, and therefore this composition is not suitable for the fabrication of microstructures on {110} wafers.

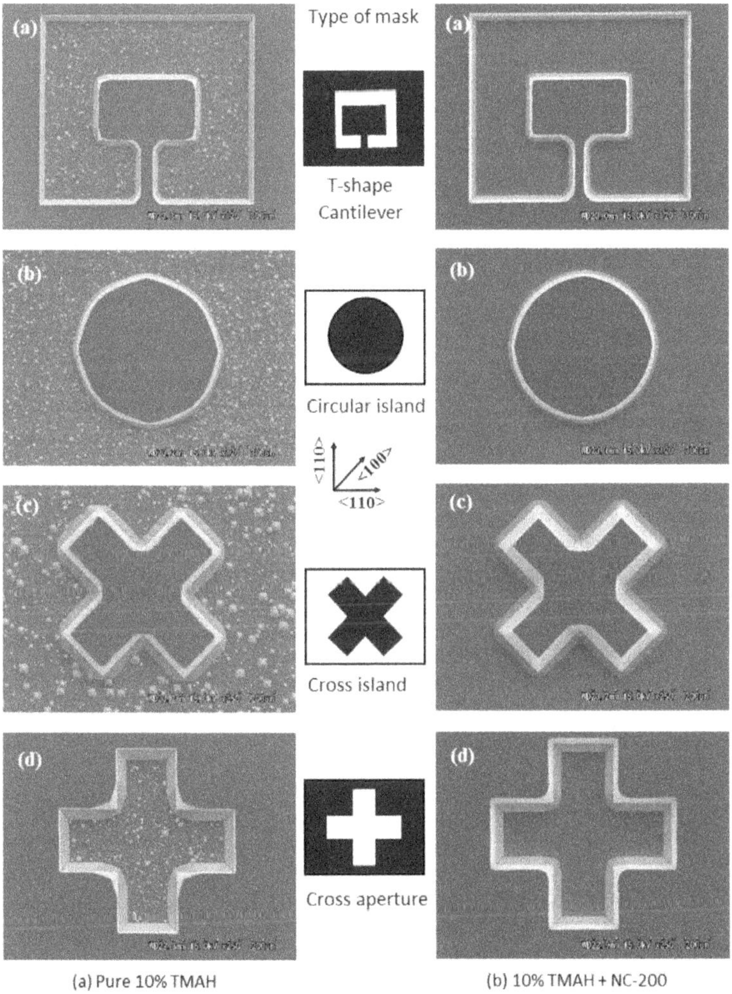

Figure 5.13 Various kinds of etched patterns in pure and surfactant (NC-200) added 10 wt% TMAH at 60°C (etch depth = 27 μm). From Ref. [31]. @ IOP Publishing. Reproduced with permission. All rights reserved.

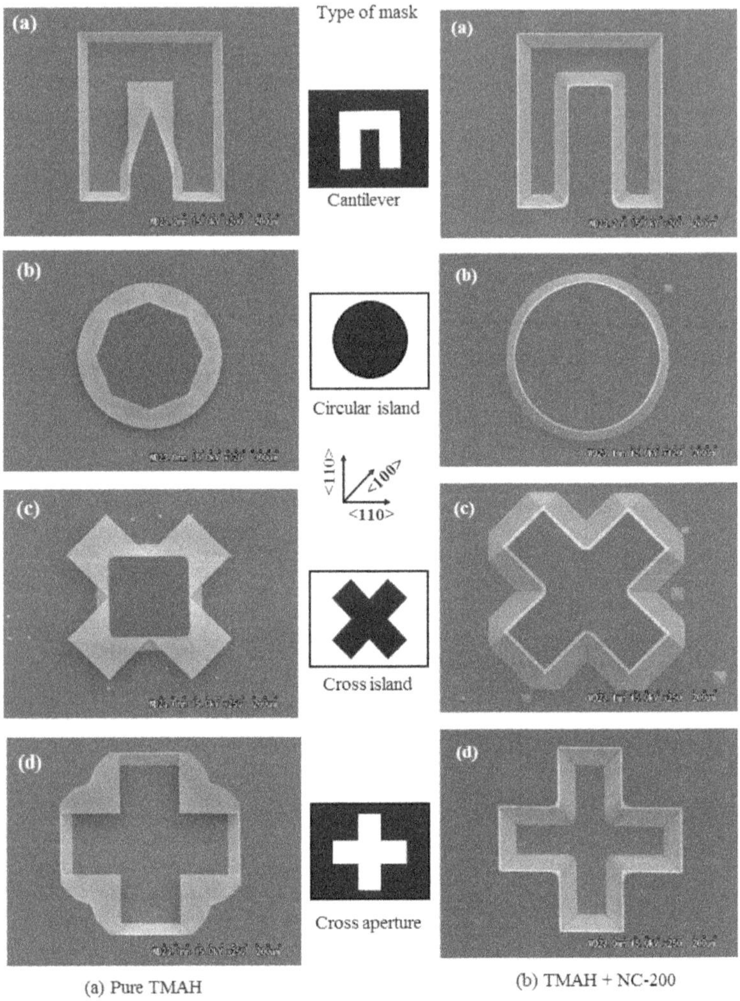

Figure 5.14 Various kinds of etched pattern in pure and surfactant (NC-200) added 25 wt% TMAH at 60°C (etch depth = 33–35 μm). From Ref. [31]. @ IOP Publishing. Reproduced with permission.

5.6 Why Are Etching Characteristics Affected When a Surfactant Is Added to TMAH?

As discussed earlier, the addition of a very small amount of a surfactant in TMAH, particularly at high concentration (i.e., 25 wt%), alters its etching characteristics considerably, especially in terms of undercutting at mask edges and corners. It significantly reduces the etch rates of Si{110} and its vicinal orientations, such as {221}, {331}, and {441}, while not strongly affecting the etch rates of Si{100} and its vicinal surfaces, such as {211}, {311}, and {411} [17, 44, 47]. It means that the effect of the surfactant is orientation dependent. To investigate the main reason behind the orientation-dependent etching characteristics, Fourier transform infrared (FTIR) spectroscopy and ellipsometry studies are performed [46, 47].

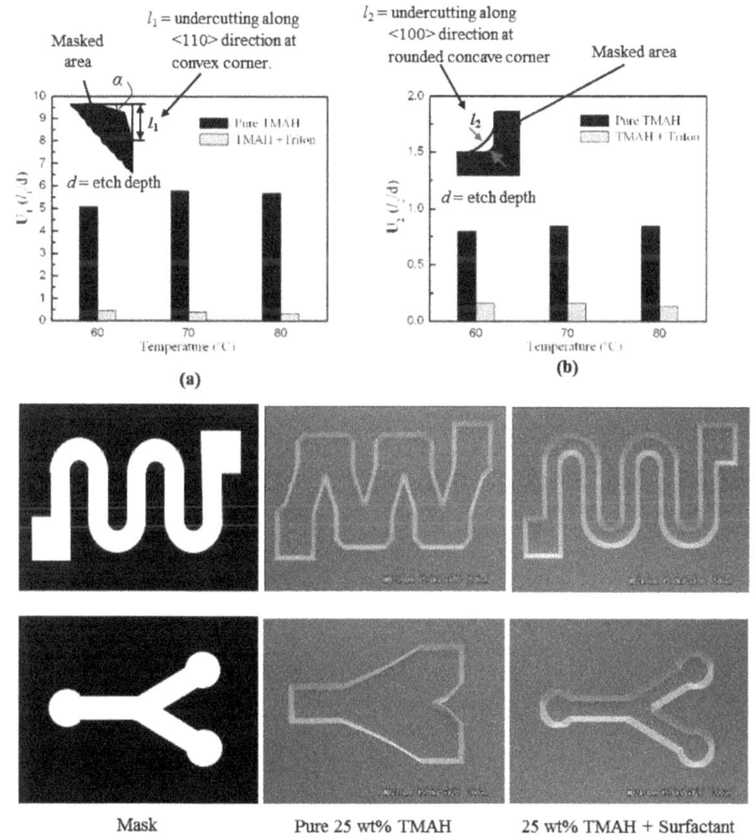

Figure 5.15 Undercutting ratio (*l*/*d*) at (a) convex and (b) rounded concave corners of the mask pattern and (c) various kinds of etched patterns on Si{100} wafer in pure and surfactant added 25 wt% TMAH. SEM pictures are taken after removal of mask layer.

5.6.1 FTIR Measurement

FTIR is an effective tool to examine the bonding and adsorption of foreign molecules at the surface under observation [78–88]. Attenuated total reflection infrared (ATR-IR) spectroscopy is a very promising method for the in situ analysis of the surface. This method has been utilized for real-time observation of silicon etching/cleaning/oxidation processes [77–88]. To investigate the orientation-dependent characteristics of surfactant-added TMAH, infrared (IR) spectroscopic measurements in multiple internal reflection (MIR) geometry are carried out on {100} and {110} surfaces to observe the Si/water, Si/TMAH, and Si/(TMAH+surfactant) interfaces before, after, and during etching [47]. Silicon prisms with 45° bevels cut at the short edges are prepared by dicing a silicon wafer with a patterned thick oxide layer followed by beveling of both ends by mechanical and chemical polishing in order to achieve MIRs, as shown in Fig. 5.16. The use of a thick oxide avoids the detection of absorption peaks from the O-ring, which is used to seal the custom-made Teflon cell in order to make it liquid-leakage free. To collect the background spectra of TMAH (or TMAH+Triton) and DI water, a very thin layer (1–2 nm) of oxide is grown by dry thermal oxidation to prevent direct contact with the silicon surface, which would be otherwise etched. The sample (or silicon prism) is mounted into a Teflon cell and fitted into the experimental system, as illustrated in Fig. 5.16c. The silicon prism provides MIRs during the measurements. When the IR spectra are collected for different states of the silicon surface, the solution is continuously supplied through a pipe and drained to another pipe connected to the outlet.

Figure 5.17a presents the IR spectra for Si-H_2 and Si-H absorption regions before etching Si{100} and Si{110} surfaces. The peaks for dihydride and monohydride termination appear dominantly at 2110 cm^{-1} and 2070 cm^{-1} for {100} and {110}, respectively. The absorbance spectra of {100} and {110} taken after every 2 min during etching in pure and Triton-added TMAH are shown in Fig. 5.17b. These spectra are obtained by subtracting the IR spectrum while TMAH flows (or TMAH+Triton) in the presence of a thin oxide (1–2 nm) layer. To collect the spectra during etching, the thin oxide layer is removed by flowing 5% HF followed by a thorough rinse using flowing DI water.

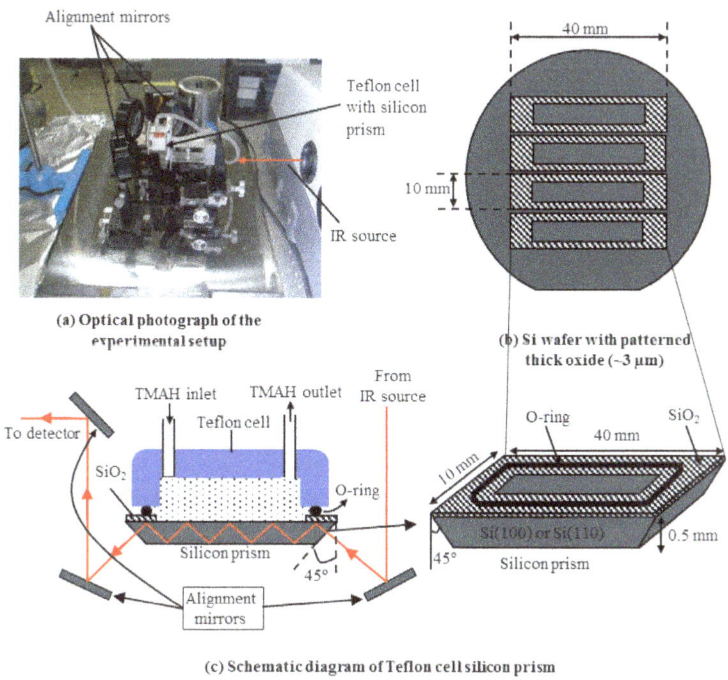

Figure 5.16 FTIR experiment setup for in situ measurements of silicon etching mechanism in pure and surfactant-added TMAH. Silicon prisms are prepared using 4-inch diameter of 500 μm thickness.

The absorption peaks before etching for dihydride and monohydride termination for Si{100} and Si{110} appear dominantly at 2110 cm^{-1} and 2070 cm^{-1}, as shown in Figs. 5.17a,i and 5.17a,ii, respectively, while these peaks during etching appear at 2077 cm^{-1} and 2045 cm^{-1}, as presented in Figs. 5.17b,i and 5.17b,ii, respectively, due to the change in the local environment from DI water only to aqueous TMAH, which contains TMA^+ and OH^- ions, in addition to water. Moreover, the peak heights for Si-H and Si-H_2 are reduced during the etching in comparison to those before etching. Hydrogen-terminated surfaces are very susceptible to the local environment and various parameters for different solvents (such as the dielectric constant, Gutmann's electron–donor and electron–acceptor numbers, etc.) are considered to be responsible for the IR frequency shifts [89, 90]. Generally, the dielectric effects are widely accepted as an explanation. The shifted peaks at 2077 cm^{-1} and 2045 cm^{-1}

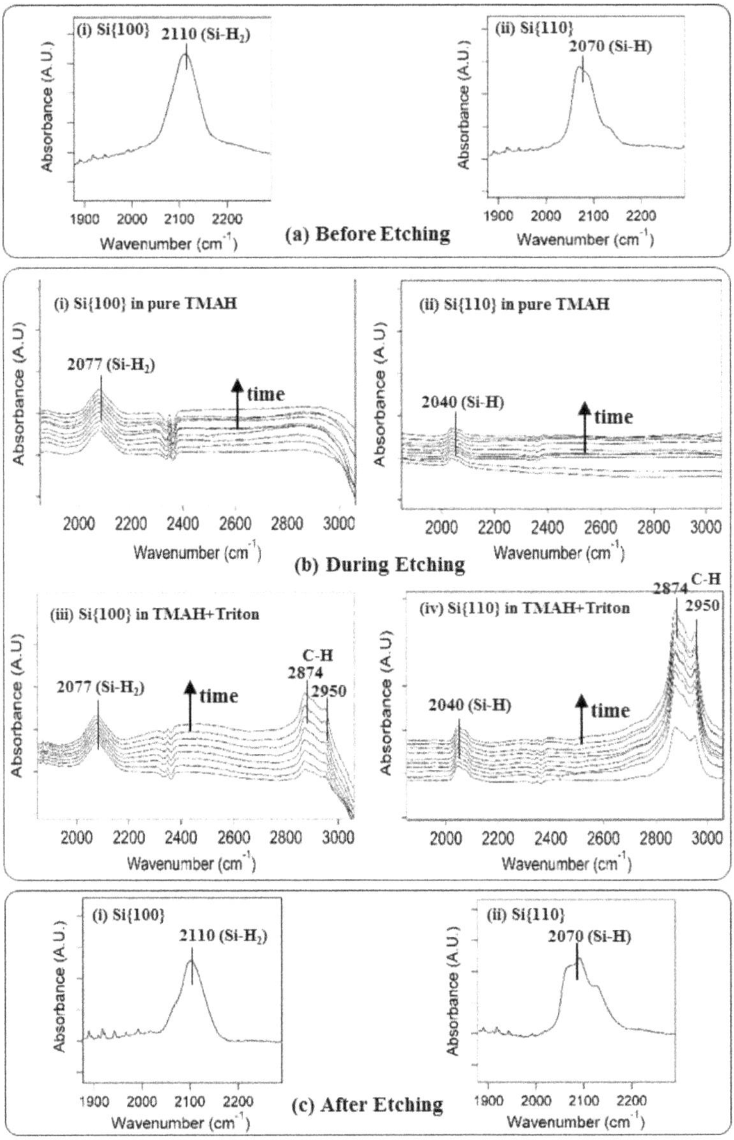

Figure 5.17 Infrared absorption spectra (a) before etching, (b) during etching and (c) after etching. In the case of (a) and (c), silicon surfaces are exposed to de-ionized (DI) water after HF treatment and spectra are collected in the Si-H_2 and S-H vibration regions, while in the case of (b) samples are exposed to etchant (i.e., pure and Triton-added TMAH) and the spectra are recorded in the Si-H and C-H stretch regions.

are attributed to strained dihydride and monohydride states of silicon surfaces, respectively [89, 91]. The spectra above 3000 cm^{-1} show only broad depressions around 3300 cm^{-1} due to absorbance by water. In the case of etching in TMAH+Triton, the same shifts in the IR frequency for the dihydride (Si-H$_2$) and monohydride (Si-H) states are detected. However, additional peaks corresponding to C-H vibrational modes are also observed, as can be seen in Fig. 5.17b,iii,iv. The peaks at 2874 cm^{-1} and 2950 cm^{-1} are attributed to the CH$_3$ symmetric and asymmetric stretch modes, respectively [92]. These additional peaks reveal unambiguously the presence of a surfactant (Triton) on the surface [78, 79, 88]. By comparing their intensities for Si{100} and Si{110}, it can be concluded that the surfactant adsorbs more strongly on Si{110}. As a result, the {110} planes are significantly prevented from being attacked by H_2O and OH^- during etching, resulting in a reduced etch rate, as shown in Fig. 5.3b. Since Triton does not adsorb as strongly on Si{100}, the corresponding reduction of the etch rate is much weaker, as can be noticed in Fig. 5.3a. Hence, the in situ observation of the silicon surface during etching using FTIR spectroscopy confirms the existence of selective adsorption of the surfactant molecules.

After etching, the surfaces treated in 5% HF and rinsed in DI water, and the IR spectra collected in the Si-H$_2$ and Si-H regions are presented in Fig. 5.17c. The dihydride and monohydride absorption peaks appear at their corresponding wave numbers, as previously observed (Fig. 5.17a), indicating that the physisorbed surfactant has been removed by the HF treatment and the surfaces are returned to the expected H termination states.

5.6.2 Ellipsometric Measurement

Ellipsometry is a nondestructive technique and widely used for thin-film characterization to measure the thickness and refractive index [46, 47, 93]. To gain further information on the orientation-dependent adsorption of surfactant molecules, the difference in the surfactant layer thickness for Si{110} and Si{100} is measured by ellipsometry after the surfactant treatment. This method can be named "ex situ ellipsometric spectroscopy measurements." For surfactant treatment of silicon samples, 1% v/v Triton (surfactant) is added to DI water. The silicon samples are cleaned properly and dipped in 5% HF to remove any trace amounts of oxide, followed by a thorough rinse in DI water. The samples are then dipped in the surfactant-added DI water for 10–15 min (bath time) to achieve a saturated thickness of surfactant molecules. Thereafter, the samples are taken out and rinsed by gentle dipping in pure DI water several times (3–4 min). Finally, the samples are dried in air and the thickness is measured using ellipsometry. Properly cleaned bare Si{100} and Si{110} samples are used as reference to measure the surfactant layer thickness on Si{100} and Si{110} surfaces after surfactant treatment, respectively. Figure 5.18 presents the surfactant layer thickness as a function of the surfactant bath time for two different surfactant bath temperatures (room temperature and 60°C). It can be easily noticed from the figure that the surfactant layer thickness is saturated on both types of surfaces and the saturated thickness on Si{110} is thicker than on Si{100}, indicating a stronger attachment to the surface.

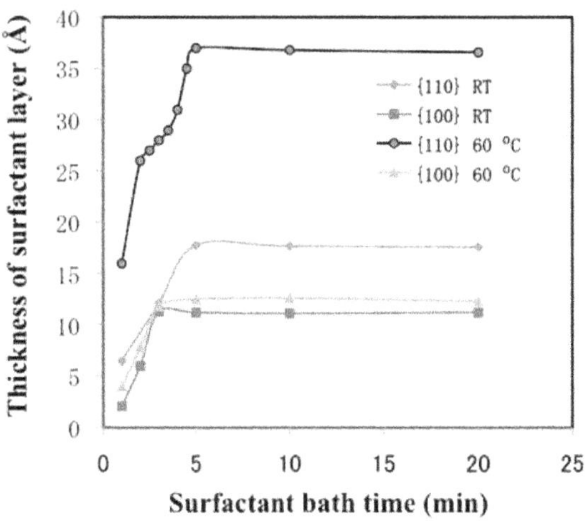

Figure 5.18 The surfactant layer thickness on Si{110} and Si{100} as a function of surfactant bath time at room temperature (RT) and 60°C.

5.6.3 The Physical Model

FTIR and ellipsometric investigations confirm that the adsorption of surfactant molecules depends on the orientation of the surface.

To explain the preferential adsorption of the surfactant and the change in etching characteristics of an etchant, a simple model is developed. The model is based on the interaction of the hydrophobic end of the surfactant molecules with the hydrophobic wafer surface. It is known that the surface of silicon is mostly H terminated during etching [57, 94] and that H-terminated surfaces are characteristically hydrophobic [80]. The relative hydrophobicity of the silicon surface can be estimated by the density of H terminations. The number of H terminations per unit area for the crystallographic cut of Si{100} and Si{110} is $4/a^2$ and $4/(a^2\sqrt{2})$, respectively, where a is the lattice parameter of the conventional unit cell. This calculation indicates that Si{110} is more hydrophobic (lower H density) than Si{100}. If the water and surfactant molecules are to compete for suitable locations over the surface, on average water will find the H-terminated sites more comfortable, while the hydrophobic ends of the surfactant molecules will find the areas between H terminations more suitable [47]. Thus the surfactant molecules form an adsorbed monolayer film on the surface with the hydrophobic part of the molecules (or head) in contact with the hydrophobic surface, while the hydrophilic part of the molecules (or tail) remains in contact with water [47, 78, 93, 95]. This argument suggests that the surfactant molecules adsorb more densely on a more hydrophobic surface. The Si{110} is more hydrophobic (lower H density) than Si{100}, resulting in the formation of a more densely packed surfactant layer [47].

The difference in the adsorption behavior of the surfactant on two differently oriented surfaces can be used to explain the effects on the etch rate and surface morphology. Figure 5.19 presents schematically the structure of the surfactant layer based on the hydrophobic nature of the surface, where the head is sketched as an ellipse and the tail is sketched as a zigzag chain. As mentioned above, Si{110} is more hydrophobic than Si{100}, resulting in the formation of a more densely packed surfactant layer. This influences more strongly the diffusion of the reactants (i.e., OH^- and H_2O) and products (essentially H_2, $Si(OH)_4$, $SiH(OH)_3$, etc.) into and out from the etched interface, as shown in Fig. 5.19. As a result, less reactants reach the surface in the presence of a thick surfactant layer and the etch rate is correspondingly reduced. Therefore, the etch rate of Si{110} and its vicinal surfaces reduces to a significantly low level due to the larger packing density of surfactant molecules on these surfaces. In the case of Si{100} surface, the layer of surfactant molecules is not able to protect the surface from the etchant and thus the etch rate is almost unaffected, as presented in Fig. 5.3a. The intense reduction of the undercutting at sharp convex (Fig. 5.15a) and rounded concave (Fig. 5.15b) corners in surfactant-added TMAH can now be explained confidently as being due to the higher packing density of the adsorbed layer of surfactant molecules on the {h h 1} and {h+2 h+2 h} high-index planes, hindering them from the action of the etchant, as presented for the {110} orientation in Fig. 5.19b.

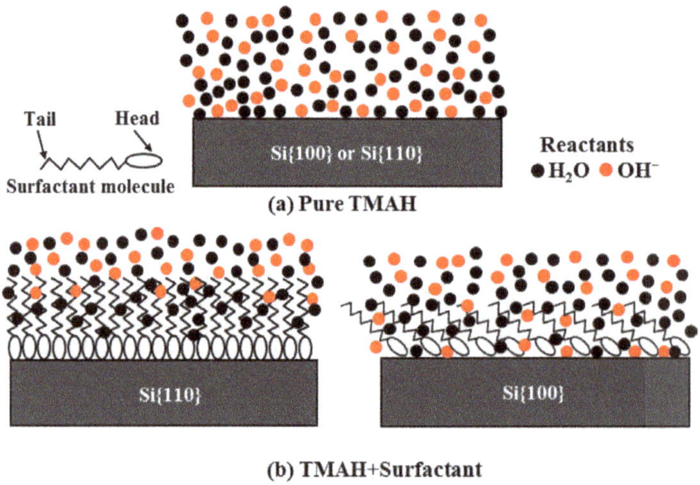

Figure 5.19 A simple model representing the wet etching on silicon surface in TMAH (a) without surfactant and (b) with surfactant. The surfactant molecules adsorb more densely on Si{110} than on Si{100}, affecting the diffusion of the reactants and products significantly and resulting in a drastic reduction in the etch rate of Si{110}.

The strong decrease in the etch rate of exact and vicinal Si{110} reduces, in turn, the production of hydrogen gas (H_2). On the other hand, the presence of the surfactant layer physically covering the surface prevents the formation of large H_2 bubbles, filtering them away before they can grow too large. As a result, the surface morphology is smoothened and the surface roughness is reduced. Similarly, the decrease in the etch rate and the physical presence of the surfactant layer result in a reduction of the amount and size of inhomogeneous regions (or inhomogeneities) in the etchant concentration and/or temperature in the close vicinity of the active regions of the surface, such as the morphological steps appearing along the zigzag morphology of {110}. These etchant inhomogeneities are the result of a time delay between the faster reaction rates at the active surface sites and the slow diffusion transport of the reactants and products in the etchant [96]. The reduction in the number and size of inhomogeneities also contributes to smoothening of the surface roughness. The zigzag structures typically observed on {110} due to limited diffusion transport become much shallower after the addition of the surfactants, as shown in Fig. 5.11. This is in agreement with the effect of other surfactants and alcohols on the morphology of {110} [96].

5.7 Etched Profile Control

In silicon wet anisotropic etching, undercutting, slow etch rate planes, and etch-stop layers play a significant role to get an etched profile of the desired shape. The selection of the method to control the profile depends on the type of structure to be fabricated. In the case of simple cavities/grooves, the etched shape is controlled by slow-etch-rate planes (e.g., {111}). In some cases, the etch-stop layer is needed, while in other cases control over the undercutting is necessary. These methods are briefly discussed in the following sections.

5.7.1 P^+ Silicon Etch Stop

All common alkaline etchants exhibit a reduction in the etch rate at high boron doping, and this feature is used to fabricate silicon MEMS components such as cantilever beams and diaphragms [4, 97]. In the case of TMAH solutions, the etch rate of silicon decreases to a significantly low level when the boron concentration reaches 2.5×10^{20} cm^{-3}, which is the boron solubility limit into silicon at 1150°C [98]. The highly doped p-type silicon is denoted by P^+-Si (or P^{++}-Si). The etch rate ratio of highly doped to lightly doped (i.e., doping concentration less than 10^{18} cm^{-3}) silicon in TMAH is about 1:40. Figure 5.20 shows the silicon etch rate as a function of boron concentration for a 25 wt% TMAH solution at 80°C and 100°C. The selectivity between heavily and lightly doped silicon decreases with increasing temperature due to the difference in activation energy, while etchant concentration does not affect it significantly. In TMAH, the etch rate of polysilicon is higher than that of monocrystalline silicon, but the selectivity (doped/undoped) is the same for both materials [99].

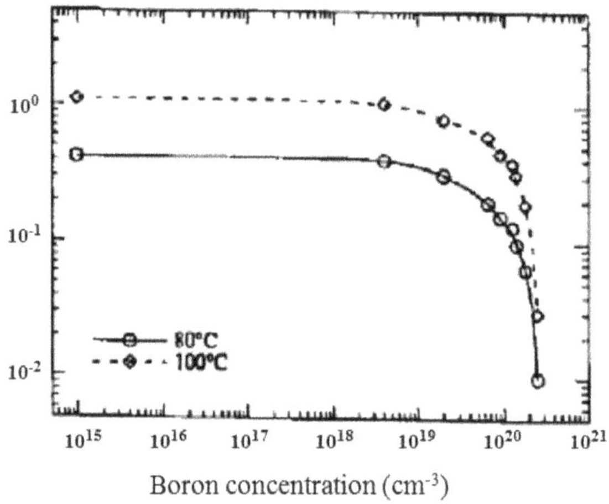

Figure 5.20 Etch rate of Si{100} at two different temperatures in 25 wt% TMAH as a function of boron concentration. Reprinted from Ref. [98], Copyright (1996), with permission from Elsevier.

5.7.2 Electrochemical Etch Stop

The electrochemical etch stop is a promising technique to achieve high etching selectivity between p- and n-type silicon in alkaline solutions to control the thickness of silicon microstructures (e.g., diaphragm, cantilever, etc.) [100–103]. This method is based on the well-known anodic passivation characteristics of silicon with a reverse-biased p–n junction. It was first proposed by Waggener in 1970 [104]. Electrochemical etching is investigated by measuring the current–voltage (*I–V*) characteristics of n- and p-type silicon using a three-electrode system in an anisotropic solution to determine the open-circuit potential (OCP) and passivation potential (PP), as shown in Fig. 5.21. A sample holder made of Teflon is used to protect the back side of the wafer from the etchant. Figure 5.21b shows the I–V characteristics of n- and p-type silicon in a 25 wt% TMAH solution [105]. The OCP, which is the potential at which $I = 0$, and the PP, where the current suddenly drops, are indicated in the graphs. If the potential is more positive than the PP, a blocking oxide layer is formed and the anodic current drops to a low value. The silicon sample etches when it is cathodic to the PP and passivated by the formation of oxide when it is anodic to the PP. To create a p–n structure, an n-type epitaxial layer is deposited on the top surface of a p-type wafer. Moreover, an n-type layer can be formed by the doping method (diffusion or ion implantation) or the direct wafer-bonding technique. In the fabrication process, an n-type layer is positively biased (above the PP) with respect to the platinum electrode and the mask layer (e.g., oxide) on a p-type surface is patterned. The wafer is mounted in a specially designed wafer holder, which seals one side of the wafer and allows exposure of p-type silicon to the etchant, as shown in Fig. 5.21a. The p-type silicon remains at the OCP as the majority of the potential drops across the reverse-biased p–n junction. The etching of p-type silicon continues until the diode is destroyed, and n-type silicon is exposed to the etchant, which is passivated by potential. Consequently the etching is stopped. This method can be used to fabricate structures of various shapes by selectively etching away p-type silicon.

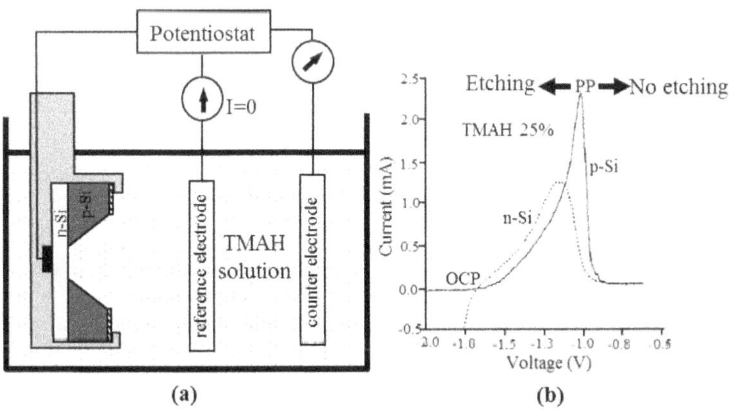

Figure 5.21 Electrochemical etch stop for the fabrication of microstructure: (a) Three electrode experimental setup and (b) I–V characteristics for p- and n-type silicon in 25 wt% TMAH. Reprinted from Ref. [105], Copyright (1995), with permission from Elsevier.

5.7.3 Protection of Convex Corners

In the design of MEMS structures many kinds of shapes such as mesa structures and proof masses for accelerometers include convex corners. The undercutting at convex corners mainly depends on the etchant concentration and its type (i.e., pure or with additive). Fabrication methods of convex corners are mainly classified into two categories, the corner compensation method and a technique based on two-mask lithography and two-step etching [106]. The most popular method is corner compensation, which relies on time-delayed etching through extra patterns at the corners. These methods are covered by large numbers of publications. Surfactant-added TMAH provides minimum undercutting and therefore can be employed for the fabrication of convex (or rounded) corners with minimum distortion. All techniques for the fabrication of convex corners using wet anisotropic etching are reviewed and discussed in the next chapter.

5.7.4 Effect of Mechanical Agitation

A wet-etched profile is strongly affected by mechanical agitation, especially in surfactant-added TMAH where surfactant adsorption on the silicon surface alters the etching characteristics [107]. The shapes of the microstructures fabricated by etching in TMAH+Triton without and with mechanical agitation are presented in Fig. 5.22. It can be easily noticed that the mechanical agitation increases the undercutting at mask edges and convex corners. The mechanical agitation directly influences the attachment/detachment of the surfactant molecules, promptly reducing the thickness of the adsorbed surfactant layer that results in increase of undercutting. The increased amount of undercutting due to agitation confirms that the nature of the surfactant adsorption on the developed silicon surfaces during etching is a physisorption process. If it is chemisorption, the agitation of the solution by mechanical agitation would minimally affect the undercutting at the mask edges. The etching solution (or silicon samples/wafers) can be agitated by various ways such as ultrasonic agitation, magnetic stirrer, etc. The amount of undercutting may vary with the strength and type of agitation. The undercutting results shown in Fig. 5.22 indicate that any type of solution and/or sample agitation should be avoided during etching in TMAH+surfactant if undercutting is to be minimized.

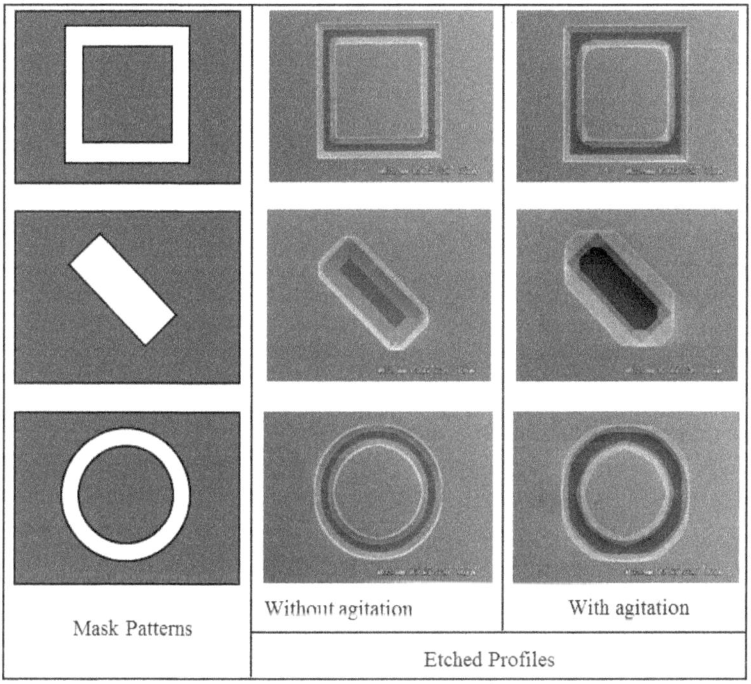

Figure 5.22 Etched shapes on Si{100} surface without and with mechanical agitation in 25 wt% TMAH + 0.1 vol% Triton at 71°C (Etch depth = 36 μm). From Ref. [107]. @ IOP Publishing. Reproduced with permission. All rights reserved.

5.7.5 Role of the Open-Circuit Potential

The different surface orientations of silicon have different electrostatic potentials and, thus, may interact galvanically, leading to the onset of similar corrosion phenomena as in metals [108]. In wet etching, the galvanic interaction between different facets of the etching front affects the etched profile [109, 110]. For instance, the OCPs of the more reactive {100} and {110} surfaces are similar (about

−1450 mV) and more negative than the one for the nobler {111} surface (about −1300 mV) [59]. If these surfaces are simultaneously immersed in an electrolyte, they may form a galvanic cell, depending on the relative areas of the various surfaces. This can result in the cathodic protection of the noble surface {111}, experiencing a reduction in the etch rate, while {100} and {110} will see an increase. As an example, the etch rate of {100} has been reported to gradually increase during the formation of V-shaped {111}-delimited trenches as the overall exposed surface continuously changes from {100} to {111} domination and the OCP changes correspondingly (gradual galvanic protection) [109].

To investigate the effect of the galvanic interaction during etching, especially on the amount of undercutting at the mask edges, different wafers with increasing size of the mask openings are utilized, as shown in Fig. 5.23 [107]. The wafer marked as 1 comprises the mask opening of width w_1. In the second wafer, extra patterns of larger opening of width w_2 are added to the previous pattern. Similarly, for the third and fourth wafers, additional patterns of larger openings of widths w_3 and w_4 are added to each previous pattern, respectively. All wafers are etched together in the same solution and for the same time. The vertical etching of the {100} surface is eventually self-stopped by the formation of {111}-delimited V-grooves. Upon going from the first to the fourth wafer, the amount of exposed {100} area increases. As a result, the moment in time when {100} disappears and {111} becomes the dominant surface in the etch front occurs earlier for wafer 1, being followed by wafers 2, 3, and finally 4. In this manner, the dependence of the amount of undercutting on the relative area ratio of {111} to {100} can be investigated. Figure 5.23 presents SEM pictures of microstructures fabricated on wafers 1–4 and the amount of undercutting in surfactant (Triton)-added TMAH. It can be easily noticed that the largest undercutting is observed on wafer 1, where the vertical etching is stopped first due to displaying the smallest opening width w_1. Then follow wafers 2 and 3, in that order, since the vertical etching also stops eventually. Finally, wafer 4 presents the smallest-measured undercutting, correlating with the fact that the vertical etching still remains active even after 8 h of etching.

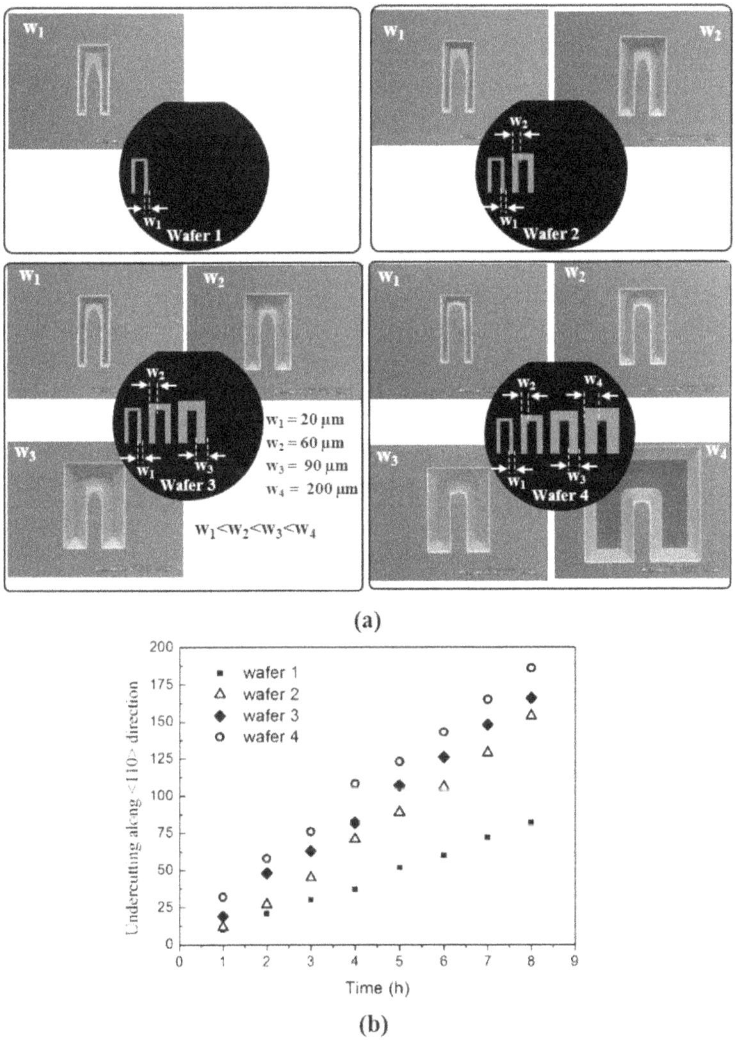

Figure 5.23 Effect of galvanic interaction on the etching behavior of TMAH+Triton, especially on the amount of undercutting: (a) SEM pictures with mask patterns, and (b) undercutting at convex corner along the <110> direction as a function of time for wafers 1, 2, 3, and 4. From Ref. [107]. @ IOP Publishing. Reproduced with permission. All rights reserved.

The etch rate of silicon is the sum of both chemical and electrochemical reactions, and the electrochemical term may become the dominant contribution, not only when an external potential is applied, but also when the galvanic interaction between the developing facets becomes large enough [111]. As the area ratio of {111} to all other orientations becomes sufficiently large during etching (i.e., the onset of the electrochemical etch rate), galvanic protection of {111} by the other orientations leads effectively to an increase in the amount of undercutting at the mask edges of straight and/or bent V-grooves increases significantly after the grooves have been formed. Surfactant-added TMAH is primarily employed to achieve minimal undercutting at the mask edges. Therefore, to reduce the undercutting at the mask edges, either etching should be stopped as soon as the grooves are formed or a large enough sacrificial {100} area must be used so that a large fraction of the etch front remains {100} oriented after the grooves are formed. If possible, the latter method is recommended [107].

5.8 Summary

The etching characteristics of TMAH strongly depend on the etchant concentration and the type and concentration of the additive. Etching temperature mainly influences the etch rates of silicon and the masking layer. Considering the effect of the different etching parameters, the selection of the etchant (i.e., without or with additive, low or high concentration) depends on the type of structure to be fabricated. The following points must be kept in mind while selecting a TMAH solution:

- If etched surface morphology is not an issue, low-concentration TMAH should be selected as it provides a high etch rate of the {100} orientation.
- To achieve smooth surface morphology, high-concentration TMAH (20–25 wt%) should be preferred.
- In the fabrication of suspended structures using one side of the wafer, high undercutting is desirable and therefore high concentration (20–25 wt%) should be preferred for the etching process as it provides significant undercutting at convex corners.
- To achieve minimum undercutting with smooth surface morphology on a Si{100} wafer, 25 wt% TMAH with a surfactant should be used.

- The etch rates of the Si{110} surface and its vicinal orientations are suppressed to a significantly low level in surfactant-added TMAH. Therefore pure TMAH should be employed for the formation of microstructures on Si{110} wafers.
- The effect of a surfactant is highly influenced by mechanical agitation, and therefore no agitation should be used in the solution and/or samples if minimum undercutting at the mask edges is desired.
- Due to galvanic protection of the developed {111} facets, the undercutting rate at the mask edges of straight and/or bent V-grooves increases significantly after the grooves have been formed. To minimize the undercutting at the mask edges, two workarounds are provided: (i) to use a large enough sacrificial {100} area so that a large fraction of the etch front remains {100} oriented after the grooves are formed and/or (ii) to stop the etching process as soon as the grooves have been formed. If possible, solution (i) is recommended.

The above-mentioned points are elaborated in Chapter 8, where the applications of wet bulk micromachining for the fabrication of simple to complex MEMS structures are explored.

References

1. Lee, D. B. (1969). Anisotropic etching of silicon, *J. Appl. Phys.*, **40**, pp. 4569–4575.
2. Bean, K. E. (1978). Anisotropic etching of silicon, *IEEE Trans. Electron Devices*, **ED-25**, pp. 1185–1193.
3. Seidel, H., Csepregi, L., Heuberger, A., and Baumgartel, H. (1990). Anisotropic etching of crystalline silicon in alkaline solutions I: orientation dependence and behavior of passivation layers, *J. Electrochem. Soc.*, **137**, pp. 3612–3626.
4. Seidel, H., Csepregi, L., Heuberger, A., and Baumgartel, H. (1990). Anisotropic etching of crystalline silicon in alkaline solutions II: influence of dopants, *J. Electrochem. Soc.*, **137**, pp. 3626–3632.
5. Sato, K., Shikida, M., Matsushima, Y., Yamashiro, T., Asaumi, K., Iriye, Y., and Yamamoto, M. (1998). Characterization of orientation-dependent etching properties of single-crystal silicon: effects of KOH concentration, *Sens. Actuators A*, **61**, pp. 87–93.

6. Pal, P., Ashok, A., Haldar, S., Xing, Y., and Sato, K. (2015). Anisotropic etching in low concentration KOH: effects of surfactant concentration, *Micro Nano Lett.*, **10**, pp. 224–228.
7. Powell, O., and Harrison, H. B. (2001). Anisotropic etching of {100} and {110} planes in (100) silicon, *J. Micromech. Microeng.*, **11**, pp. 217–220.
8. Tanaka, H., Yamashita, S., Abe, Y., Shikida, M., and Sato, K. (2004). Fast etching of silicon with a smooth surface in high temperature ranges near the boiling point of KOH solution, *Sens. Actuators A*, **114**, pp. 516–520.
9. Matsuoka, M., Yoshida, Y., and Moronuki, M. (1992). Preparation of silicon thin diaphragms free from micropyramids using anisotropic etching in KOH solution, *J. Chem. Eng.*, **25**, pp. 735–740.
10. Baryeka, I., and Zubel, I. (1995). Silicon anisotropic etching in KOH-isopropanol etchant, *Sens. Actuators A*, **48**, pp. 229–238.
11. Shikida, M., Sato, K., Tokoro, K., and Uchikawa, D. (2000). Differences in anisotropic etching properties of KOH and TMAH solutions, *Sens. Actuators A*, **80**, pp. 179–188.
12. Backlund, Y., and Rosengren, L. (1992). New shapes in (100) Si using KOH and EDP etches, *J. Micromech. Microeng.*, **27**, pp. 5–9.
13. Tabata, O., Asahi, R., Funabashi, H., Shimaoka, K., and Sugiyama, S. (1992). Anisotropic etching of silicon in TMAH solutions, *Sens. Actuators A*, **34**, pp. 51–57.
14. Sato, K., Shikida, M., Yamashiro, T., Asaumi, K., Iriye, Y., and Yamamoto, M. (1999). Anisotropic etching rates of single-crystal silicon for TMAH water solution as a function of crystallographic orientation, *Sens. Actuators A*, **73**, pp. 131–137.
15. Pal, P., Sato, K., Gosalvez, M. A., Tang, B., Hida, H., and Shikida, M. (2011). Fabrication of novel microstructures based on orientation dependent adsorption of surfactant molecules in TMAH solution, *J. Micromech. Microeng.*, **21**, p. 015008.
16. Pal, P., and Sato, K. (2010). Fabrication methods based on wet etching process for the realization of silicon MEMS structures with new shapes, *Microsyst. Technol.*, **16**, pp. 1165–1174.
17. Gosalvez, M. A., Tang, B., Pal, P., Sato, K., Kimura, Y., and Ishibashi, K. (2009). Orientation and concentration dependent surfactant adsorption on silicon in aqueous alkaline solutions: explaining the changes in the etch rate, roughness and undercutting for MEMS applications, *J. Micromech. Microeng.*, **19**, p. 125011.

18. Tellier C. R., and Charbonnieras, A. R. (2003). Characterization of the anisotropic chemical attack of (hhl) silicon plates in a TMAH 25 wt% solution: micromachining and adequacy of the dissolution slowness surface, *Sens. Actuators A*, **105**, pp. 62–75.

19. Steinsland, E., Finstad, T., and Hanneborg, A. (2000). Etch rates of (100), (111), and (110) single-crystal silicon in TMAH measured in situ by laser reflectance interferometry, *Sens. Actuators A*, **86**, pp. 73–80.

20. Tang, B., Yao, M. Q., Tan, G., Pal, P., Sato, K., and Su, W. (2014). Smoothness control of wet etched Si{100} surfaces in TMAH+Triton, *Key Eng. Mater.*, **609**, pp. 536–541.

21. Shikida, M., Masuda, T., Uchikawa, D., and Sato K. (2001). Surface roughness of single-crystal silicon etched by TMAH solution, *Sens. Actuators A*, **90**, pp. 223–231.

22. Tang, B., Sato, K., Zhang, D., and Cheng, Y. (2014). Fast Si (100) etching with a smooth surface near the boiling temperature in surfactant modified tetramethylammonium hydroxide solutions, *Micro Nano Lett.*, **9**, pp. 582–584.

23. Chung, G. S. (2005). Anisotropic etching characteristics of Si in tetramethylammonium hydroxide: isopropyl alcohol: pyrazine solutions, *J. Korean Phys. Soc.*, **46**, pp. 1152–1156.

24. Choi, W. K., Thong, J. T. L., Luo, P., Tan, C. M., Chua, T. H., and Bai, Y. (1998). Characterisation of pyramid formation arising from the TMAH etching of silicon, *Sens. Actuators A*, **71**, pp. 238–243.

25. Resnik, D., Vrtacnik, D., Aljancic, U., and Amon, S. (2003). Effective roughness reduction of {100} and {311} planes in anisotropic etching of {100} silicon in 5% TMAH, *J. Micromech. Microeng.*, **13**, pp. 26–34.

26. Sakaino, K., and Adachi, S. (2001). Study of Si(100) surfaces etched in TMAH solution, *Sens. Actuators A*, **88**, pp. 71–78.

27. Resnik, D., Vrtacnik, D., Aljancic, U., and Amon, S. (2000). Wet etching of silicon structures bounded by (311) sidewalls, *Microelectron. Eng.*, **51–52**, pp. 555–566.

28. Holke, A., and Henderson, H. T. (1999). Ultra-deep anisotropic etching of (110) silicon, *J. Micromech. Microeng.*, **9**, pp. 51–57.

29. Thong, J. T. L., Choi, W. K., and Chong, C. W. (1997). TMAH etching of silicon and the interaction of etching parameters, *Sens. Actuators A*, **63**, pp. 243–249.

30. Zhang, J., Hon, W. C., Leung, L. L. W., and Chen, K. J. (2005). CMOS-compatible micromachining techniques for fabricating high-

performance edge-suspended RF/microwave passive components on silicon substrates, *J. Micromech. Microeng.*, **15**, pp. 328–335.

31. Pal, P., Sato, K., Gosalvez, M. A., and Shikida, M. (2007). Study of rounded concave and sharp edge convex corners undercutting in CMOS compatible anisotropic etchants, *J. Micromech. Microeng.*, **17**, pp. 2299–2307.
32. Mukhiya, R., Bagolini, A., Bhattacharyya, T. K., Lorenzelli, L., and Zen, M. (2011). Experimental study and analysis of corner compensation structures for CMOS compatible bulk micromachining using 25 wt% TMAH, *Microelectron. J.*, **42**, pp. 127–134.
33. Chen, P. H., Peng, H. Y., Hsieh, C. M., and Chyu, M. K. (2001). The characteristic behavior of TMAH water solution for anisotropic etching on both silicon substrate and SiO_2 layer, *Sens. Actuators A*, **93**, pp. 132–137.
34. Yan, G., Chan Philip, C. H., Hsing, I. M., Sharma, R. K., Sin, J. K. O., and Wang, Y. (2001). An improved TMAH Si-etching solution without attacking exposed aluminum, *Sens. Actuators A*, **89**, pp. 135–141.
35. Brida, S., Faes, A., Guarnieri, V., Giacomozzi, F., Margesin, B., Paranjape, M., Pignatel, G. U., and Zen, M. (2000). Microstructures etched in doped TMAH solutions, *Microelectron. Eng.*, **53**, pp. 547–551.
36. Fujitsuka, N., Hamaguchi, K., Funabashi, H., Kawasaki, E., and Fukada, T. (2004). Silicon anisotropic etching without attacking aluminum with Si and oxidizing agent dissolved in TMAH solution, *Sens. Actuators A*, **114**, pp. 510–515.
37. Tasur, J. J., Du, C. H., and Lee, C. (2001). Investigation of TMAH for front-side bulk micromachining process from manufacturing aspect, *Sens. Actuators A*, **92**, pp. 375–383.
38. Paranjape, M., Pandy, A., Brida, S., Landsberger, L., Kahrizi, M., and Zen, M. (2000). Dual-doped TMAH silicon etchant for microelectromechanical structures and systems applications, *J. Vac. Sci. Technol. A*, **18**, pp. 738–742
39. Pandy, A., Landsberger, L. M., Nikpour, B., Paranjape, M., and Kahrizi, M. (1998). Experimental investigation of high Si/Al selectivity during anisotropic etching in tetra-methyl ammonium hydroxide, *J. Vac. Sci. Technol. A*, **16**, pp. 868–872.
40. Sundaram, K. B., Vijayakumar, A., and Subramanian, G. (2005). Smooth etching of silicon using TMAH and isopropyl alcohol for MEMS applications, *Microelectron. Eng.*, **77**, pp. 230–241.

41. Zubel, I., and Kramkowska, M. (2001). The effect of isopropyl alcohol on etching rate and roughness of (100) Si surface etched in KOH and TMAH solutions, *Sens. Actuators A*, **93**, pp. 138–147.

42. Merlos, A., Acero, M., Bao, M. H., Bausells, J., and Esteve, J. (1993). TMAH/IPA anisotropic etching characteristics, *Sens. Actuators A*, **37–38**, pp. 737–743.

43. Cheng, D., Gosalvez, M. A., Hori, T., Sato, K., and Shikida, M. (2006). Improvement in smoothness of anisotropically etched silicon surfaces: effects of surfactant and TMAH concentrations, *Sens. Actuators A*, **125**, pp. 415–421.

44. Pal, P., Gosalvez, M. A., and Sato, K. (2010). Silicon micromachining based on surfactant-added tetramethylammonium hydroxide: etching mechanism and advanced application, *Jpn. J. Appl. Phys.*, **49**, p. 056702.

45. Gosalvez, M. A., Pal, P., Tang, B., and Sato, K. (2010). Atomistic mechanism for the macroscopic effects induced by small additions of surfactants to alkaline etching solutions, *Sens. Actuators A*, **157**, pp. 91–95.

46. Tang, B., Pal, P., Gosalvez, M. A., Shikida, M., Sato, K., Amakawa, H., and Itoh, S. (2009). Ellipsometry study of the adsorbed surfactant thickness on Si{110} and Si{100} and the effect of pre-adsorbed surfactant layer on etching characteristics in TMAH, *Sens. Actuators A*, **156**, pp. 334–341.

47. Pal, P., Sato, K., Gosalvez, M. A., Kimura, Y., Ishibashi, K., Niwano, M., Hida, H., Tang, B., and Itoh, S. (2009). Surfactant adsorption on single crystal silicon surfaces in TMAH solution: orientation-dependent adsorption detected by in-situ infra-red spectroscopy, *J. Microelectromech. Syst.*, **18**, pp. 1345–1356.

48. Resnik, D., Vrtacnik, D., Aljancic, U., Mozek, M., and Amon, S. (2005). The role of Triton surfactant in anisotropic etching of {110} reflective planes on (100) silicon, *J. Micromech. Microeng.*, **15**, pp. 1174–1183.

49. Ashok, A., and Pal, P. (2016). Silicon micromachining in 25 wt% TMAH without and with surfactant concentrations ranging from ppb to ppm, *Microsyst. Technol.*, pp. 1–8.

50. Yang, C. R., Chen, P. Y., Yang, C. H., Chiou, Y. C., and Lee, R. T. (2005). Effects of various ion-typed surfactants on silicon anisotropic etching properties in KOH and TMAH solutions, *Sens. Actuators A*, **119**, pp. 271–281.

51. Yang, C. R., Yang, C. H., and Chen, P. Y. (2005). Study on anisotropic silicon etching characteristics in various surfactant-added tetramethyl

ammonium hydroxide water solutions, *J. Micromech. Microeng.*, **15**, pp. 2028–2037.

52. Sato, K., Uchikawa, D., and Shikida, M. (2001). Change in orientation-dependent etching properties of single-crystal silicon caused by a surfactant added to TMAH solution, *Sens. Mater.*, **13**, pp. 285–291.
53. Sarro, P. M., Brida, D., van der Vlist, W., and Brida, S. (2000). Effect of surfactant on surface quality of silicon microstructures etched in saturated TMAHW solutions, *Sens. Actuators A*, **85**, pp. 340–345.
54. Sekimura, M. (1999). Λnisotropic ctching of surfactant-added TMAII solution, *Proc. IEEE Microelectromech. Syst.*, pp. 17–21.
55. Zubel, I., Kramkowska, M., and Rola, K. (2012). Silicon anisotropic etching in TMAH solutions containing alcohol and surfactant additives, *Sens. Actuators A*, **178**, pp. 126–135.
56. Fruhauf, J. (2005). *Shape and Functional Elements of the Bulk Silicon Microtechnique: A Manual of Wet-Etched Silicon Structures* (Springer, Berlin).
57. Wind, R. A., and Hines, M. A. (2000). Macroscopic etch anisotropies and microscopic reaction mechanisms: a micromachined structure for the rapid assay of etchant anisotropy, *Surf. Sci.*, **460**, pp. 21–38.
58. Wind, R. A., Jones, H., Little, M. J., and Hines, M. A. (2002). Orientation-resolved chemical kinetics: using microfabrication to unravel the complicated chemistry of KOH/Si etching, *J. Phys. Chem. B*, **106**, pp. 1557–1569.
59. Nguyen, Q. D. (2007). Electrochemistry in anisotropic etching of silicon in alkaline solutions: a kinematic wave analysis, *PhD Thesis*, University of Twente, The Netherlands.
60. Gosalvez, M. A., Pal, P., Ferrando, N., Hida, H., and Sato, K. (2011). Experimental procurement of the complete 3D etch rate distribution of Si in anisotropic etchants based on vertically micromachined wagon wheel samples, *J. Micromech. Microeng.*, **21**, p. 125007.
61. Gosalvez, M. A., Pal, P., Hida, H., and Sato, K. (2011). Reliability assessment of the complete 3D etch rate distribution of Si in anisotropic etchants based on vertically micromachined wagon wheel samples, *J. Micromech. Microeng.*, **21**, p. 125008.
62. Gosalvez, M. A., Pal, P., and Sato, K. (2011). Reconstructing the 3D etch rate distribution of silicon in anisotropic etchants using data from vicinal {100}, {110} and {111} surfaces, *J. Micromech. Microeng.*, **21**, p. 105018.

63. Xu, Y. W., Michael, A., and Kwok, C. Y. (2011). Formation of ultra-smooth 45° micromirror on (100) silicon with low concentration TMAH and surfactant: techniques for enlarging the truly 45° portion, *Sens. Actuators A*, **166**, pp. 164–171.

64. Pal, P., and Singh, S. S. (2013). A simple and robust model to explain convex corner undercutting in wet bulk micromachining, *Micro Nano Syst. Lett.*, **1**, pp. 1–6.

65. Pal, P., and Singh, S. S. (2013). A new model for the etching characteristics of corners formed by Si{111} planes on Si{110} wafer surface, *Engineering*, **5**, pp. 1–8.

66. Pal, P., Haldar, S., Singh, S. S., Ashok, A., Xing, Y., and Sato, K. (2014). A detailed investigation and explanation to the appearance of different undercut profiles in KOH and TMAH, *J. Micromech. Microeng.*, **24**, p. 095026.

67. Trieu, H. K., and Mokwa, W. (1998). A generalized model describing corner undercutting by the experimental analysis of TMAH/IPA, *J. Micromech. Microeng.*, **8**, pp. 80–83.

68. Schroder, H., and Obermeier, E. (2000). A new model for Si{100} convex corner undercutting in anisotropic KOH etching, *J. Micromech. Microeng.*, **10**, pp. 163–170.

69. Shikida, M., Nanbara, K., Koizumi, T., Sasaki, H., Sato, K., Odagaki, M., Ando, M., Furuta, S., and Asaumi, K. (2000). A model explaining mask-corner undercut phenomena in anisotropic silicon etching: a saddle point in the etching-rate diagram, *Sens. Actuators A*, **97–98**, pp. 758–763.

70. Chang Chien, W. T., Chang, C. O., Lo, Y. C., Li, Z. W., and Chou, C. S. (2005). On the Miller-indices determination of Si{100} convex corner undercut planes, *J. Micromech. Microeng.*, **15**, pp. 833–842.

71. Merlos, A., Acero, M. C., Bao, M. H., Bausells, J., and Esteve, J. (1992). A study of the undercutting characteristics in the TMAH:IPA system, *J. Micromech. Microeng.*, **2**, pp. 181–183.

72. Dong, W., Zhang, X., Liu, C., Li, M., Xu, B., and Chen, W. (2004). Mechanism for convex corner undercutting of (110) silicon in KOH, *Microelectron. J.*, **35**, pp. 417–419.

73. Pal, P., and Sato, K. (2009). Various shapes of silicon freestanding microfluidic channels and microstructures in one step lithography, *J. Micromech. Microeng.*, **19**, p. 055003.

74. Pal, P., and Chandra, S. (2004). Bulk-micromachined structures inside anisotropically etched cavities, *Smart Mater. Struct.*, **13**, pp. 1424–1429.

75. Hui, W. C. (1991). Technique for protecting chip corners in wet chemical etching of silicon wafers, *Technical Report Lawrence Livermore National Lab.*, CA, United States.

76. Wacogne, B., Zeggari, R., Sadani, Z., and Gharbi, T. (2006). A very simple compensation technique for bent V-grooves in KOH etched (100) silicon when thin structures or deep etching are required, *Sens. Actuators A*, **126**, pp. 264–269.

77. Pal, P., Sato, K., Shikida, M., and Gosalvez, M. A. (2009). Study of corner compensating structures and fabrication of various shapes of MEMS structures in pure and surfactant added TMAH, *Sens. Actuators A*, **154**, pp. 192–203.

78. Jeon, J. S., Raghavan, S., and Carrejo, J. P. (1996). Effect of temperature on the interaction of silicon with nonionic surfactants in alkaline solutions, *J. Electrochem. Soc.*, **143**, pp. 277–283.

79. Jeon, J. S., Raghavan, S., and Sperline, R. P. (1995). Behavior of polyethylene oxide based nonionic surfactants in silicon processing using alkaline solutions, *J. Electrochem. Soc.*, **142**, pp. 621–627.

80. Haiss, W., Raisch, P., Bitsch, L., Nichols, R. J., Xia, X., Kelly J. J., and Schiffrin, D. J. (2006). Surface termination and hydrogen bubble adhesion on Si(100) surfaces during anisotropic dissolution in aqueous KOH, *J. Electroanal. Chem.*, **597**, pp. 1–12.

81. Haiss, W., Raisch, P., Schiffrin, D. J., Bitsch, L., and Nichols, R. J. (2002). An FTIR study of the surface chemistry of the dynamic Si(100) surface during etching in alkaline solution, *Faraday Discuss.*, **121**, pp. 167–180.

82. Niwano, M., Terashi, M., and Kuge, J. (1999). Hydrogen adsorption and desorption on Si(100) and Si(111) surfaces investigated by in situ surface infrared spectroscopy, *Surf. Sci.*, **420**, pp. 6–16.

83. Shinoharaa, M., Katagiri, T., Iwatsuji, K., Matsuda, Y., Fujiyama, H., Kimura, Y., and Niwano, M. (200). Oxidation of the hydrogen terminated silicon surfaces by oxygen plasma investigated by in-situ infrared spectroscopy, *Thin Solid Films*, **475**, pp. 128–132.

84. Boonekamp, E. P., Kelly, J. J., van de Ven, J., and Sondag, A. H. M. (1994). The chemical oxidation of hydrogen-terminated silicon (111) surfaces in water studied in situ with Fourier transform infrared spectroscopy, *J. Appl. Phys.*, **75**, pp. 8121–8127.

85. Pietsch, G. J., Chabal, Y. J., and Higashi, G. S. (1995). The atomic-scale removal mechanism during chemo-mechanical polishing of Si(100) and Si(111), *Surf. Sci.*, **331–333**, pp. 395–401.

86. Pietsch, G. J., Chabal, Y. J., and Higashi, G. S. (1995). Infrared-absorption spectroscopy of Si(100) and Si(111) surfaces after chemomechanical polishing, *J. Appl. Phys.*, **78**, pp. 1650–1658.

87. Workman, A. M. A., Raghavan, S., and Sperline, R. P. (2000). In situ ATR-FTIR analysis of surfactant adsorption onto silicon from buffered hydrofluoric acid solutions, *Langmuir*, **16**, pp. 3636–3640.

88. Imanishi, A., Omoda, R., and Nakato, Y. (2006). In-situ FTIR studies on self-assembled monolayers of surfactant molecules adsorbed on H-terminated Si(111) surfaces in aqueous solutions, *Langmuir*, **22**, pp. 1706–1710.

89. Ozanam, F., Djebri, A., and Chazalviel, J. N. (1996). The hydrogenated silicon surface in organic electrolytes probed through in-situ IR spectroscopy in the ATR geometry, *Electrochim. Acta*, **41**, pp. 687–692.

90. Gutmann, V. (1997). *The Donor-Acceptor Approach to Molecular Interactions* (Plenum, New York).

91. Jakob, P., and Chabal, Y. J. (1991). Chemical etching of vicinal Si(111): dependence of the surface structure and the hydrogen termination on the pH of the etching solutions, *J. Chem. Phys.*, **95**, pp. 2897–2909.

92. Smith Brian, C. (1998). *Infrared Spectral Interpretation: A Systematic Approach* (CRC Press, Boca Raton).

93. Tiberg, F. (1996). Physical characterization of non-ionic surfactant layers adsorbed at hydrophilic and hydrophobic solid surfaces by time-resolved ellipsometry, *J. Chem. Soc. Faraday Trans.*, **92**, pp. 531–538.

94. Allongue, P., Costa-Kieling, V., and Gerischer, H. (1993). Etching of silicon in NaOH solutions II. electrochemical studies of n-Si(111) and (100) and mechanism of the dissolution, *J. Electrochem. Soc.*, **140**, pp. 1018–1026.

95. Paria, S., and Khilar, K. C. (2004). A review on experimental studies of surfactant adsorption at the hydrophilic solid-water interface, *Adv. Colloid Interface Sci.*, **110**, pp. 75–95.

96. Gosálvez, M. A., Sato, K., Foster, A. S., Nieminen, R. M., and Tanaka, H. (2007). An atomistic introduction to anisotropic etching, *J. Micromech. Microeng.*, **17**, pp. S1–S26.

97. Tatic-Lucic, S., Zhang, W.-Y., and Navneet, N. (2005). Etch-stop characteristics of heavily B/Ge-doped silicon epilayer in KOH and TMAH, *Sens. Actuators A*, **123–124**, pp. 640–645.

98. Steinsland, E., Nese, M., Hanneborg, A., Bernstein, R. W., Sandmo, H., and Kittilsland, G. (1996). Boron etch-stop in TMAH solutions, *Sens. Actuators A*, **54**, pp. 728–732.

99. Charavel, R., Laconte, J., and Raskin, J. P. (2003). Advantages of p^{++} polysilicon etchstop layer versus p^{++} silicon, *Proc. SPIE*, **5116**, pp. 699–709.

100. Wallman, L., Bengtsson, J., Danielsen, N., and Laurell, T. (2002). Electrochemical etch-stop technique for silicon membranes with p- and n-type regions and its application to neural sieve electrodes, *J. Micromech. Microeng.*, **12**, pp. 265–270.

101. Kloeck, B., Collins, S. D., Rooij, N. F. D., and Smith, R. L. (1989). Study of electrochemical etch-stop for high precision thickness control of silicon membranes, *IEEE Trans. Electron Devices*, **36**, pp. 663–669.

102. Ashruf, C. M. A., French, P. J., Sarro, P. M., Bressers, P. M. M. C., and Kelly, J. J. (1998). Electrochemical etch stop engineering for bulk micromachining, *Mechatronics*, **8**, pp. 585–612.

103. Chung, G.-S. (2008). Study of electrochemical etch-stop for high-precision thickness control of single-crystal Si in aqueous TMAH:IPA:pyrazine solutions, *Microelectron. Eng.*, **85**, pp. 271–277.

104. Waggener, H. A. (1970). Electrochemically controlled thinning of silicon, *Bell Labs Tech. J.*, **49**, pp. 473–475.

105. Acero, M. C., Esteve, J., Burrer, C., and Gotz, A. (1995). Electrochemical etch-stop characteristics of TMAH:IPA solutions, *Sens. Actuators A*, **46**, pp. 22–26.

106. Pal, P., and Sato, K. (2015). A comprehensive review on convex and concave corners in silicon bulk micromachining based on anisotropic wet chemical etching, *Micro Nano Syst. Lett.*, **3**, pp. 1–42.

107. Pal, P., Sato, K., and Gosalvez, M. A. (2012). Etched profile control in anisotropic etching of silicon by TMAH+Triton, *J. Micromech. Microeng.*, **22**, p. 065013.

108. Schweitzer, P. A. (1989). *Corrosion and Corrosion Protection Handbook* (M. Dekker, New York).

109. Kretschmer, H. R., Xia, X. H., Kelly, J. J., and Steckenborn, A. (2004). Anisotropic etching of three-dimensional shapes in silicon: the important role of galvanic interaction, *J. Electrochem. Soc.*, **151**, pp. C633–C636.

110. Philipsen, H. G. G., and Kelly, J. J. (2005). Anisotropy in the anodic oxidation of silicon in KOH solution, *J. Phys. Chem. B*, **109**, pp. 17245–17253.

111. Allongue, P., Henry de Villeneuve, C., Morin, S., Boukherroub, R., and Wayner, D. D. M. (2000). The preparation of flat H-Si(111) surfaces in 40% NH_4F revisited, *Electrochim. Acta*, **45**, pp. 4591–4598.

Chapter 6

Convex and Concave Corners in Silicon Wet Bulk Micromachining

6.1 Introduction

In wet anisotropic etching-based bulk micromachining, stable etched profiles are shaped by slowest-etch-rate planes (i.e., {111}) [1–9]. In the case of a Si{100} wafer, the {111} planes appear at the edges aligned along the <110> directions, as shown in Fig. 6.1 [10]. These planes form a square/rectangular cavity/groove. In the case of a Si{110} wafer, six {111} planes are exposed during the etching process. The intersection of the {111} planes at the {110} surface forms a polygon, as shown in Fig. 6.2 [11]. The edges of the polygon structure are oriented along the <112> and <110> directions. Using the <112> and <110> directions, different kinds of patterns can be formed. The <112> direction contains vertical {111} planes, whereas the <110> direction comprises slanted {111} planes. The <112> direction can be used to form a parallelogramic structure. The intersection of {111} planes forms either concave corners (<180°, that is, corners turning inside) or convex corners (>180°, that is, corners turning outside), as shown in Fig. 6.3. Although both types of corners are bounded by {111} planes, their etching characteristics are exactly opposite to each other. The concave corner remains intact during the etching, while convex corners are immediately attacked by the etchant, leading to undercutting. The rate of undercutting depends on the type of etchant employed for etching. Hence the

Silicon Wet Bulk Micromachining for MEMS
Prem Pal and Kazuo Sato

ISBN 978-981-4613-72-9 (Hardcover), 978-1-315-36492-6 (eBook)
www.panstanford.com

fabrication of structures with protected convex corners is a tedious task for both {100} and {110} silicon wafers. The mechanism of undercutting [10–21] and the fabrication of convex corners [22–65] are widely investigated. In this chapter, the fabrication methods of convex corners on {100} and {110} silicon wafers using silicon bulk micromachining based on anisotropic wet chemical etching are discussed. The main reason behind undercutting, protection of convex corners, the advantages and disadvantages of undercutting in silicon-based microelectromechanical system (MEMS) fabrication, etc., are explained in detail.

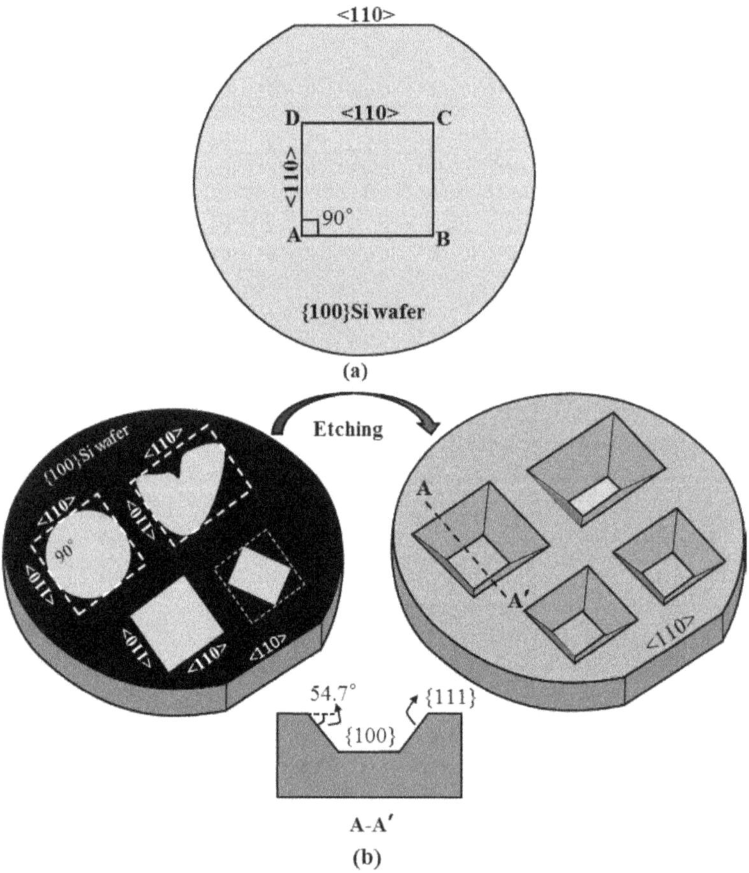

Figure 6.1 Schematic illustration of a square-shaped pattern formed on a Si{100} wafer by the directions at which the {111} planes appear during etching: (a) <110> directions on the wafer surface and (b) etched profiles of mask geometries of different shapes.

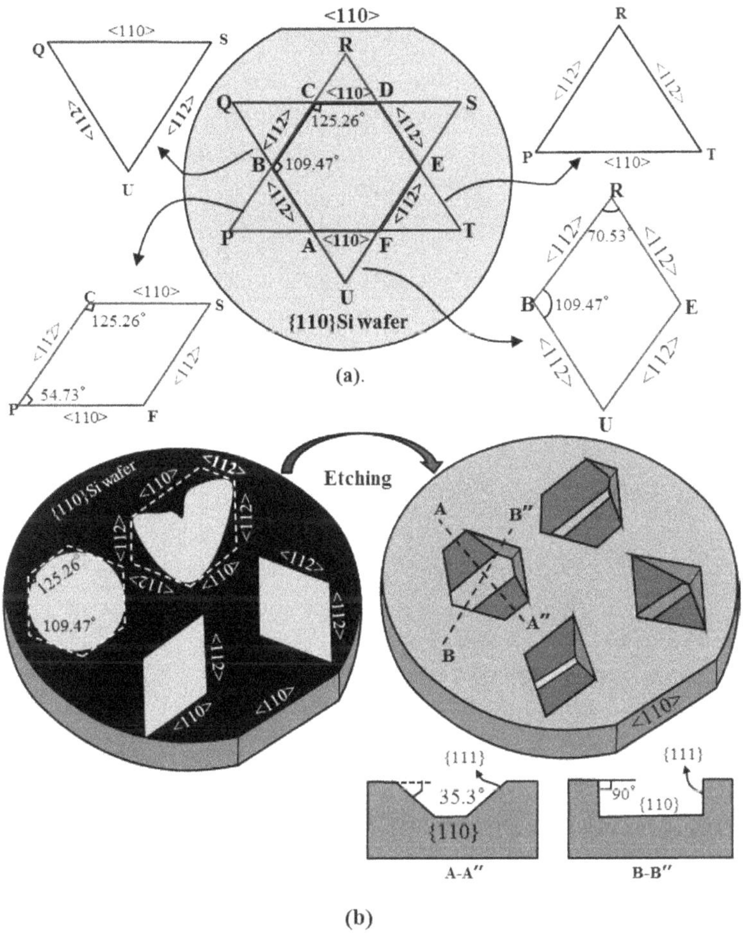

Figure 6.2 Schematic representation of different shapes formed on Si{110} by the directions at which the {111} planes appear during etching: (a) <112> and <110> directions on the wafer surface and (b) etched profile of mask geometries of different shapes.

6.2 Etch Rate, Underetching, and Undercutting

To select an etchant for silicon wet anisotropic etching, various parameters such as etch rate, anisotropy, handling, etching selectivity over other materials (especially with masking material), complementary metal-oxide-semiconductor (CMOS) compatibility, undercutting, etc., are considered. The etch rate and undercutting are primarily used to determine the etching time and the etched shape of a structure. The etch rate is a measure of the thickness removed per unit time. It is measured along the direction perpendicular to the surface being etched. The etch rate of a crystallographic plane $\{hkl\}$ (i.e., $R_{\{hkl\}}$) can be expressed by the following relation:

$$R_{\{hkl\}} = \frac{d_{\{hkl\}}}{t}$$

where $d_{\{hkl\}}$ and t are the etch depth measured perpendicular to the $\{hkl\}$ plane and the etching time, respectively.

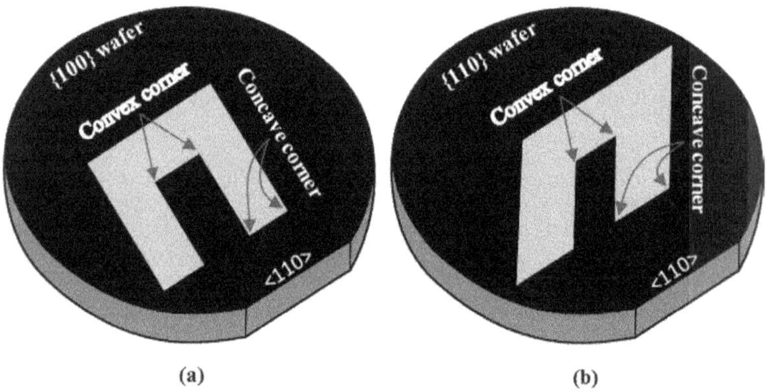

Figure 6.3 A mask pattern comprising concave and convex corners on (a) {100} and (b) {110} silicon wafers.

The etch rate is a very useful parameter of an anisotropic etchant as it is used to estimate other parameters such as throughput, anisotropy ratio between two planes (particularly amongst {100}, {110}, and {111} planes), undercutting, etc. The absolute and relative etch rates of different crystallographic planes are usually a strong function of the etchant type, concentration, and etching temperature. Moreover, the etch rates are significantly affected by the impurities (or additives) in the etchant. A low concentration of isopropyl alcohol (IPA) in KOH [66–71] and a very small concentration of surfactant (e.g., 0.1% v/v) in tetramethylammonium hydroxide (TMAH) [72–85] ensures a significant reduction in the etch rates of {110} and its vicinal planes (or the planes between {110} and

{111} surfaces, such as {441}, {331}, {221}, etc.). The etching characteristics of low-concentration KOH are also altered when a small amount of surfactant is added to it [86–88]. The etch rates of principal orientations (i.e., {100}, {110}, and {111}) can easily be determined using {100}, {110}, and {111} silicon wafers, which are commonly available in the market. As discussed in previous chapters, the detailed data on the crystal orientation dependence of the etch rate, either a wagon-wheel structure or a silicon hemisphere, is used. The widest application of the wagon-wheel pattern is to know the relation between the lateral etch rate and the crystallographic direction on different orientation surfaces (e.g., {100}, {110}). As the Si{111} planes have the lowest etch rate in anisotropic wet chemical etching, the direction containing {111} planes (e.g., <110> and <112>) exhibits the minimum lateral etching.

Undercutting and underetching are the lateral etching which occurs under the masking layer. Mostly, the words "undercutting" and "underetching" are used interchangeably. In Ref. [1], underetching is specifically used to define the etching under the mask edges which do not contain extruded/convex corners. This kind of underetching takes place due to the misalignment of mask edges or/and owing to finite etching of the {111} planes. If the mask patterns include convex corners, the underetching at convex corners is termed "undercutting." Hereafter, mainly "undercutting" is used for the lateral etching at any type of mask edge and corner. In the case of {100} wafers, significant undercutting takes place at the edges aligned along non-<110> directions if an etchant is used without any additives (i.e., pure KOH or TMAH). To fabricate a square/rectangular cavity of controlled dimensions in {100} silicon wafers, the sides of the square/rectangular mask opening are aligned along the <110> directions as the mask edges aligned with these directions exhibit minimum undercutting owing to the appearance of {111} planes. This type of cavity comprises four concave corners which are formed by well-defined {111} planes. Thus, the limiting shape at the concave corner is bounded by the slowest-etching planes; as a result these corners exhibit no undercutting in wet anisotropic etchants. If the sides are slightly misaligned from the <110> direction, the undercutting starts at the edges, resulting in a cavity of bigger dimensions than the requirement, as shown in Fig. 6.1.

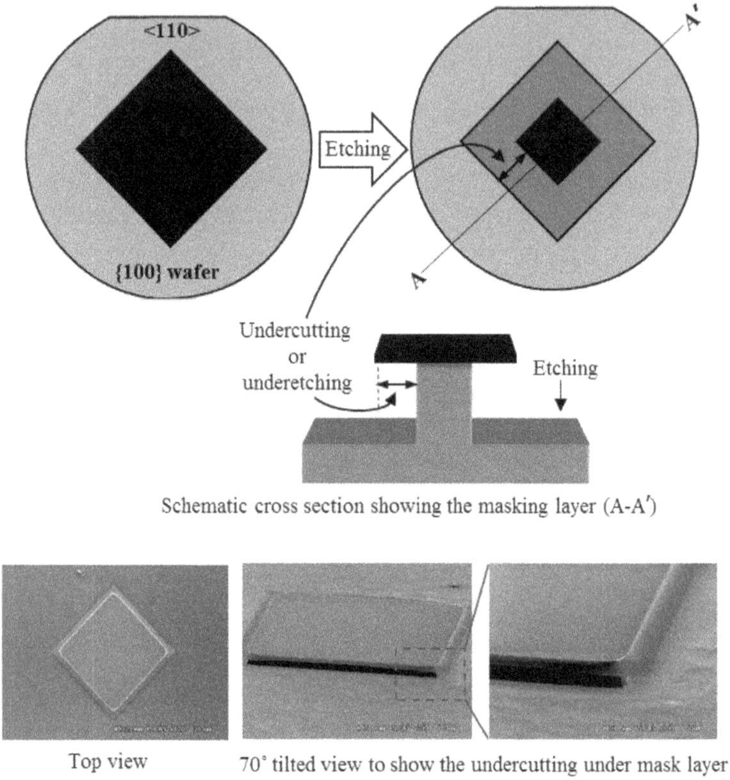

Figure 6.4 Schematic diagram with SEM pictures illustrating undercutting and etching.

Figure 6.4 illustrates the undercutting at the mask edges aligned along the <100> direction on a {100} wafer. The undercutting at sharp convex corners on {100} and {110} surfaces is shown in Figs. 6.5 and 6.6, respectively. In this book, the sharp convex and concave corners are always referred to as the corners bounded by only {111} planes. These corners are bounded by two mutually perpendicular <110> directions on a {100} silicon wafer, as shown in Fig. 6.5. Like the undercutting, the emergence of a facet at convex corners depends on several parameters such as etchant type, concentration, temperature, and the types of additive. Due to undercutting, a mask opening of any arbitrary shape on a {100} wafer finally acquires a square/rectangular cavity after a sufficient time (or prolonged time) of wet etching, as shown in Fig. 6.1b. The sidewalls of the cavity are the slowest {111} planes. In the case of a {110} surface, except the parallelogramic structures formed by <112> and <110> directions, any kind of mask opening become a polygon-shaped cavity after a sufficiently long etching time, as illustrated in Fig. 6.2b. The final shape and size can be determined using etching parameters and trigonometric relations. The ratio of the undercut length (l) to the etch depth (d) is defined as the undercutting ratio ($U_r = l/d$).

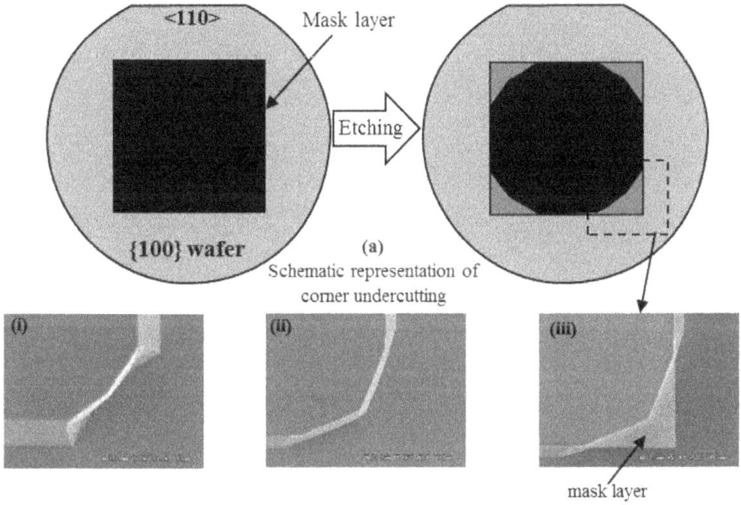

(b) SEM micrographs of the undercut convex corners without and with mask layer fabricated in (i) 40 wt% KOH and (ii)-(iii) 25 wt% TMAH

Figure 6.5 Undercutting at the convex corners of the mask pattern on a Si{100} wafer.

Alignment of mask edges along crystallographic directions on the wafer surface plays a significant role in controlling the shape and size of etched profiles. To align the mask patterns with respect to the crystallographic directions on the wafer surface, a primary flat is commonly employed as reference. In this case, any degree of misorientation in the primary flat leads to the misalignment of mask patterns with respect to crystallographic directions. A small degree of misalignment of the mask edge with the crystallographic direction results in an oversized microstructure due to the undercutting at the misaligned mask edges. Hence, in the fabrication of silicon-based MEMS structures using wet etching, a high-precision alignment of the mask pattern to the crystal orientation is desirable in order to control the dimensions of fabricated structures. Several studies have been performed for the precise alignment of mask patterns with respect to crystallographic directions on {110} and {100} silicon wafers [89–95]. All these techniques are based on the development of mask patterns to create a pre-etched pattern for the identification of crystallographic directions, for instance, <110> directions on a {100}-oriented wafer surface. A detailed description of these methods has been presented in the next chapter.

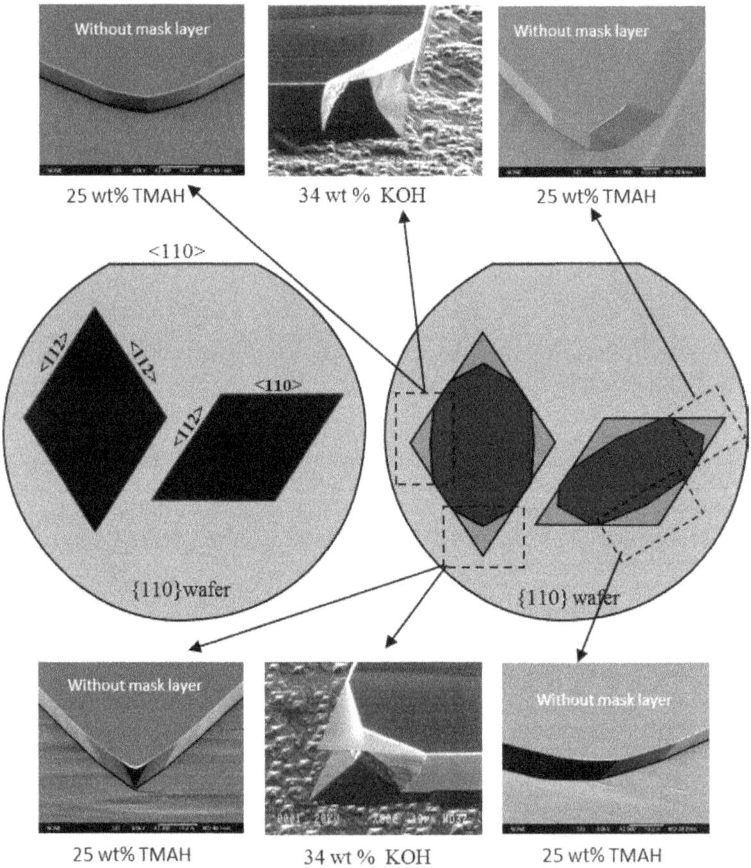

Figure 6.6 Undercutting at the convex corners of the mask pattern on a Si{110} wafer in 25 wt% TMAH and 34 wt% KOH.

6.3 Role of Corner Undercutting in MEMS Fabrication

In the fabrication of wet-bulk-micromachined structures, undercutting at convex corners plays a significant role. It has its own advantages and disadvantages in the fabrication process. For the realization of suspended/freestanding microstructures made of materials such as P^+-Si, SiO_2, Si_3N_4, etc., silicon beneath a structural component needs to be removed [1–8, 96–99]. If the process is done only on one side of the wafer, removal of the underneath material is only possible by an undercutting process [97–99]. To minimize the release time as fast as possible, undercutting is desirable. Consequently, a high undercutting rate is advantageous for the formation of suspended structures. Figure 6.7 illustrates the application of undercutting for the fabrication of an overhanging cantilever beam.

In many cases, undercutting is detrimental as it distorts the desired shape of the structure, for example, a proof mass for an accelerometer, mesa structures, chip isolation grooves, bent V-grooves, etc., as shown in Fig. 6.8 [1, 2, 27, 30, 41, 48, 49, 99–101]. Consequently, the undercutting for the fabrication of these types of microstructures is unwanted and it is highly anticipated to eliminate this problem by some means. Considering the pros and cons of undercutting, special attention must be given while designing mask patterns. Several methods which are employed to circumvent the problem of undercutting are explained in this chapter.

6.4 Why Does Undercutting Start at Convex Corners?

The etching characteristics of KOH with different alcohol additives are presented in Fig. 6.9. The undercutting in pure KOH is significantly high, whereas it is dramatically suppressed in alcohol-added solutions. In the case of TMAH, the surfactants are the best additives to minimize the undercutting at all types of corners and edges, as shown in Fig. 6.10. In pure alkaline solutions, the convex corners encounter severe undercutting, while no undercutting is observed at concave corners, as illustrated in Fig. 6.7. Now the question is,

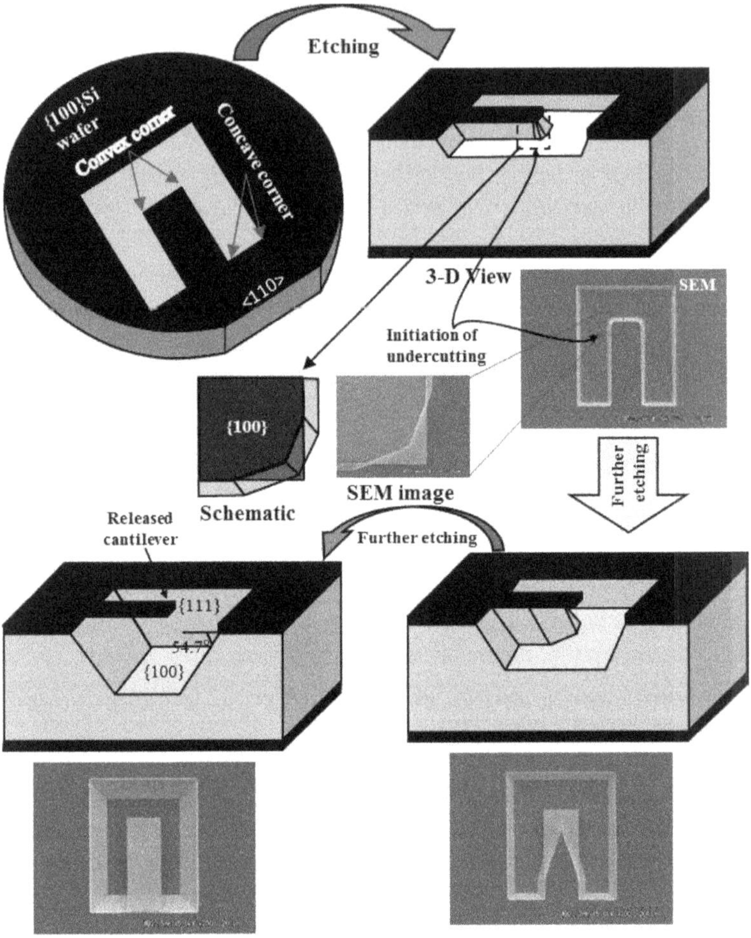

Figure 6.7 Schematic diagrams with SEM pictures representing the application of convex-corner undercutting in the releasing of a microcantilever (or microstructure) for MEMS.

Why does the undercutting start at the convex corner and not at the concave corner, even though both types of corners are formed by the intersection of {111} planes? Moreover, why is the undercutting suppressed when the etching is done in alcohol-added KOH (or surfactant-added TMAH), as shown in Figs. 6.9 and 6.10? Several models have been developed to explain the mechanism behind the undercutting at convex corners [12–19]. Most of the models talk about Si{100} wafers as this orientation is most widely employed in microfabrication. In general, they explain that the undercutting occurs due to the appearance of high-etch-rate planes during the etching process. At the atomistic scale, the convex corner exposes

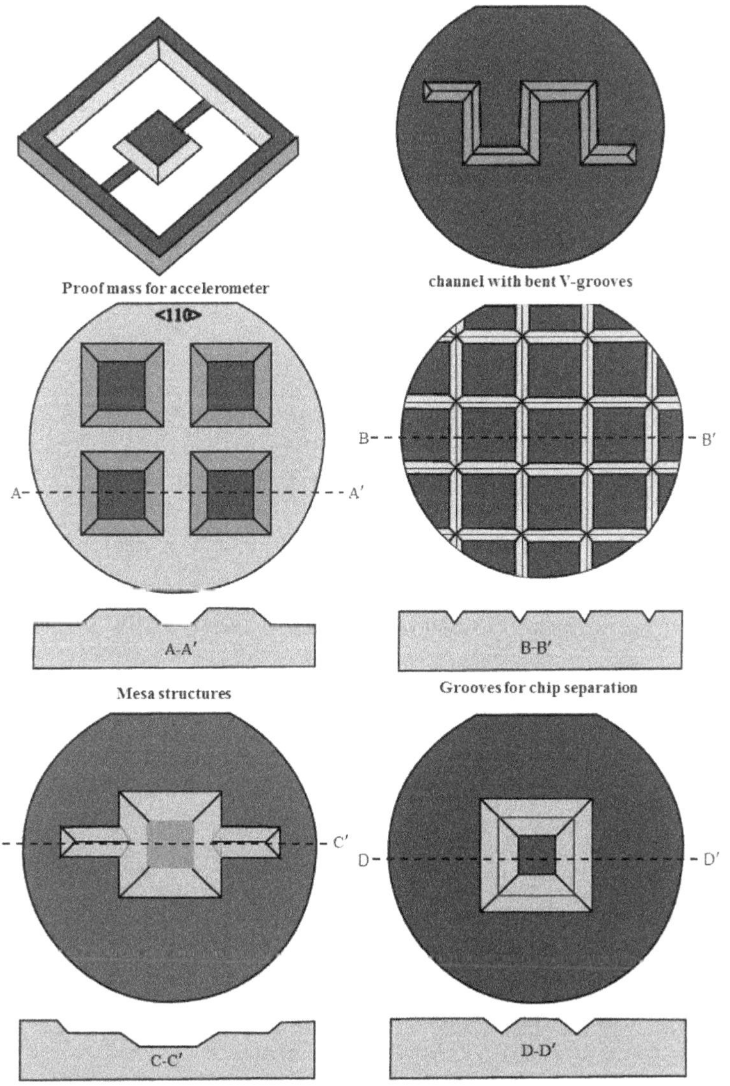

Figure 6.8 A few examples of microstructures in a Si{100} wafer where undercutting at convex corners is absolutely undesirable.

low-coordination atoms with high removal rates, hence inducing the emergence of fast-etching planes as the sloping facets or bevels [102]. Wei Dong et al. explained the undercutting mechanism on a {110} surface by comparing the density of break bonds (i.e., dangling bonds) at the convex corner and the {111} surface [18]. The break-bond density of silicon atoms at the convex corner is higher than that of the {111} planes [18]. This fact results in undercutting at the convex corner as the removal rate of the atoms belonging to the convex corner is higher than that of the atoms pertaining to the {111} planes. The silicon atoms at concave corners do not contain any break bond and therefore the shape of the concave corner is not distorted.

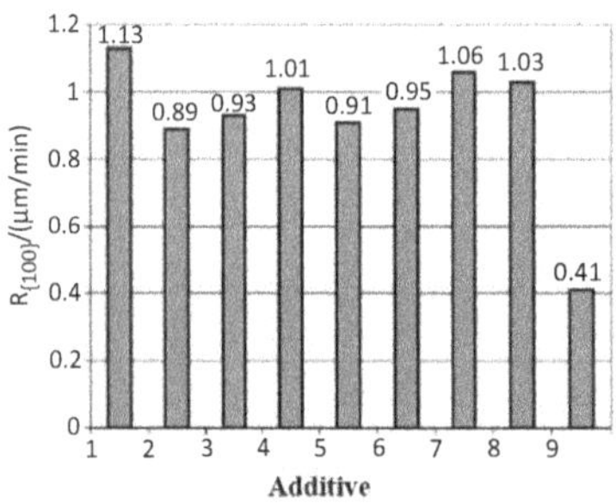

(a) Etch rates of Si{100} (1) 10M KOH; 5M KOH + (2) propanol-1; (3) isopropanol; (4) butanol-1; (5) isobutanol; (6) *sec*-butanol; (7) *tert*-butanol; (8) glycerol; (9) ethylene glycol.

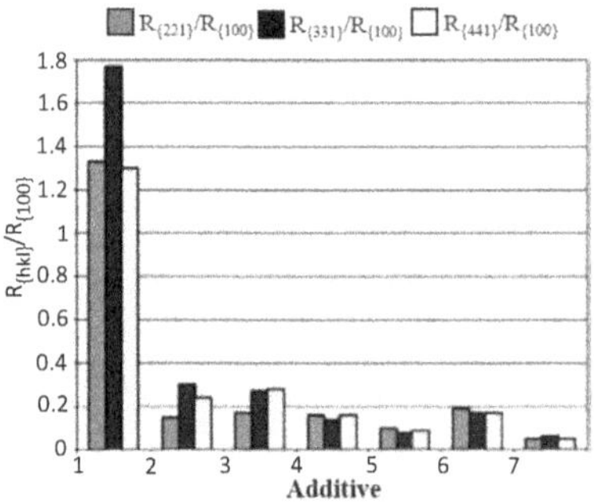

(b) Etch rate ratios $R_{(hkl)}/R_{(100)}$ in (1) 10M KOH; 5M KOH + (2) propanol-1; (3) isopropanol; (4) butanol-1; (5) isobutanol; (6) *sec*-butanol; (7) *tert*-butanol.

(c) Undercutting at convex corners in (a) 10M KOH (Etch time: 30 min), (b) 5M KOH + IPA (Etch time = 60 min), and (c) 5M KOH + *tert*-butanol (Etch time = 60 min).

Figure 6.9 Etching characteristics of KOH solutions at 80°C with and without different alcohol additives. Reprinted from Ref. [71], Copyright (2002), with permission from Elsevier.

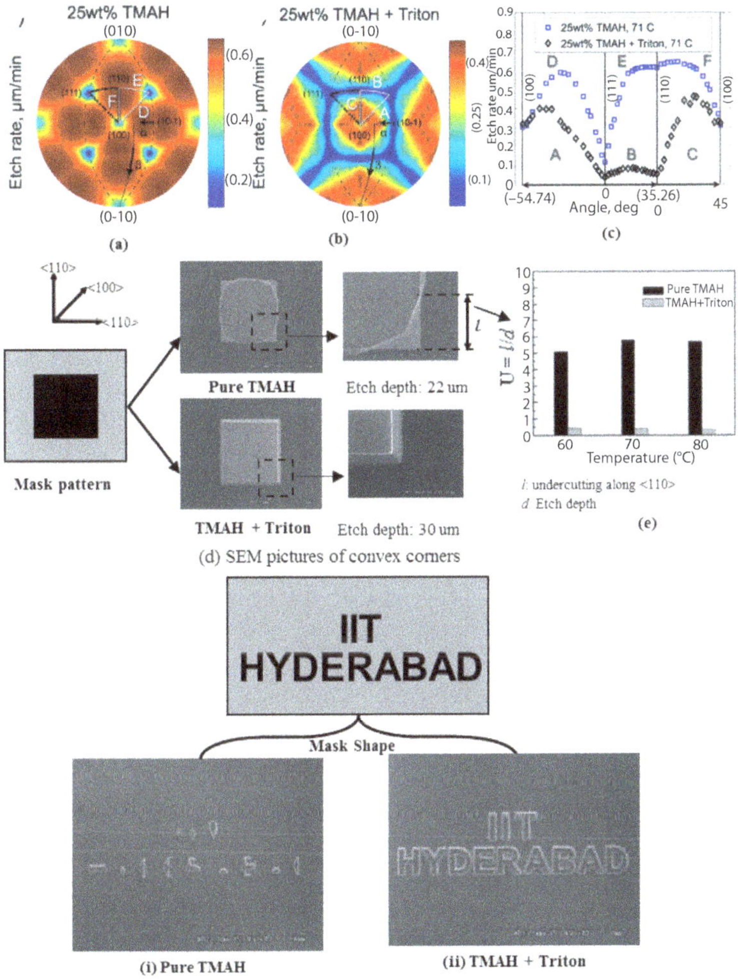

(f) SEM pictures of wet-bulk-micromachined alphabet letters exhibiting significant reduction in undercutting at different types of corners and edges in surfactant-added 25 wt% TMAH (etch depth: 30 µm).

Figure 6.10 Etching characteristics of pure and surfactant (Triton X-100)-added 25 wt% TMAH: (a–c) 3D distribution and 2D plot of the etch rates, (d) etched profiles of convex corners, (e) undercutting ($U = l/d$) at convex corners, and (f) micromachining of alphabet letters in Si{100} using (i) pure TMAH and (ii) TMAH+Triton. From Ref. [103]. @ IOP Publishing. Reproduced with permission. All rights reserved.

A simple model developed by Pal and Singh describes the etching characteristics of concave and convex corners on {100} and {110} surfaces in all kinds of wet anisotropic etchants [10, 11]. This model is based on the etching behavior of the tangent planes at the convex edge and the role of dangling bonds in the etching process. There are an infinite number of planes that can pass through the convex edge of the mesa structure, but the silicon atoms of the convex edge belong fully to only the tangent plane at that edge, as can quite simply be noticed in Fig. 6.11. The orientation of tangent planes depends on the angle of the convex corners (i.e., acute, right, and obtuse). The tangent planes at the acute and obtuse convex corners are {100} and {110}, respectively. In the case of a right-angled convex corner on a {100} wafer, the orientation of the tangent plane is {110}. In anisotropic etchants, the etch rates of the {110} and {100} planes is higher than that of the {111} planes. Therefore, the undercutting starts at the convex edge as the tangent planes containing the atoms of the convex ridge exhibit a much higher etch rate than that of the {111} planes which form the corners.

As presented in Fig. 6.10, undercutting at the convex corner is significantly reduced in surfactant-added TMAH. The etch rate of Si{100} is almost the same in pure and surfactant-added 25 wt% TMAH (Fig. 6.10c), while the etch rates of {110} and its vicinal planes are decreased to a considerably low level when a very small amount of surfactant is added to the etchant. The main cause behind the change in etching rates is attributed to the orientation-dependent adsorption of surfactant molecules during the etching process, as discussed in the previous chapter [73–75, 103, 104]. The etch rate of {100} is almost unaffected because the layer of surfactant molecules on the {100} surface is not able to protect the surface from the etchant, owing to its lower adsorptivity. In the case of Si{110} and its vicinal planes, significant reduction in the etch rate indicates that the layer of adsorbed surfactant molecules is sufficient enough to shield the surface from the etchant. Since the tangent plane at the convex corner on a {100} surface is {110}, the surfactant molecules form a dense layer on the convex corners, as illustrated in Fig. 6.12. This dense layer inhibits the etchant from reacting chemically with the silicon atoms at the corner, which results in a dramatic reduction in the undercutting.

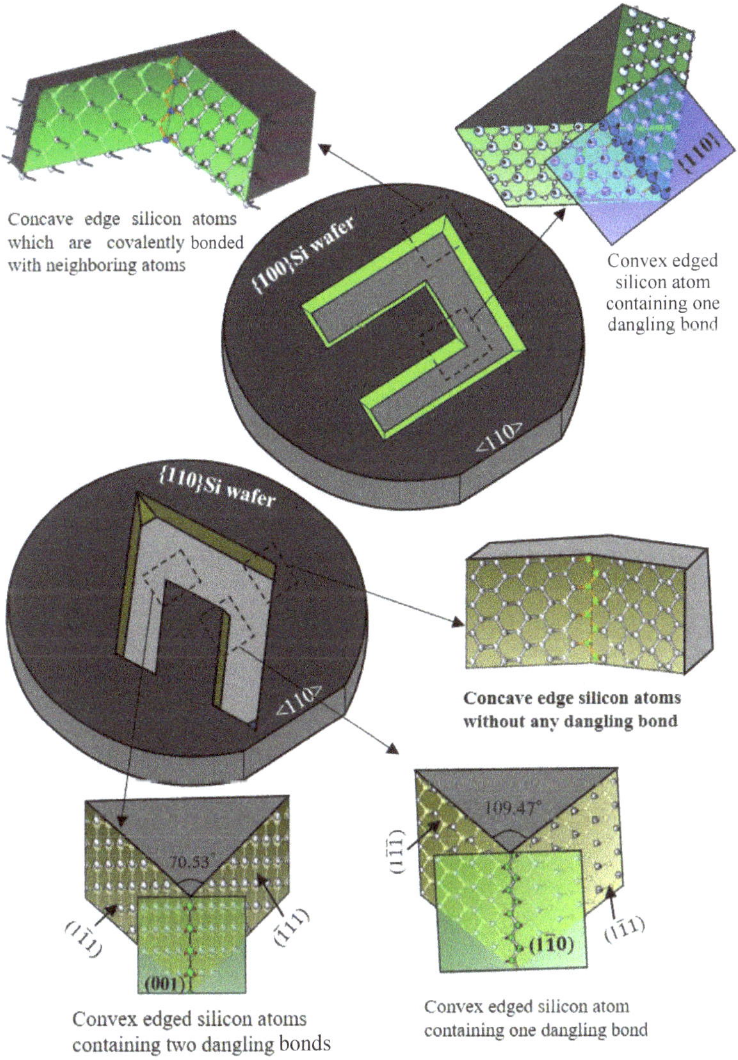

Figure 6.11 Schematic representation of concave and convex corners in {100} and {110} wafers. The close-up view of the concave corner exhibits silicon atoms of a {111} surface with one dangling bond, while the atoms belonging to the edge of the concave corner do not contain any dangling bond. The close-up view of the convex corner shows the tangent planes.

In the case of a concave corner, regardless of its shape, the concave ridge consists of atoms with no dangling bonds (or unsatisfied bond). In other words, all the bonds of concave-edge atoms are engaged with neighboring atoms. Owing to this fact, no undercutting is initiated at the concave corners and they remain intact and firmly defined by the intersection of the {111} planes regardless of the etching time, etch depth, etchant concentration, and etching temperature, as schematically shown in Fig. 6.11.

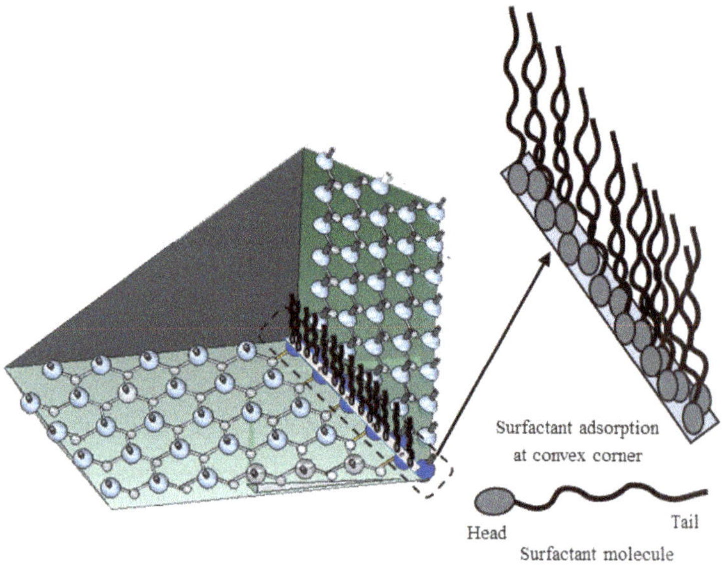

Figure 6.12 Schematic representation of surfactant adsorption at silicon atoms of convex corners, which belong to the {110} planes, in a surfactant-added TMAH solution during the etching process. The dense layer of surfactant molecules protects the convex corners from the etchant, which results in the reduction of undercutting.

6.5 Etched Profiles at Sidewalls and Corners

The initiation of undercutting at a convex corner exposes other types of high-etch-rate planes. In the beginning, these planes are vicinal {110} (or vicinal {100}) planes, depending on the type of corner, and have high etch rates. The contour of an etched corner is analyzed by the lateral undercutting etch rates in different directions, which are estimated by the etching of a wagon-wheel structure, as discussed in previous chapters. The lateral etch rate (or underetching) at different crystallographic directions on a plane surface depends on the type of etchant, concentration, and temperature [9, 105]. Consequently the amount of undercutting and the geometry of the developed etch front are typically different for different etchants. The final etch front of the undercut convex corners are constructed by the intersection of the directions which are parallel to the maximum lateral etch rate directions nearest to the edges forming convex corners, as shown in Figs. 6.13 and 6.14 for {100} and {110} surfaces, respectively. This leads to differently shaped etched profiles at different types of corners, as can be seen in Figs. 6.5, 6.6, 6.13, and 6.14.

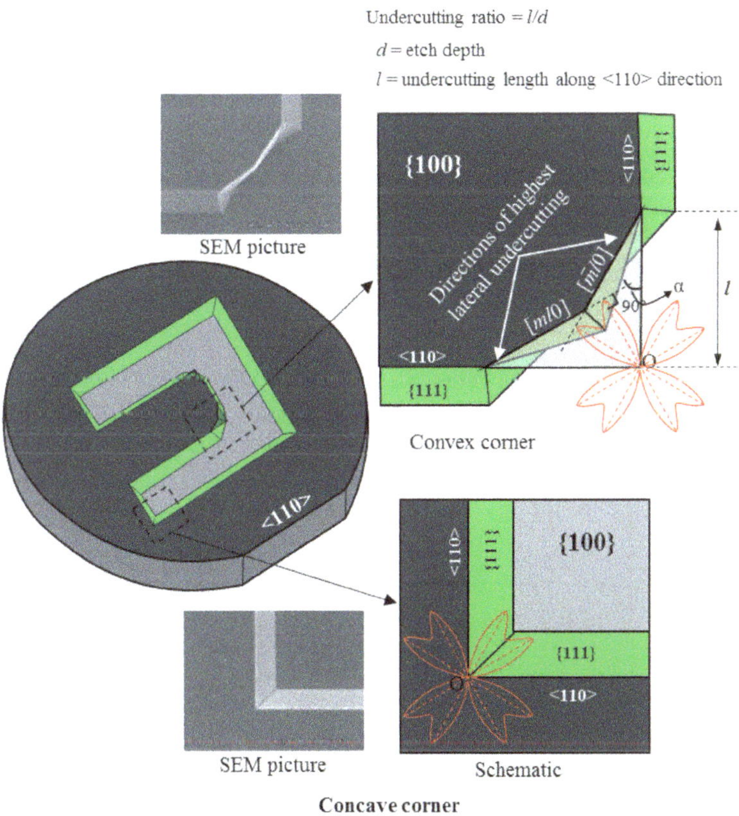

Figure 6.13 Schematic representation of corner undercutting using a lateral underetching diagram.

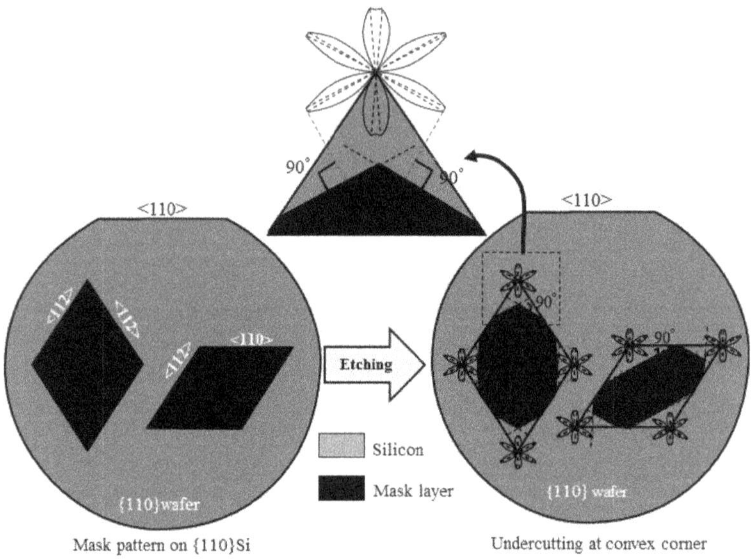

Figure 6.14 Determination of the etch front of convex corners on a Si{110} surface using a lateral undercutting diagram.

The etched profile of the sidewalls appearing at masked edges can easily be determined using the Wulff–Jaccodine method [9, 103, 106]. In this method, the etched profile is estimated by the distribution of the etch rate vectors existing in the planes belonging to the mask edge from where the etched sidewall profile appears. The mask edge (or line) is called the zone axis of the planes passing through it. A zone is a set of {*hkl*} planes that mutually intersect along (or are parallel to) a common direction (or axis) [*uvw*] with respect to a given crystal, as illustrated in Fig. 6.15. The line of intersection [*uvw*] is called a zone axis. The etch rates of the planes of any zone axis can be determined by the etch rate data of the hemisphere [103, 107–109]. The sidewall profile of the etched pattern at the mask edge is the minimal envelop of all intersecting lines which are perpendicular to the etch rate vectors of the different planes belonging to the mask edge (zone axis), as demonstrated in Figs. 6.16 and 6.17. The planes appearing at the mask edge are the minimum etch rate planes between the horizontal plane and vertical plane at that edge. Similarly the etched profile of an undercut structure are estimated. In this case, the sidewalls are the minimum-etch-rate planes at the maximum lateral underetch rate directions.

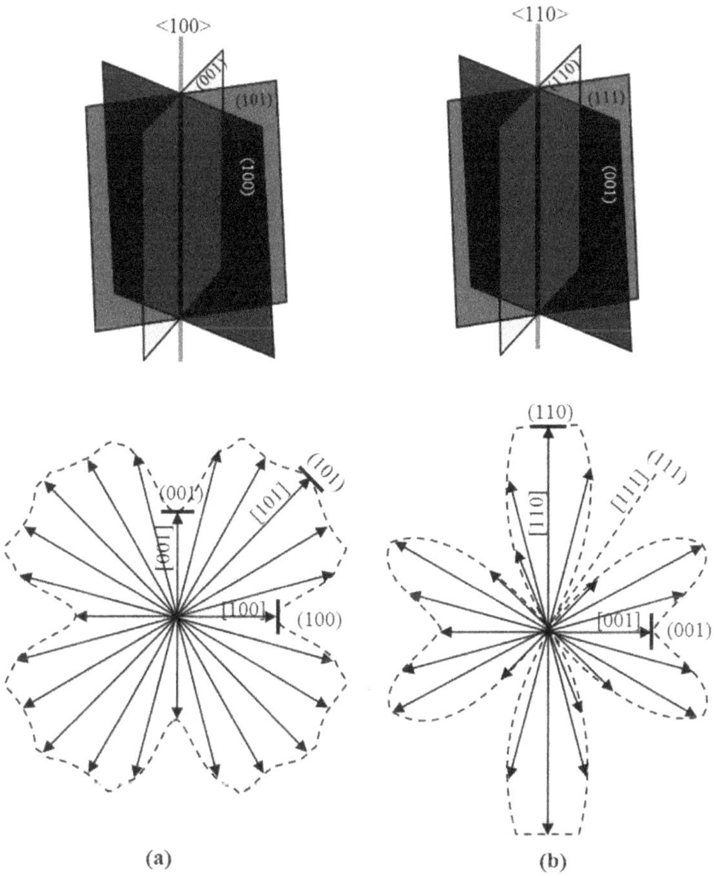

Figure 6.15 Schematic representation of the zone axis and its planes, and the diagram of the etch rates of different planes in high-concentration KOH-type etchant for (a) <100> and (b) <110> zone axes which are utilized to predict the shape of the etched profile using the Wulff–Jaccodine method.

The orientation of the facets appearing at the convex corners mainly depends on the type of etchant, concentration, and additives [10–20]. The etched shapes of an undercut convex corner in pure TMAH and KOH are schematically represented in Figs. 6.18a and 6.18b, respectively. In the case of TMAH, one type of planes appears at the undercut convex corner (i.e., plane A_1), while in the case of KOH-based etching two differently orientated planes are exposed after undercutting (i.e., planes B_1 and C_1) [15, 16, 110]. The etching time and temperature also affect the shape and orientation of beveled planes. In general, these facets in the case of a {100} surface are typically {311}, {211}, {331}, {411}, {212}, {772}, etc., [12–17, 25]. The orientation of the undercutting planes is significantly affected by the additives in the etchant, for example, IPA in KOH (Fig. 6.9) and the surfactant in TMAH (Fig. 6.10). However, the orientation of beveled planes at undercut corners depends on the choice of etchant

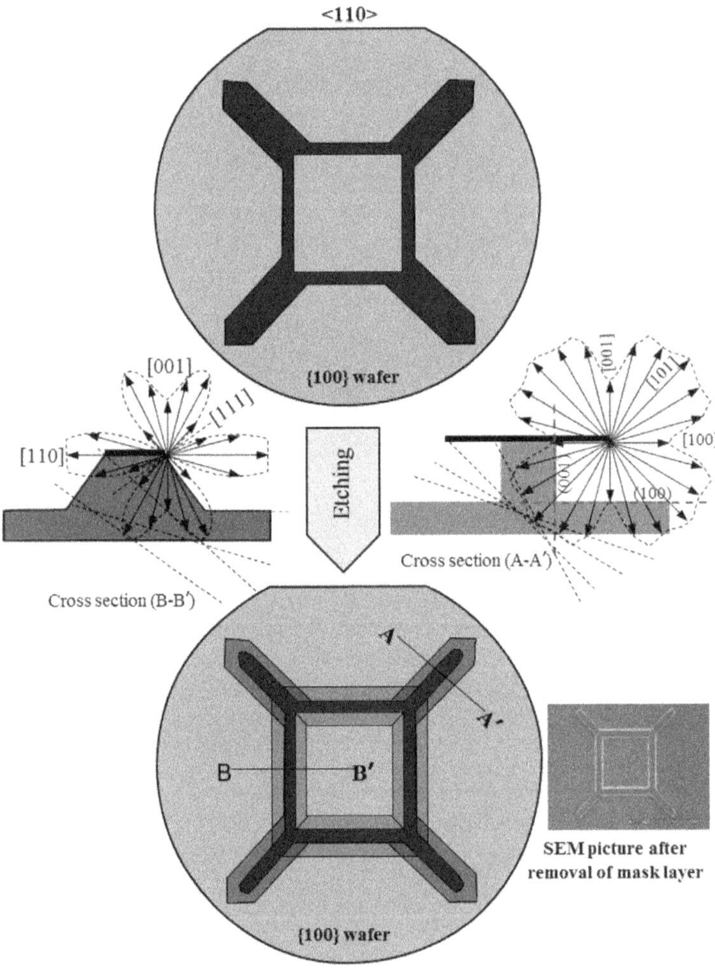

Figure 6.16 Schematic demonstration of the Wulff–Jaccodine method to estimate the etched profile at the mask edges aligned along the <110> and <100> directions on a Si{100} surface (etchant: KOH type).

and its concentration; different research groups have reported different indices for the same etchant. For instance, in a KOH solution, the beveling planes at the convex corners reported by Shikida et al., Chang Chien et al., and Mayer et al. are {311}, {772}, and {411}, respectively [15, 16, 27]. Hence, there is some disagreement in the literature about the exact orientation of planes that emerge at the convex corners during the etching process. Shikida et al. explained that the fast-etching planes emerging at convex corners are located at the saddle point in the etch rate diagram [15]. The location of the saddle point in the etch rate diagram depends on the etching parameters, for instance, it is located around the {311} plane for 34 wt% KOH. In the case of {110} wafers, the etch front planes of the mesas structures formed by the <112> direction estimated by Kim and Cho are {311} and {771} planes for acute and obtuse corners, respectively [57]. The beveled angles (α) at an undercut convex corner on a {100} surface in 25 wt% TMAH and 35 wt% KOH at 76 ± 1°C are presented in Fig. 6.19 [110]. In the case of KOH etching, the undercutting direction makes an angle of 30 ± 1° with the <110> direction (i.e., beveled angle), while it makes an angle of 25 ± 1° when etching is performed in TMAH. This kind of variation is also reported by Trieu et al. [12]. Thus the undercutting direction is not unique as the included angle of the undercut edges varies in the band of 4° for both TMAH and KOH at different structures throughout the wafer. As the undercutting direction is not unique, the absolute indexing of the standing plane is not possible and correct. However, the undercut planes are located around the saddle point in the etch rate contour [15].

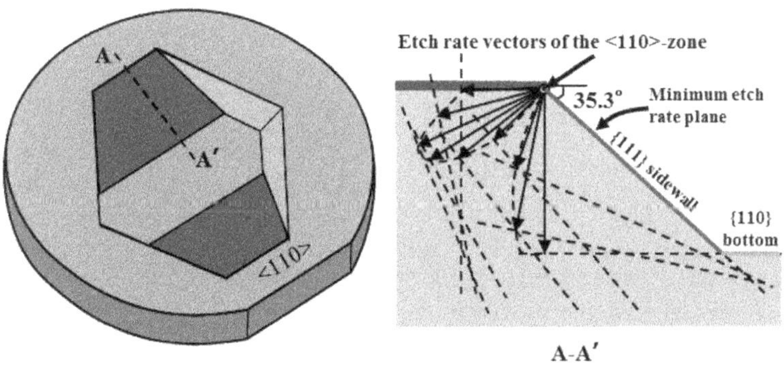

Figure 6.17 Schematic illustration of the estimation of the sidewall profile at a <110> mask edge on a Si{110} surface using the Wulff–Jaccodine method.

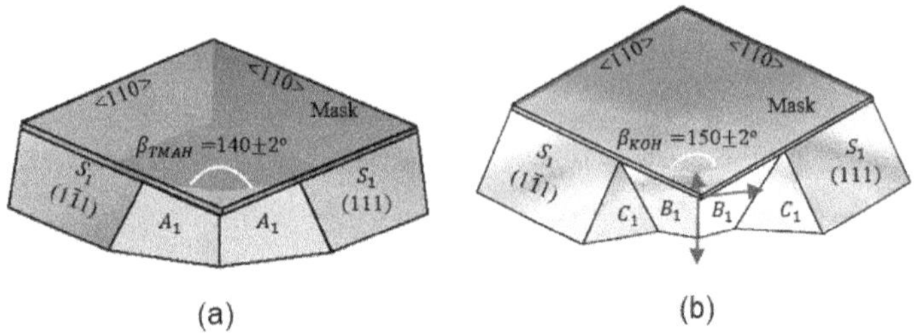

Figure 6.18 Schematic view of convex-corner undercutting in (a) TMAH and (b) KOH solutions. A_1, B_1, and C_1 are high-indexed planes.

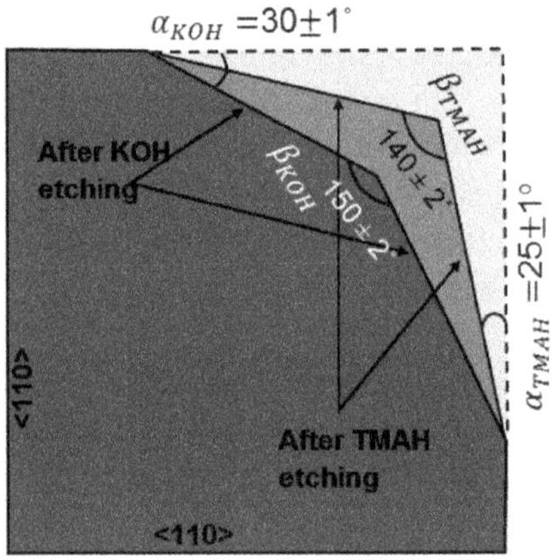

Figure 6.19 Schematic view of undercutting at a convex corner in TMAH and KOH solutions.

As discussed the appearance of differently shaped corners in anisotropic etchants is generally explained using the lateral etch rate diagram, as shown in Fig. 6.13 [10, 12]. It shows that the undercutting direction is along the maximum lateral etch rate direction. However, this description does not explain why the maximum lateral etch rate occurs along that particular direction only and not along any other direction. Figure 6.20 shows SEM pictures and schematic diagrams of convex corners of different shapes etched in TMAH and KOH etchants. One is with a 90° corner (corner B) and the others are the corners formed after undercutting of a 90° corner (corner A). It can be seen in Figs. 6.20a and 6.20c that the sidewall at the edge of the 150° corner in KOH is formed by more than one type of planes, that is, a smooth plane and a rugged surface toward the bottom. In the case of TMAH, mainly one type of plane appears at the edges of the 142° corner; however, a closer look at the zoomed scanning electron microscopy (SEM) image of the sidewalls shown in Fig. 6.20c reveals the appearance of an additional plane toward the top edge, which is appearing as a white line. However, this is a recessive plane and its appearance is overshadowed by one kind of plane. With prolonged etching this recessive plane may slightly extend, but its appearance will continue to fade out in comparison to the widely extending and strongly dominating other type of plane.

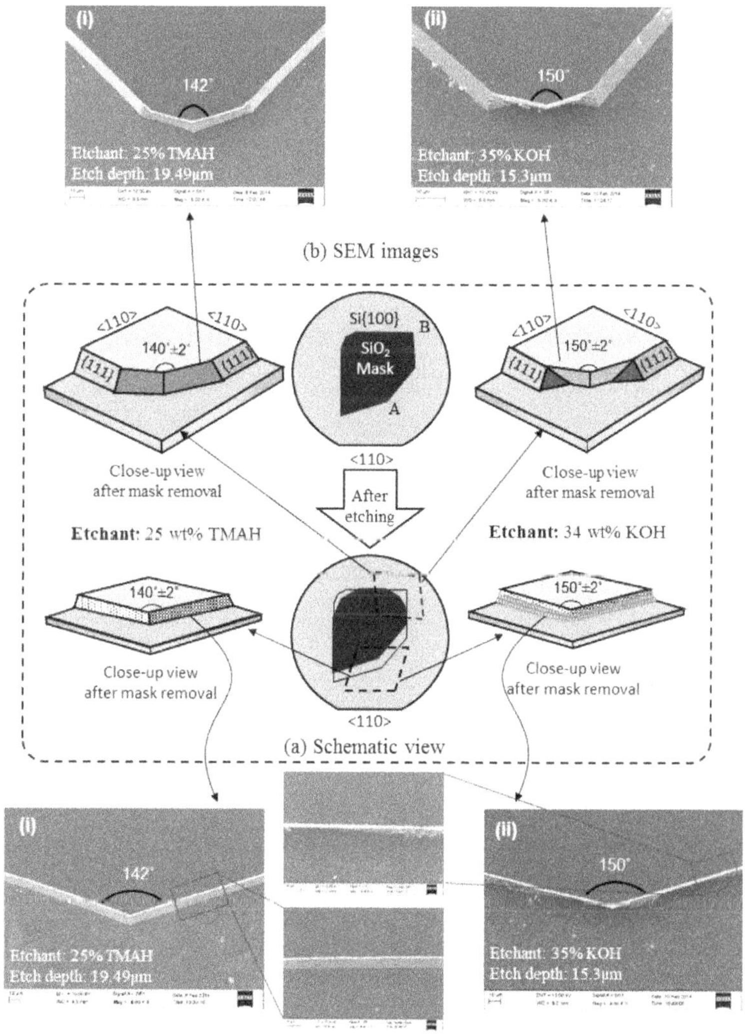

Figure 6.20 Schematic illustration and SEM pictures of convex corners of different shapes on a Si{100} wafer.

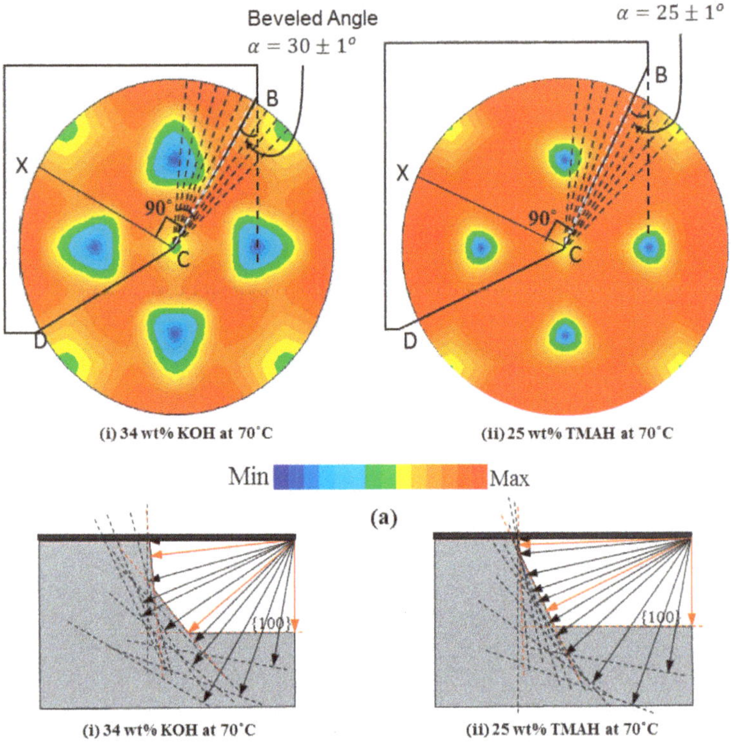

(b) Etched profiles at the mask edges aligned along the beveled directions (i.e., CB)

Figure 6.21 Explanation of corner undercutting using etch rate contours in (i) KOH and (ii) TMAH.

Figrue 6.21a presents contour plots of the etching rates of different planes in 34 wt% KOH and 25 wt% TMAH solutions. The undercutting directions for KOH and TMAH are shown by dotted white lines. The etched profiles of the sidewalls at the beveling edges of the undercut convex corners are schematically presented in Fig. 6.21b. The minimal envelope is indicated by red-colored lines, which is the envelope of the planes perpendicular to the red-colored etch rate vectors. The minimal envelope for KOH encompasses more than one dominating plane. However, in the case of TMAH an additional plane does appear at the top edge, but its appearance is overshadowed by another plane, which is primarily one dominating plane. Therefore we can say that the sidewall profile at the beveled direction in the case of TMAH is predominantly formed by one type of plane. Undercutting at a convex corner propagates mainly along three directions marked by arrows in Fig. 6.18. One is the downward propagation (i.e., <100> direction perpendicular to the wafer surface), which leads to the formation of a mesa structure; another is the lateral undercutting direction, which distorts the 90° corner to eventually become obtuse angled; and the third is in the direction perpendicular to both <100> and the lateral direction. The last two undercutting directions primarily define the etch front of the convex corner. As the sidewall of the 148°–152° corner in KOH contains more than one plane, undercut 90° convex corners are formed by more than one plane, as shown in Fig. 6.20b-ii.

From the etch rate contour maps shown in Fig. 6.21, we notice that any zone axis lying in the white band of 2° (i.e., the beveled direction which the corner finally takes) has an etch rate which is the local maximum-etch-rate plane. This leads to the conclusion that the beveled line is that line along which the etch rate of the slowest-etching plane (i.e., the standing plane) is larger than the minimum etch rates of any other plane along any other line (or zone axis). Thus the beveled direction comprises a local maximum-etch-rate plane which the convex corner will finally seek after the initiation of undercutting from the 90° corner. As the undercutting begins at the corner, many high-etch-rate planes appear, which increases the rate of undercutting. This haphazard advancement of undercutting comes to a stable profile when it encounters a zone axis which contains a local maximum-etch-rate plane. Thereafter, the undercutting propagates parallel to this zone axis and does not try to seek any other direction as it has already encountered that particular direction along which the planes have a local maximum etch rate. It answers the question, Why does the 90° convex corner take the shape of 140 ± 2° (for TMAH) and 150 ± 2° (for KOH) and why not some other angle? Moreover, it can be concluded that the standing planes at the undercut convex corners are not the highest-etch-rate planes. Instead, they are local maximum-etch-rate planes which lie in the vicinity of the saddle point in the etch rate contour map [15]. In surfactant-added TMAH and IPA-added KOH, the etch rates of {110} and its vicinal planes are suppressed to a considerably low level. Due to this factor, the undercutting at extruded corners, curved and non-<110> edges on a {100} surface, is significantly reduced [66–84]. It may be emphasized here that surfactant-added TMAH is not suitable for the fabrication of microstructures on a {110} wafer as the etch rate of the {110} orientation is very low, as presented in Fig. 6.10. It is discussed in the next section that the knowledge of the beveled planes is not so important to fabricate protected convex corners. The undercutting ratio (l/d) and beveled angle (α, an angle between the direction of the maximum lateral undercutting and mask edges), as illustrated in Fig. 6.13, are primarily required to develop a method to realize well-shaped convex corners.

6.6 Fabrication Techniques of Convex Corners

Most of the studies for the fabrication of convex and concave corners are reported for the {100} surface (i.e., Si{100} wafer) as this orientation is most extensively employed in MEMS fabrication [23–55, 61–65]. Very less is reported for Si{110} as this orientation is used only for specific applications [56–60]. In this chapter, the fabrication of convex corners is explained for both types of orientation (i.e., {100} and {110}). There are several different methods that have been employed for the formation of protected convex corners using wet anisotropic etching. Each fabrication technique has its own set of advantages and disadvantages in terms of process flexibility, time, cost, and the quality of the fabricated corner. These methods are described in the following sections.

6.6.1 The Corner Compensation Method

To realize well-defined convex corners in the wet bulk micromachining process, corner compensation is a most popular method. This method is based on the addition of extra geometry at the convex corner in order to delay the etching, so no undercutting can start at the corner. The extra features at convex corners are called compensating structures. This technique in wet anisotropic etching is covered by a large number of publications [22–59]. Most of the studies are carried out for pure and IPA-added KOH solutions. Some of them are performed for ethylenediamine pyrocatechol (EDP) [24, 25] and TMAH [43, 48, 51]. Various types of compensation patterns employed for the realization of convex corners on {100} and {110} silicon wafers are shown in Figs. 6.22

and 6.23, respectively. The design methodology of these structures is elaborated in the subsequent sections. The size and shape of the compensating geometries strongly depend on the etch depth and the etching characteristics of the etchant, such as etch rate, beveled angle (α), and undercutting ratio ($U_r = l/d$).

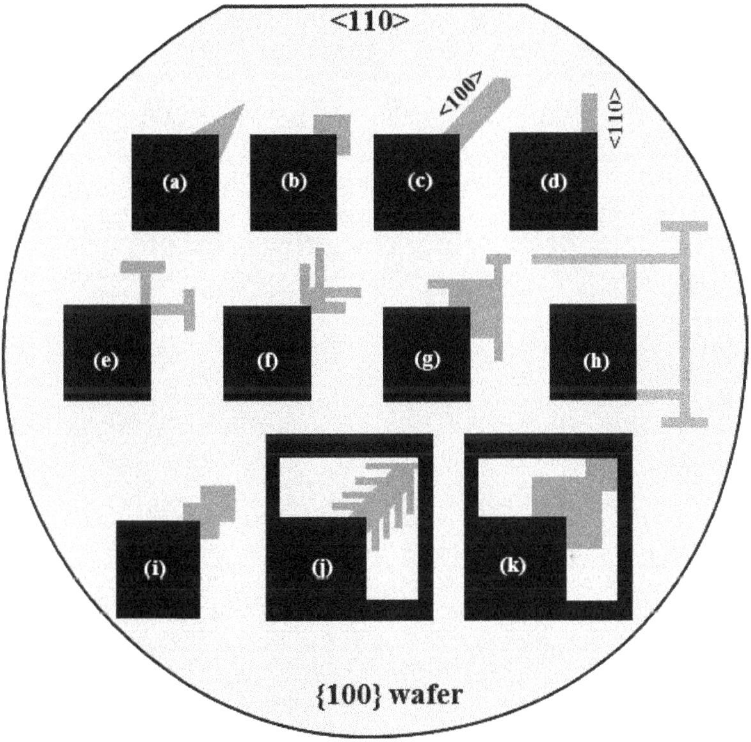

Figure 6.22 Various types of corner compensation geometries used to realize the convex corners on Si{100} wafers.

6.6.1.1 Corner compensation geometries for a Si{100} wafer

The design methodology of corner compensation geometry is the same for all types of anisotropic etchants. The dimensions and shapes are analyzed by relative etch rates of crystallographic planes. However, different kinds of compensation geometries, as presented in Fig. 6.22, are proposed: four types of designs, namely triangle, square, <110> band, and <100>-oriented beam are the basic structures. Other shapes are derivatives of these structures.

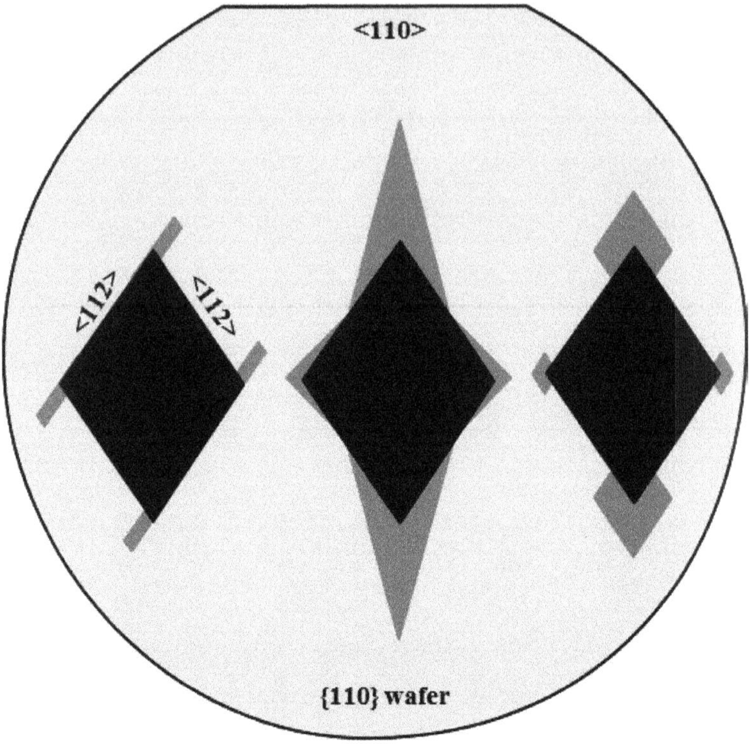

Figure 6.23 Different kinds of corner compensation designs employed to fabricate the convex corners on Si{110} wafers.

- **Triangle**: The triangular-shaped geometry, as shown in Fig. 6.24, is a simplest compensating design in terms of determining its shape and dimensions [25, 26, 48]. To calculate the dimensions of the compensating triangle, the undercutting length along the <110> direction (OP or OC or l in Fig. 6.24) and the beveled angle α (i.e., $\angle$OPB or $\angle$OCB) are needed. The lines PB and CB are representing the intersection of the beveling planes with the wafer surface if no compensating pattern is used. The sides of the triangle AP and AC are chosen to coincide with the <lm0> family of lines corresponding to PB and CB (i.e., $\angle$OPB = $\angle$OPA = $\angle$OCA = $\angle$OCB. In other words, the sides of the triangle are the highest lateral etch rate directions on the {100} surface, as illustrated using a red-colored diagram of lateral etch rates whose center is matched with the convex corner. The length of the sides of the compensating triangle (i.e., AP or AC) can be determined using the law of sines (or sine rule) as follows:

$$\frac{\mathrm{AP}}{\sin 135} = \frac{l}{\sin\left(\dfrac{90-2\alpha}{2}\right)} \Rightarrow \mathrm{AP(or\,AC)} = \frac{l}{\sqrt{2}\sin(45-\alpha)}$$

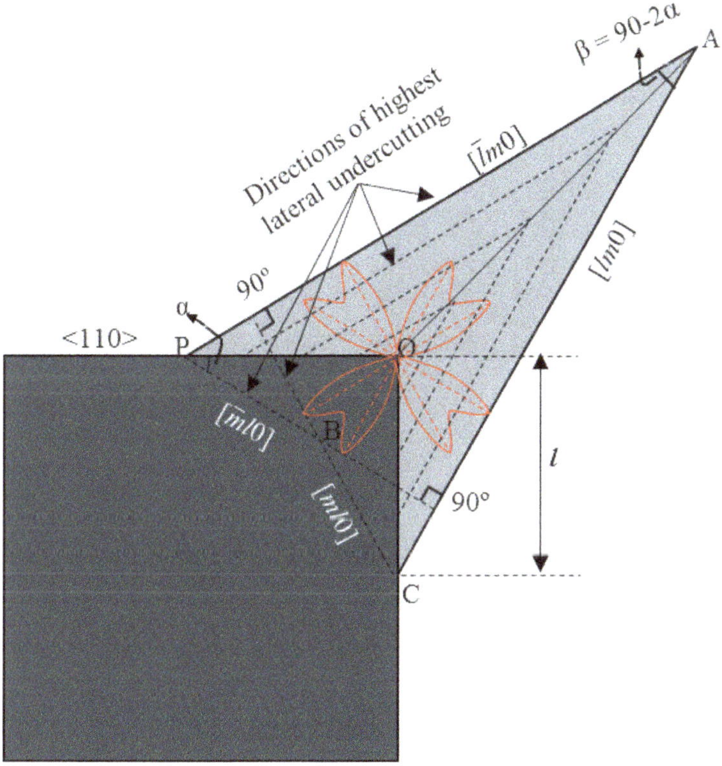

Figure 6.24 Schematic representation of triangular-shaped compensation geometry.

It is clear from this formula that the length of the sides of the compensating triangle depends on the undercutting length l and the beveled angle α. The angle α and the undercutting ratio ($U_r = l/d$) depend on the type of etchant. Thus, to determine the dimensions of the compensating triangle to fabricate a protected convex corner for etch depth d, the length l and angle α for an etchant must be known.

The successive consumption of the compensating design during etching is shown by dotted lines. As the sides of this geometry are the maximum lateral etch rate directions on the wafer surface, the successive etch pattern follows its original shape.

The SEM pictures of mesa structures fabricated in 10 wt% KOH, 40 wt% KOH, and 25 wt% TMAH using triangular patterns are shown in Figs. 6.25a, 6.25b, and 6.25c, respectively. In the case of 25 wt% TMAH, the convex corners of the mesa structure are formed with slight beveling at the top side, while in 40 wt% KOH extrusion at the fabricated corner is clearly visible. Hence the triangular compensation geometry does not ensure complete compensation of the convex corners.

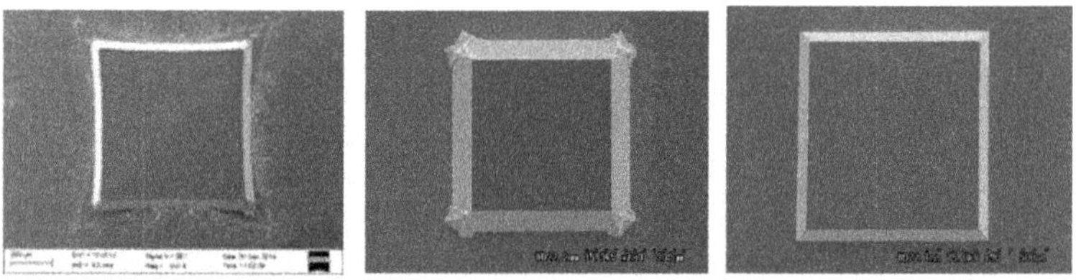

(a) Etch depth: 43 μm (b) Etch depth: 52 μm (c) Etch depth: 55 μm

Figure 6.25 SEM pictures of mesa structures fabricated in (a) 10 wt% KOH, (b) 40 wt% KOH, and (c) 25 wt% TMAH using triangular-shaped compensating geometry shown in Fig. 6.24.

In several studies, it is found that 25% TMAH is the most favorable choice for minimizing undercutting at sharp convex corners and achieving a smooth surface finish with a reasonable etch rate of Si{100} by adding a small amount of surfactant [72–80, 82, 84]. Figure 6.26 shows the SEM image of consecutive etch shapes of a mesa structure with a triangular compensation structure. It can be noticed in Fig. 6.26b that the compensating structures at the top part of the corners have been consumed, but the convex corners have not acquired the desired shape as a lot of lump mass is left at the edges of the corners. Although further etching has removed the extruded structure, the upper side shape of the corner is distorted. The etched profile of the triangular-shaped compensating pattern is different from the estimated one according to the analysis discussed in Fig. 6.24. As can be seen in Fig. 6.26, the planes appearing at the free end of the compensating geometry exhibit a fast etch rate in comparison to that emerging at the edges. The etch rates of the planes appearing at the edges are very low in comparison to that of {100} ones. Therefore, the edges and corners of the compensating pattern advance outward with a fast rate due to this etch rate difference that results in an extrusion at the desired convex corners. Thus, it can be concluded that triangular-shaped geometry cannot provide a sharp convex corner in a surfactant-added TMAH solution.

- **Square**: The triangular-shaped geometry requires more space due to its long sides. In this scheme, a square-shaped design is proposed to reduce the spatial requirement to fit in lesser space at the target corner in comparison to a triangular shape, as shown in Fig. 6.27. In this case, a square whose center coincides with the apex of the convex corner is used for the time-delayed etching to protect the convex corner [22–24, 26, 48, 51, 53]. It can be noticed from Fig. 6.27, the square-shaped compensation design contains three convex corners (m, T, and n) and therefore this structure is consumed by the undercutting that starts from these corners, as illustrated by dotted lines. This structure is completely consumed by fast-etching planes that intersect the {100} surface at <*lm*0> directions. To determine the side length of the compensation geometry (a) for etch depth d, a simple formula can be derived using geometrical relations. The calculation is based on the time required for the beveled edge [$\bar{l}m0$] (or <*lm*0>) to evolve from point m to point O (or n to O), as illustrated in Fig. 6.27.

(a) Etching time: t_1 (b) Etching time: $t_2 > t_1$ (c) Etching time: $t_3 > t_2 > t_1$

Figure 6.26 Sequentially etched shape for a mesa structure with triangular-shaped compensating design in surfactant-added 25 wt% TMAH. Reprinted from Ref. [48], Copyright (2009), with permission from Elsevier.

The perpendicular distance r from a point m $(-a/2, a/2)$ to a line OS which is parallel to <*lm*0> and passing through point O (i.e., line $y = \tan\alpha.x$)

$$r = \frac{\frac{a}{2} - \tan\alpha.\left(-\frac{a}{2}\right)}{\sqrt{1+\tan^2\alpha}} \quad \Rightarrow r = 0.5a(\cos\alpha + \sin\alpha)$$

The relation between r and l is $l = \frac{r}{\sin\alpha}$.

Using these two equations, we can get a relation between a, r, and the undercutting ratio ($U_r = l/d$)

$$U_r = \frac{l}{d} \quad \Rightarrow d = \frac{0.5a(\cos\alpha + \sin\alpha)}{U_r \sin\alpha}$$

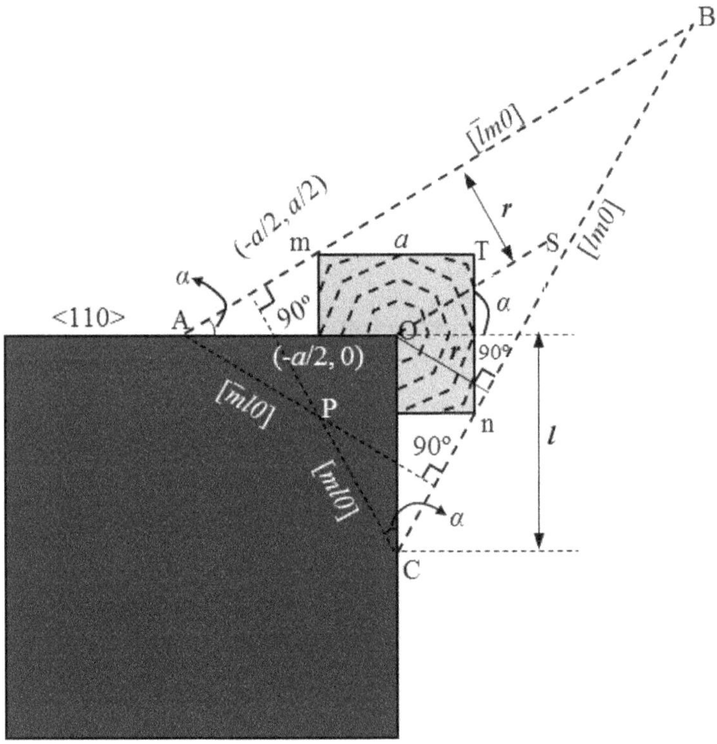

Figure 6.27 Schematic view of square-shaped compensation design.

The mesa structures fabricated in 10 wt% KOH, 40 wt% KOH, and 25 wt% TMAH using square compensation geometry are presented in Figs. 6.28a, 6.28b, and 6.28c, respectively. In all three cases, significant beveling occurs at the convex corners, which alters their shapes. The SEM images of a mesa structure etched in surfactant-added 25 wt% TMAH for different durations are shown in Fig. 6.29. In this case also, the fabricated convex corners are not well shaped. If the etching is done till the compensating geometry is consumed, a lot of extruded silicon remains at the corners, as can be seen in Fig. 6.29b. Similar to triangular-shaped compensation, further etching to remove the extra mass at the convex corner alters the desired shape of the corner, as shown in Fig. 6.29c.

(a) Etch depth: 44 μm (b) Etch depth: 50 μm (c) Etch depth: 50 μm

Figure 6.28 SEM pictures of mesa structures fabricated in (a) 10 wt % KOH, (b) 40 wt% KOH, and (c) 25 wt% TMAH using square-shaped compensating geometry, as shown in Fig. 6.27.

(a) Etching time: t_1 (b) Etching time: $t_2>t_1$ (c) Etching time: $t_3>t_2>t_1$

Figure 6.29 Sequentially etched shape for a mesa structure with square compensation patterns in surfactant-added 25 wt% TMAH. Reprinted from Ref. [48], Copyright (2009), with permission from Elsevier.

- **The <110>-oriented beam**: The main objective behind the design of a compensation structure is to reduce the spatial requirement and to achieve a well-shaped convex corner. In this case, a simple <110>-oriented beam (or rectangle), as illustrated in Figs. 6.22d and 6.30, or a combination of <110>-oriented beams and squares, as shown in Fig. 6.22e–h, is added at the convex corner [24, 29–34, 44, 48, 49, 54]. The progressive etched profiles of the simple and asymmetric

<110> beams are shown by dotted lines. The consumption of the <110> compensation beam takes place by the initiation of undercutting at its convex corners, which is illustrated using the lateral etch rate diagrams shown by red-colored lines. It can be observed from the etched front indicated by dotted lines that the <110> beam-type compensation design exhibits significant beveling, depending on the beam's width. To reduce the beveling, the width of the beam should be as small as possible. The beveling can also be minimized using an asymmetric-shaped beam, as shown in Fig. 6.30b. Let W and $L_{<110>}$ be the width and length of the compensation beam, respectively. A mathematical relation between these dimensions and the etch depth d can be determined using simple geometrical formulae as employed for the square compensation geometry in the previous section. In this case, the perpendicular distance r is calculated from point B to a line OG, which passes through point O (i.e., $y = \cot\alpha . x$). The following relations are obtained for two differently shaped beams (i.e., symmetric and asymmetric):

(a) Simple beam (Fig. 6.30a)

$$d = \frac{0.5W\cos\alpha + L_{<110>}\sin\alpha}{U_{\mathrm{r}}} \qquad \text{for } L_{<110>} \geq 0.5W$$

(b) Asymmetric beam (Fig. 6.30b)

$$d = \frac{W\cos\alpha + L_{<110>}\sin\alpha}{U_{\mathrm{r}}\sin\alpha}$$

SEM pictures of mesa structures fabricated using an asymmetric <110> beam (Fig. 6.30b) in 40 wt% KOH and 25% TMAH are presented in Figs. 6.31a and 6.31b, respectively. It can be easily noticed from the SEM images that the corners are in a very bad shape in both etchants. A lump mass extruded along the <110> direction is clearly visible in the case of the structure realized in 25 wt% TMAH. Therefore, this compensating design is not suitable for achieving sharp corners in any kind of etchant. As discussed earlier, the triangular- and square-shaped compensation geometries are not appropriate for the formation of sharp corners in surfactant-added TMAH. In this case also, it can be easily estimated that the etched shape of the convex corner using a <110> beam will also result in an extruded mass around the convex corner but along the <110> directions only.

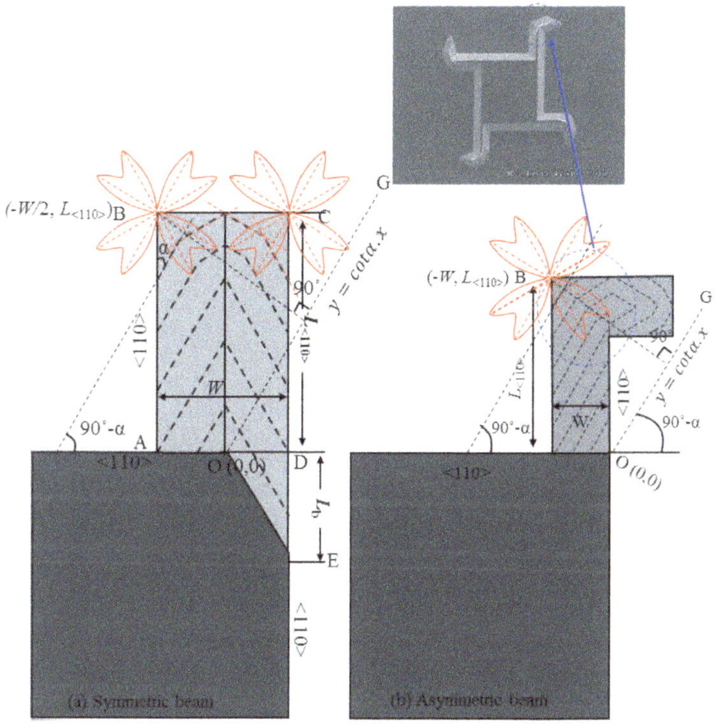

Figure 6.30 Schematic illustration of <110>-oriented compensation geometry.

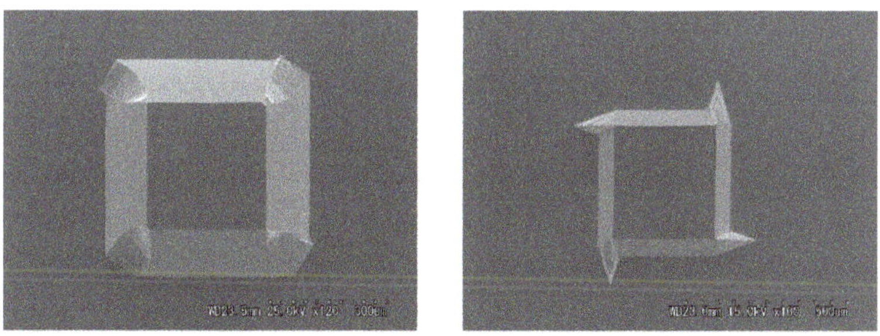

(a) Etch depth: 150 μm

(b) Etch depth: 61 μm

Figure 6.31 SEM pictures of mesa structures fabricated in (a) 40 wt% KOH and (b) 25 wt% TMAH at 70°C using the asymmetric <110> beam compensation geometry shown in Fig. 6.30b.

- **Simple <100>-oriented beam**: None of the compensation structures discussed so far (i.e., triangle, square, <110> beam) provides a sharp-edge convex corner. In continuation of the efforts to fabricate a sharp convex corner, a simple <100>-oriented beam is proposed, as presented in Fig. 6.32 [27, 35, 38, 40, 43, 45, 48, 49, 51, 53]. As shown by dotted lines, the beam is consumed by undercutting initiated from the free end and the lateral undercutting starts at the long edges of the beam.

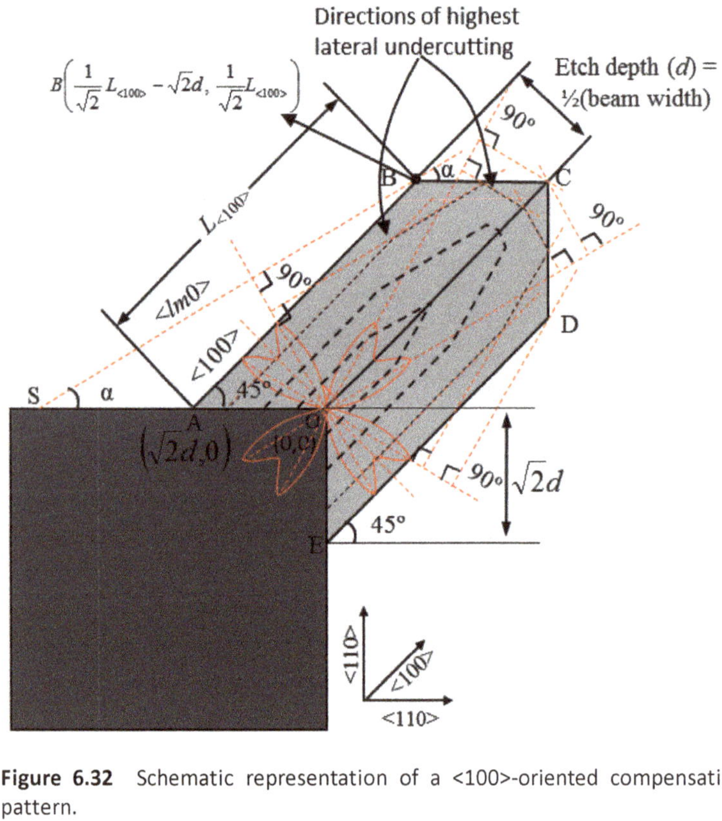

Figure 6.32 Schematic representation of a <100>-oriented compensation pattern.

The etched profile of a <100> beam depends on the etching characteristics of different planes appearing at the <100> mask edge. These planes are either {100} or {*ij*0}, as illustrated in Figs.

6.33 and 6.34. To predict the etched profile of a <100> beam, different conditions based on the etch rate ratio of {*ij*0} and {100} (i.e., $R_{\{ij0\}}/R_{\{100\}}$) can be defined using Fig 6.33 as follows:

(a) If $\frac{R_{\{ij0\}}}{R_{\{100\}}} \leq \sin\theta$, only {*ij*0} planes will be developed at <100> edges.

(b) If $\sin\theta < \frac{R_{\{ij0\}}}{R_{\{100\}}} < \frac{1}{\cos\theta}$, both (100) and {*ij*0} will expose.

(c) If $\frac{R_{\{ij0\}}}{R_{\{100\}}} \geq \frac{1}{\cos\theta}$, only vertical (100) planes emerge.

The etched shapes corresponding to the conditions (a), (b), and (c) are illustrated in Figs. 6.34a, 6.34b, and 6.34c, respectively. It can be easily observed in Fig. 6.34 that the <100> compensation geometry provides sharp convex corners only if the condition $\frac{R_{\{ij0\}}}{R_{\{100\}}} \geq \frac{1}{\cos\theta}$ is satisfied, that is, the sidewalls possess only vertical {100} planes during etching, as demonstrated by the SEM picture and the schematic drawing in Fig. 6.34c.

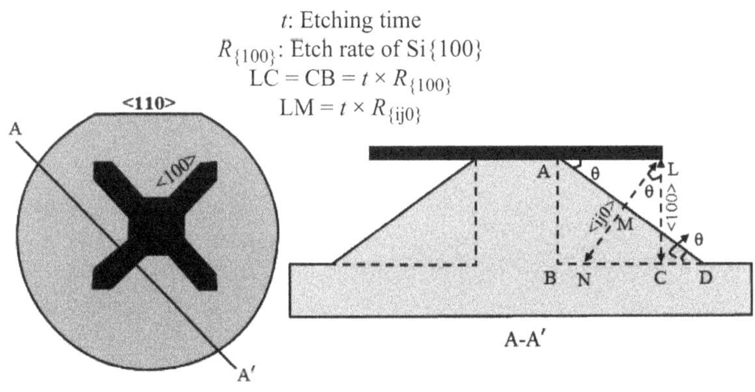

Figure 6.33 A square-shaped mask pattern with a <100> compensation design. The cross-sectional view of the etched profile of the compensating structure is utilized for analyzing the conditions for the exposure of {ij0} planes during etching, which are presented in Fig. 6.34.

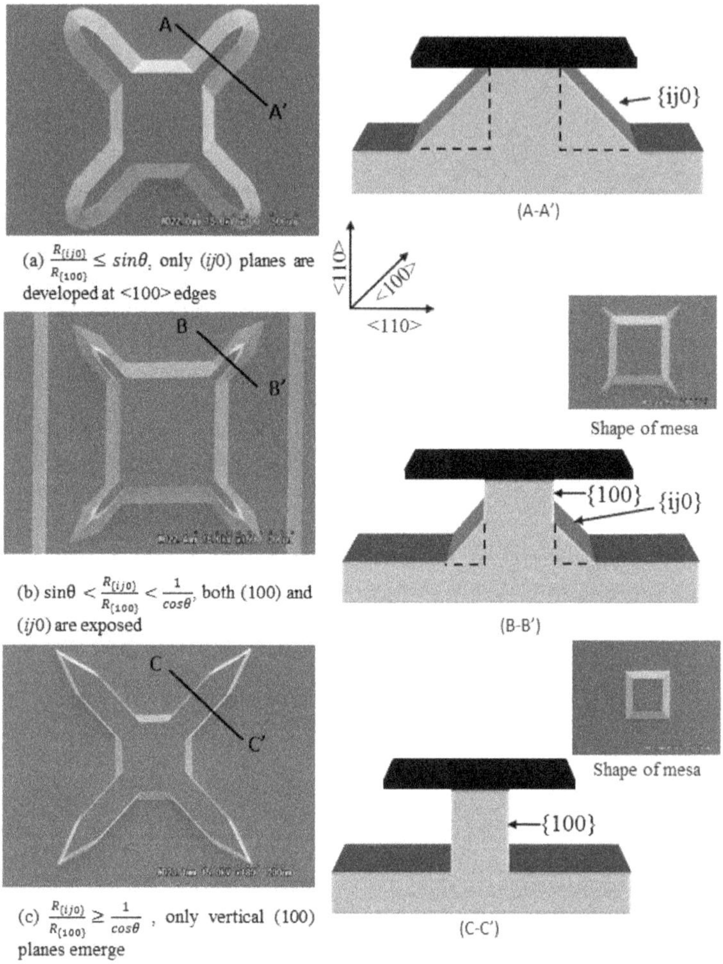

Figure 6.34 Etched profiles of a <100>-oriented beam compensation structure on a {100} wafer surface for three different conditions between the angle θ and the etch rates of $R_{\{100\}}$ and $R_{\{ij0\}}$, as illustrated in Fig. 6.33.

To get a relation to determine the length of a <100> beam (i.e., $L_{<100>}$), the successive etched profile, as shown by black-colored dotted lines in Fig. 6.32, should be analyzed. The lateral etch rate diagram is used to illustrate the directions of undercutting at the free end and the <100> mask edges. The progressive etched profile in Fig. 6.32 is presented for the condition $\frac{R_{\{ij0\}}}{R_{\{100\}}} \geq \frac{1}{\cos\theta}$ as it provides a sharp-edge convex corner. The undercutting at the free end is caused by fast-etching high-index planes, while lateral undercutting at <100> edges occurs mainly due to the etching of {100} planes. This design aims to realize the convex corner by the lateral undercutting of {100} planes which appear at <100> edges. For the same reason, the width of the beam (W) should be twice the required etch depth and the beam should be sufficiently long so that the final shape of the convex corner is formed by only the lateral undercutting of {100} planes. To determine the beam length (i.e., $L_{<100>}$) to realize a microstructure with a well-protected convex corner up to etch depth d (i.e., $\frac{1}{2}W$), a relation between $L_{<110>}$ and d can be developed by formulating the equation of various lines.

The equation of a straight line SB (i.e., <*lm0*>, a fast-etching front line) passing through a fixed point B $\left(\frac{1}{\sqrt{2}}L_{<100>} - \sqrt{2}d, \frac{1}{\sqrt{2}}L_{<100>}\right)$ and having slope tanα with a <110> edge is

$$y - \frac{1}{\sqrt{2}}L_{<100>} = \tan\alpha\left(x - \frac{1}{\sqrt{2}}L_{<100>} + \sqrt{2}d\right)$$

$$\Rightarrow \quad \sin\alpha.x - \cos\alpha.y + \frac{L_{<100>}}{\sqrt{2}}(\cos\alpha - \sin\alpha) + \sqrt{2}d\sin\alpha = 0$$

The perpendicular distance from a point O (0, 0) to a line SB (i.e., <*lm*0> direction) is

$$d_1 = \frac{L_{<100>}}{\sqrt{2}}(\cos\alpha - \sin\alpha) + \sqrt{2}d\sin\alpha$$

The equation of a straight line AB (i.e., <100> line) passing through a point A($\sqrt{2}d$,0) and making an angle of 45° with the <110> edge is

$$x - y - \sqrt{2}d = 0$$

It can be noticed in Fig. 6.32 that the perpendicular distance from point O to line AB is d (i.e., etch depth). To achieve a sharp-edge convex corner the beam should be consumed by lateral etching of {100} planes at <100> edges, and therefore following condition must be satisfied:

$$d_1 R_{<lm0>} = dR_{<100>}$$

where $R_{<100>}$ and $R_{<lm0>}$ are the etch rates along the directions perpendicular to them.

On substitution of value d_1 and rearranging the terms we get

$$d = \frac{R_{<lm0>}}{R_{<100>}}\left(\frac{L_{<100>}}{\sqrt{2}}(\cos\alpha - \sin\alpha) + \sqrt{2}d\sin\alpha\right)$$

The microstructures fabricated using a <100> beam in 25 wt% TMAH and 40 wt% KOH are shown in Figs. 6.35a and 6.35b, respectively. It can be noted from SEM pictures that the microstructures contain well-shaped convex corners. As stated earlier, a <100> beam provides a sharp convex corner only if the condition $\frac{R_{\{ij0\}}}{R_{\{100\}}} \geq \frac{1}{\cos\theta}$ is satisfied. If an etchant (or etching conditions) does not satisfy this relation, a triangular (or square) shape can be a better choice. The etching in surfactant-added TMAH (or IPA-added KOH) exposes {110} planes at <100> edges, as shown in Fig. 6.34a, and therefore this type of etchant is not suitable to fabricate sharp convex corners using a <100> beam as compensation geometry [48, 49, 70, 84].

- **Superimposed squares**: The major part of the research on the corner compensation method has been focused on reducing the spatial requirement without compromising on the quality of convex corners. The spatial requirement for a particular etchant is primarily determined by two factors: the shape of the compensating pattern and the amount of corner undercutting in that etchant. Although a <100>-oriented beam compensation design provides a very sharp corner, a high spatial requirement along its length is a serious concern. To reduce the spatial requirement, a superimposed square-shaped design is proposed [42]. In this geometry, one square of side a and two squares of side $a/2$ are superimposed at the apex of the convex corner, as illustrated in Fig. 6.36. The shape of the compensating structure at different steps during etching is designated by dotted lines. In this case, the structure is first consumed by the fast-etching planes appearing at its convex corners (i.e., B, D, E, F, and H). After a finite interval of time, depending on the dimensions and etching parameters, the etch front is transformed into a <100>-oriented beam whose free end is formed by the <*lm*0> etch front lines, as shown in Fig. 6.36b. After this stage, etching proceeds by both lateral undercutting of the {100} sides and fast propagation of the planes at the free end. Similar to the <100> beam, the shape of the convex corner should be taken by the lateral etching of vertical {100} planes, as shown in Fig. 6.36c. To get the sharp-edge convex corners (i.e., criterion to get the convex corner by lateral etching of vertical {100} planes), the following condition must be satisfied [42]:

$$V\left(=\frac{R_{<lm0>}}{R_{<100>}}\right)<\frac{(3\cos\alpha-2\sin\alpha)(\tan\alpha+1)}{\sqrt{2}}$$

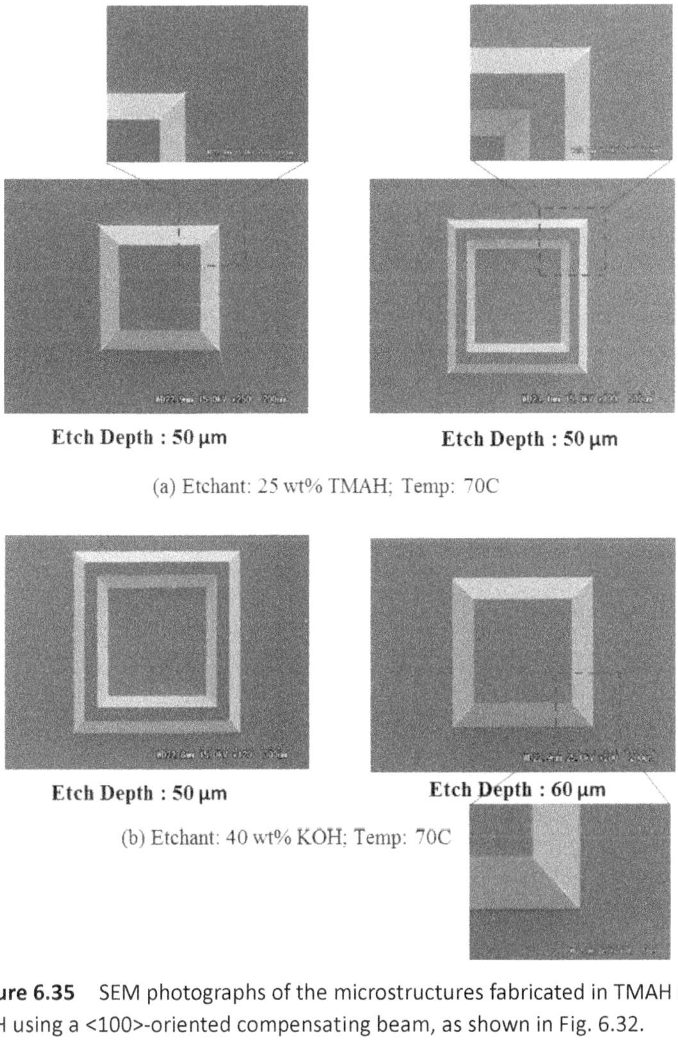

Figure 6.35 SEM photographs of the microstructures fabricated in TMAH and KOH using a <100>-oriented compensating beam, as shown in Fig. 6.32.

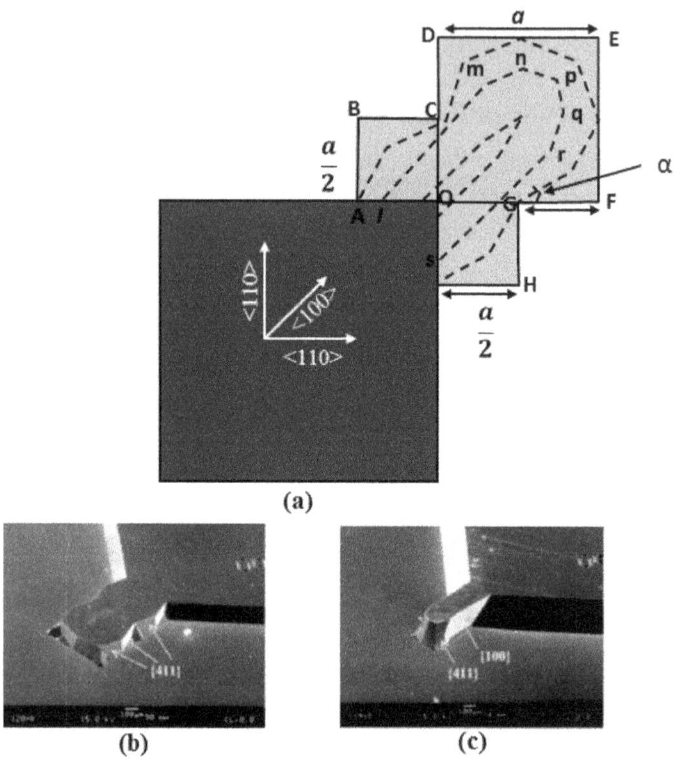

Figure 6.36 Superimposed squares geometry for corner compensation: (a) schematic and etched profile after (b) time t_1 and (c) time t_2 ($>t_1$) when the structure transformed to a <100> beam. From Ref. [42]. @ IOP Publishing. Reproduced with permission. All rights reserved.

A detailed derivation of the relation between the square side *a* and the etch depth *d* gives [42]

$$d = \left[\frac{2\sin\alpha + \cos\alpha}{4V} + \frac{\sqrt{2}}{4(\tan\alpha + 1)} \right] a$$

SEM pictures of mesa structures fabricated in 40 wt% KOH and 25 wt% TMAH are shown in Figs. 6.37a and 6.37b, respectively. In both cases, the convex corners of mesa structures are well defined.

As mentioned, this design relies on the transformation of the etch profile into a <100> beam, and the <100> beam is not suitable for the fabrication of convex corners using surfactant-added TMAH. Therefore the compensation geometry formed by superimposed squares is not appropriate to achieve sharp corners in surfactant-added TMAH. On the basis of the above discussion, it can be stated that none of the compensating designs provides any improvement over the already nearly sharp convex corners obtained in surfactant-added 25 wt% TMAH without compensation, as presented in Fig. 6.10d [48].

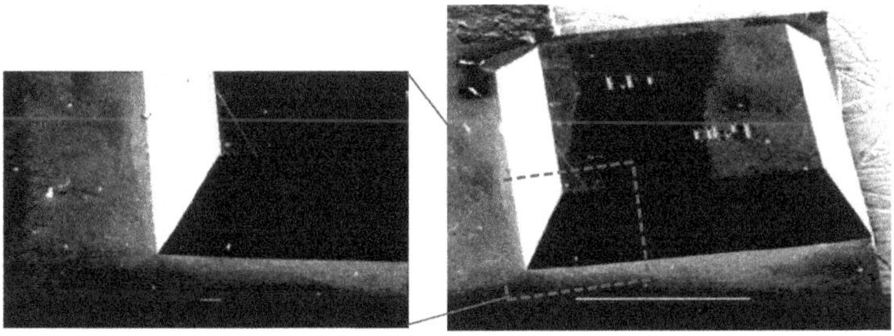

(a) Etch depth : 379 μm [42]

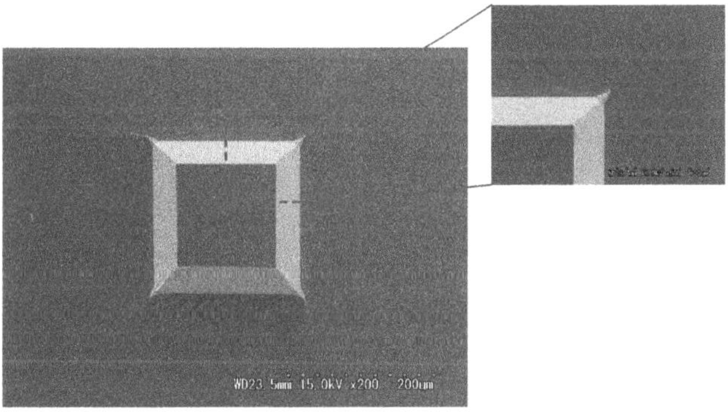

(b) Etch depth : 50 μm

Figure 6.37 SEM pictures of mesa structures fabricated in (a) 40 wt% KOH and (b) 25 wt% TMAH at 70°C using superimposed compensating structures shown in Fig. 6.36.

- **Corner compensation design for bent V-grooves**: The basic problem associated with corner compensation design is the large spatial requirement around the convex corner. In several kinds of microstructures, for instance, the mesa structure surrounded by bent V-grooves and the crossed V-grooves for chip isolation grooves, as presented in Fig. 6.38, the space available around the convex corner is less than that required for the incorporation of the compensating structure.

In these cases, simple compensation structures (e.g., square, triangle, <100> and <110> beams) cannot be employed as they cannot be fitted in the existing space around the corner. As mentioned earlier, the spatial requirement for a particular etchant is primarily determined by two factors: the shape of the compensating pattern and the amount of corner undercutting in that etchant. Various designs have been reported to fabricate bent V-grooves.

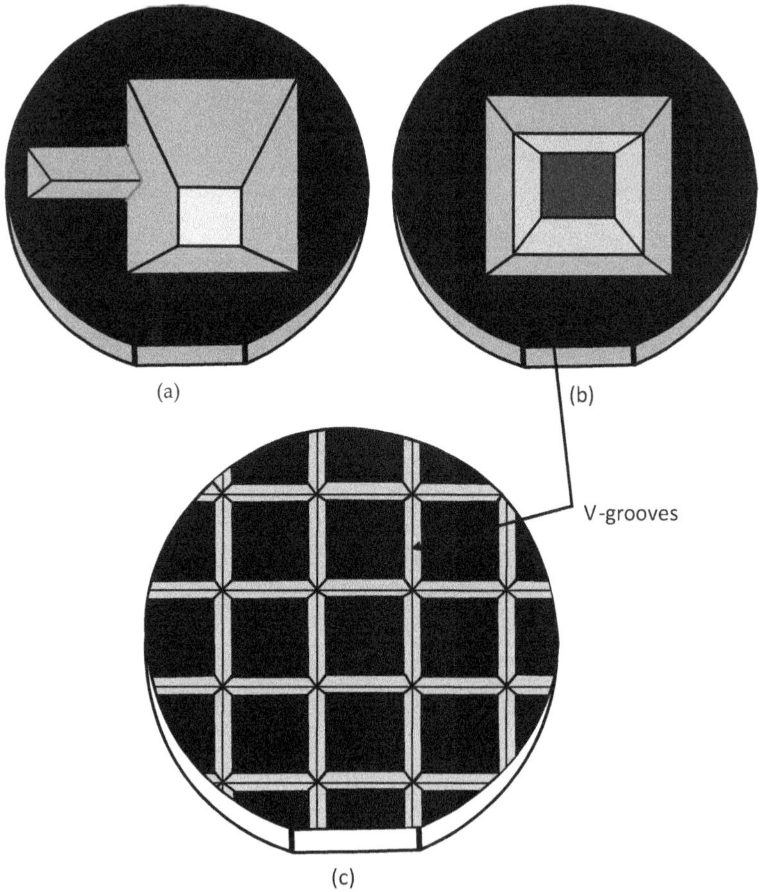

Figure 6.38 Schematic representation of the wet-bulk-micromachined structures which have limited space around the convex corners: (a) V-groove connected to a square aperture, (b) mesa structure surrounded by bent V-grooves, and (c) V-grooves for chip isolation.

Mayer et al. proposed a design in which a <100>-oriented beam is fanned out into narrow <110> strips at both sides and its free end is connected to an external mask (or frame) by a narrow beam just to eliminate the convex corners, as presented in Fig. 6.39 [27]. In this case, the etching of the compensation structure starts only from the convex corners at the free end of the <110> narrow beam. As soon as the undercutting from the free end of <110> strips reaches the main <100> beam, it proceeds by lateral etching at both sides of the beam. The dimensions of the compensating pattern can easily be calculated using the schematic drawing shown in Fig. 6.39b. The consumption of the main <100> beam through lateral undercutting by vertical {100} planes gives $\frac{W_{<100>}}{2}$ etch depth, where $W_{<100>}$ is the width of the <100> beam. The length of the <110>-oriented narrow strips (i.e., $L_{<110>}$) for etch depth d can be calculated as follows:

$$L_{<110>} = \left(d - \frac{W_{<100>}}{2} \right) \times V \times \frac{1}{\sin\alpha}$$

where $V \left(= \frac{R_{<lm0>}}{R_{<100>}} \right)$ is the etch rate ratio between <*lm*0> (or the undercut direction at the convex corner) and the <100> directions.

It is obvious from the relation that the length of the <110> strips is independent of their width. It depends on the etch depth d, width of the main <100> beam (i.e., $\frac{W_{<100>}}{2}$), and the beveled angle α at the convex corner. Hence the width of the <110> strips can be chosen according to convenience. This type of compensating structure can provide a sharp-edge convex corner only if the compensation geometry at the end is consumed by only the lateral etching of vertical {100} planes. However, practically the undercut profile at the edges of the main <100> beam is distorted by the <110> strips, which results in the appearance of a protruded mass at the bottom of the convex corner, as can be observed in Fig. 6.39c. If the etching is continued to remove this extra mass at the bottom, severe undercutting starts at the main convex corners, which degrades the shape of the mesa, as can be seen in Fig. 6.39d. Moreover, this method requires more space along the length of the V-grooves and therefore is suitable only if the grooves are long enough.

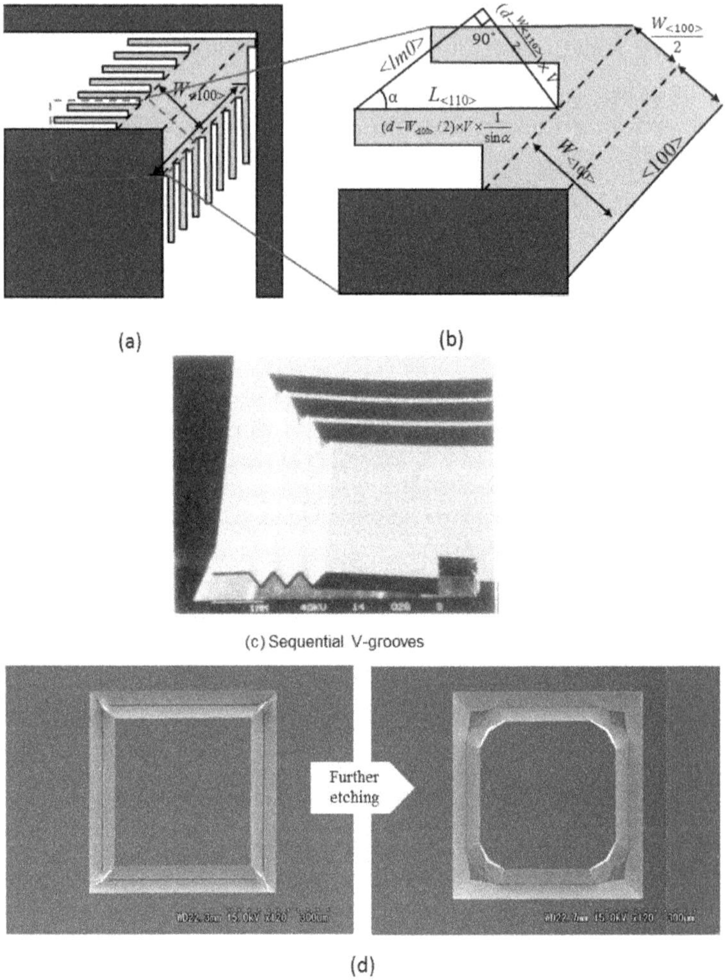

Figure 6.39 A <100>-oriented beam where several <110>-oriented narrow strips are added on both sides, (b) dimensioning of the narrow strips, (c) sequential V-grooves, and (d) mesa structures surrounded by bent V-grooves fabricated in 40 wt% KOH. From Ref. [27]. Reproduced by permission of The Electrochemical Society.

Enoksson made some modifications in the <100> beam, as presented in Fig. 6.40, to minimize the problem of high spatial requirement [36]. In this design a narrow slit is incorporated in the middle from the concave corner. The length of the slit must be slightly longer than twice the required etch depth to ensure that the compensating structure is completely underetched. The progressive change in the shape of the compensating structure during the etching process is illustrated by dotted lines. The first half of the beam (i.e., up to point C) is consumed by the undercutting of {100} vertical planes. The undercutting at A–A′ and B–B′ crosses each other at point C and subsequently the undercutting also start from point C. Similar to a simple <100> beam (Fig. 6.32), the width of the beam is chosen as twice of the etching depth. Figure 6.40b presents the SEM picture of a convex corner fabricated using a modified <100> beam, as shown in Fig. 6.40a. Although the design presented in Fig. 6.40a requires less space in comparison to a simple <100> beam (Fig. 6.32), it is not suitable for the fabrication of bent V-grooves. Additionally, it does not provide a sharp-edge corner, as some residue left at the lower part is clearly visible in Fig. 6.40b.

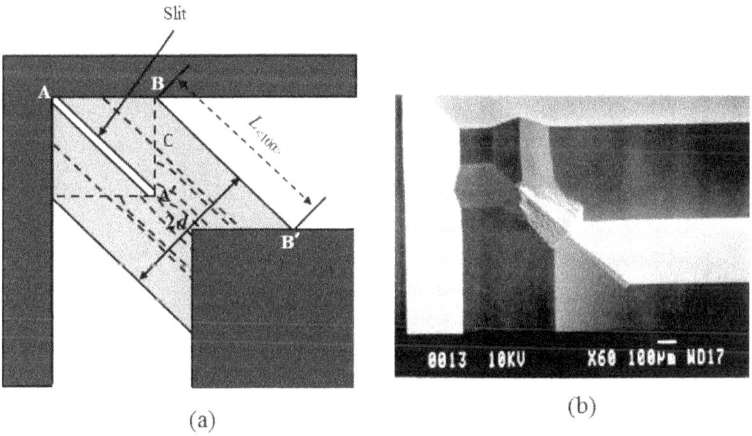

Figure 6.40 Modified <100> beam: (a) schematic view and (b) SEM picture of fabricated convex corner using this design. From Ref. [36]. @ IOP Publishing. Reproduced with permission. All rights reserved.

To realize bent V-grooves, Wacogne et al. proposed a design composed of a superimposed rectangle and square patterns, as shown in Fig. 6.41 [46]. The evolution of the etched profile, initiating from corners B and H, is shown by dotted lines. In this design, the protection time mainly depends on the distance of the beveled direction <*lm*0> at point B (or H) from point O, as indicated by *r* in Fig. 6.41. This can be formulated in terms of the beveled angle (α) and the dimensions of the rectangular strip (i.e., length l_1 and width [g_1–g_2]), as done for square and <110> compensation geometries.

Furthermore, the distance r can be correlated with etch depth d and the etch rates of <*lm*0> and <100> (i.e., $R_{<lm0>}$ and $R_{<100>}$) as follows:

$$rR_{<100>} = dR_{<lm0>} \Rightarrow r = d\frac{R_{<lm0>}}{R_{<100>}} \Rightarrow r = dV$$

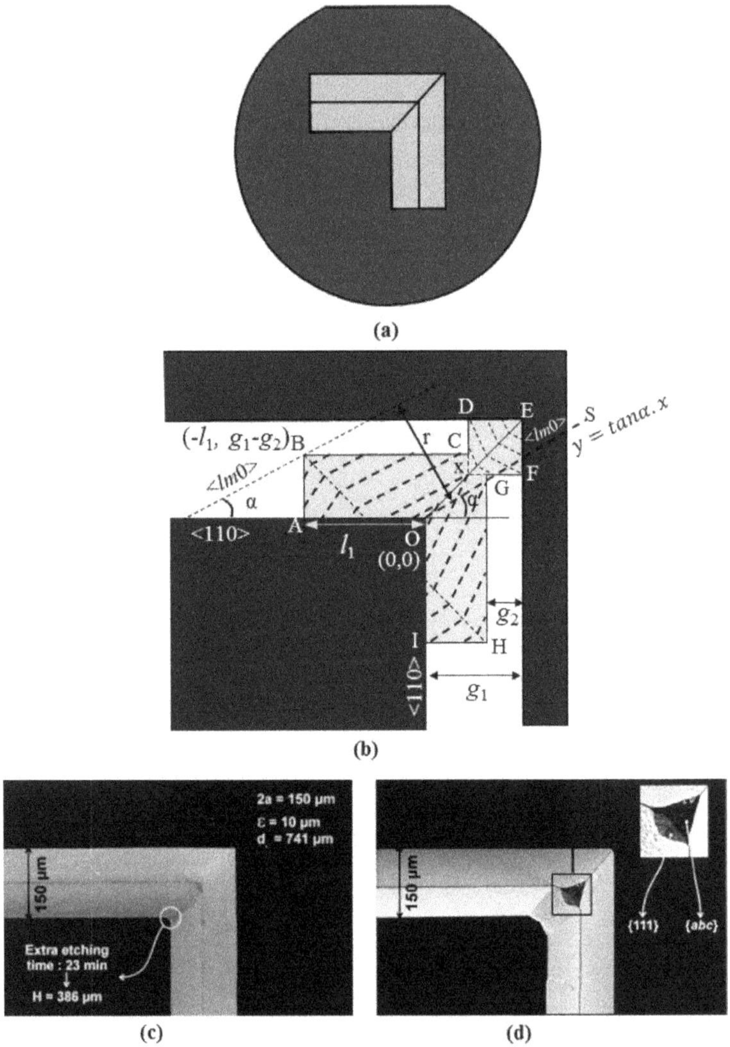

Figure 6.41 Fabrication of microstructures with bent V-grooves: (a) schematic view of a bent V-groove in a {100} wafer, (b) a mask pattern with a corner compensation structure, and (c and d) fabricated structures. Reprinted from Ref. [46], Copyright (2006), with permission from Elsevier.

The etch rate of a direction means the velocity of the edge during etching along a direction perpendicular to it. The length of the compensation strip can be defined as follows:

$$l_1 = \frac{R_{<lm0>}}{R_{<100>}} \frac{d}{\sin\alpha} - \frac{g_1 - g_2}{\tan\alpha}$$

This equation is valid only if the length l_1 is greater than the width (g_1 - g_2). This is the same case as considered for the <110> compensation geometry. Furthermore, the size of square DEFX (or width g_2) should be as small as possible to ensure that it is consumed before the beveling directions <*lm*0> from point B and H reach the main convex corner O. Figures 6.41c and 6.41d present the shape of bent V-grooves fabricated in 40 wt% KOH at 55°C using the compensation design shown in Fig. 6.41b. It can be noticed that the perfect compensation is not possible using this design as the deformation of the convex corner and some residues of the compensation structure at the bottom of the corner is clearly visible. Moreover this design can be used only if the length of the groove is sufficient for the compensation design to fit in the pattern.

As mentioned previously, the design of a compensation pattern has been a major concern when the space is limited around the corner. In continuation of the efforts to overcome the problem of high spatial requirement and to fabricate well-shaped corners, Scheibe et al. proposed superimposed <110>-oriented beams, as shown in Fig 6.42, for the formation of crossed V-grooves for chip isolation [33]. As can be noted from the SEM picture of crossed V-grooves in Fig. 6.42b, the corners are damaged and some residual mass remains at the bottom surface.

Figure 6.43a shows a compensation scheme for the fabrication of a pyramidal aperture connected to the V-groove, as presented in Fig. 6.38a [41]. The dotted line designates the shape of the structure as etching proceeds. The evolving etched profile of the compensation pattern clearly indicates that the significant beveling will take place at the convex corners A and B. This beveling can be reduced if the square-shaped compensation geometry (Fig. 6.43a) is replaced by asymmetric <110> beams, as illustrated in Fig. 6.43b. However, this design will also not be able to provide the well-defined convex corners, as discussed previously.

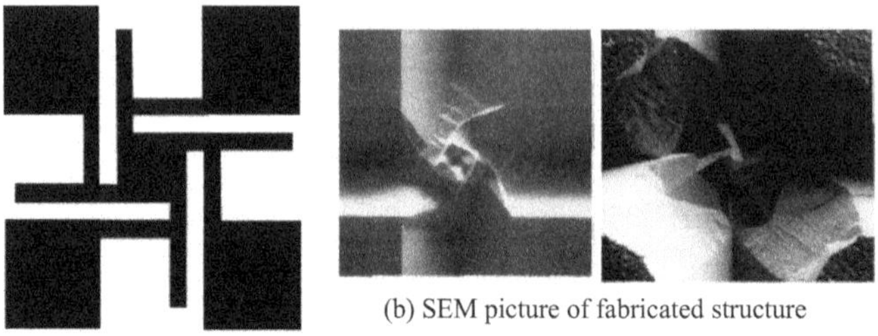

Figure 6.42 Corner compensation structure for the fabrication of grooves for chip isolation. From Ref. [33]. @ IOP Publishing. Reproduced with permission.

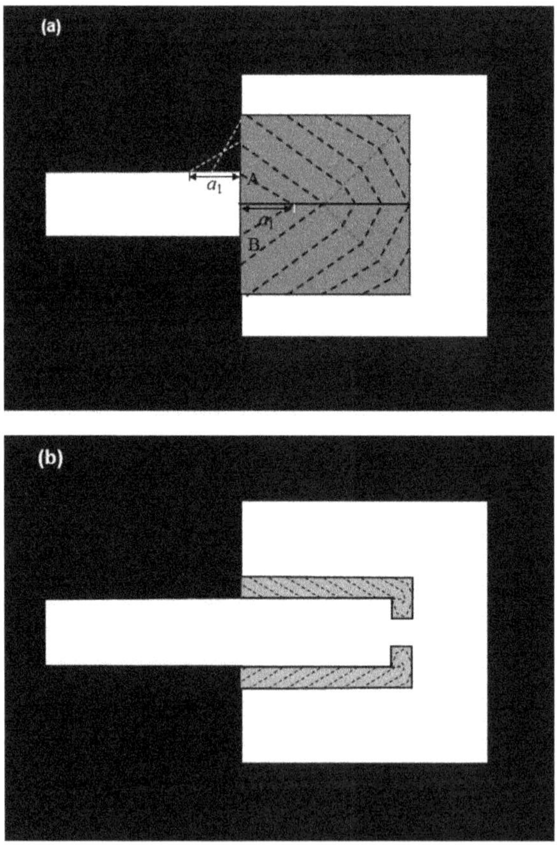

Figure 6.43 Schematic view of a mask pattern with compensation geometry to fabricate a microstructure with a V-groove connected to a square aperture, as shown in Fig. 6.38a.

In the case of a convex corner surrounded by bent V-grooves, it can be concluded that no compensating geometry provides a well-protected convex corner if etching is done in pure KOH and TMAH solutions. Nevertheless, the investigation of new etchants has improved the situation [48, 49, 72–84]. Surfactant-added TMAH is known to provide minimum undercutting at all kinds of corners and edges, as can be seen in Fig. 6.10. In different concentrations of TMAH, 25% TMAH is the most favorable choice for minimizing the undercutting while achieving a smooth surface finish with a reasonable etch rate of {100} planes [48, 49, 72–74, 78–80, 83, 84]. In this etchant, any shape of the compensating design can conveniently be fitted in the accessible space for the formation of grooves for chip isolation and bent V-grooves. It means that the surfactant-added TMAH eliminates the problem of large spatial requirement. As discussed earlier, the <100>-beam-shaped compensation geometry is not appropriate for the fabrication of convex corners in surfactant-added TMAH (Figs 6.34a). If triangular (or square) compensation geometry is employed, the extruded corners toward the bottom side will encounter the opposite {111} planes as a barrier, as shown in Fig. 6.44a. Therefore, extrusion due to extra silicon at the convex corners can be reduced in such structures.

In square and triangular compensation structures, the triangular shape is more convenient due to its simple design. During the etching process, it is consumed while maintaining its original shape. The mesa structure surrounded by bent V-grooves and the grooves for chip isolation fabricated using the triangular shape are shown in Figs. 6.44a and 6.44b, respectively. It can be observed in SEM photographs that the convex corners are well protected. It can be emphasized here that surfactant-added TMAH is the best choice for the realization of a convex corner surrounded with bent V-grooves. As discussed previously, pure KOH and TMAH are not suitable to fabricate a pyramidal aperture connected to a V-groove with well-defined convex corners. The surfactant-added solution is an appropriate choice to fabricate such kind of pyramidal aperture, as presented in Fig. 6.45. This kind of structure is frequently required for a wide range of applications, especially in the field of microfluidics and microbiology.

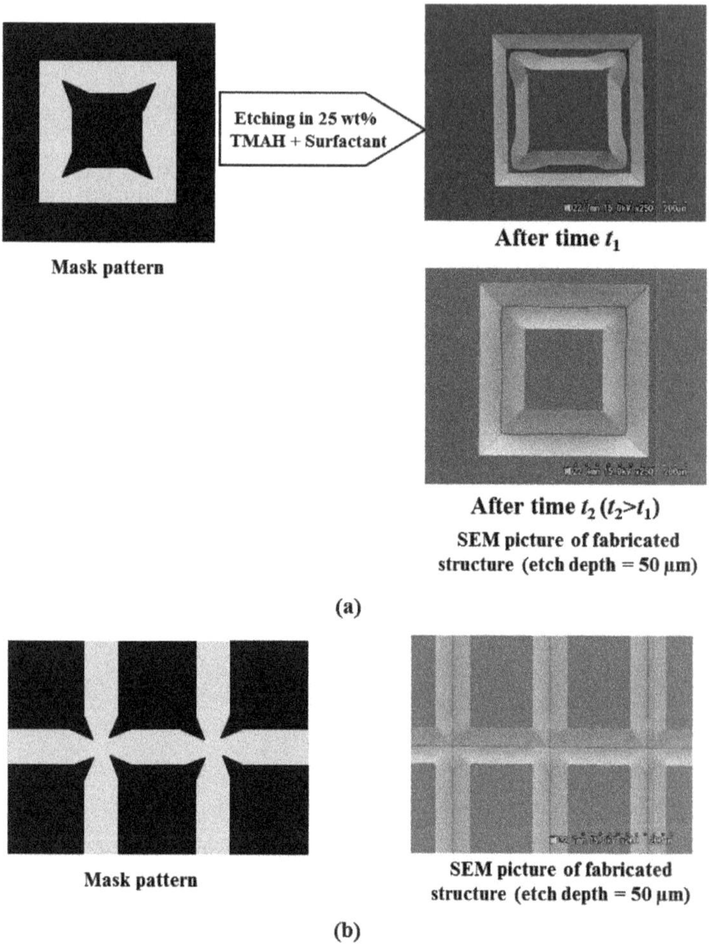

Figure 6.44 Fabrication of (a) microstructure with bent V-grooves and (b) grooves for chip isolation in surfactant-added 25 wt% TMAH using a triangular-shaped compensation structure.

6.6.1.2 Corner compensation geometries for a Si{110} wafer

The {110} silicon surface has twofold symmetry and therefore the angle at the sharp corners depends upon the direction at which the {111} planes appear, as presented in Figs. 6.2 and 6.3b. Analogous to the {100} surface, the undercutting takes place at the convex corner on the {110} surface. Therefore, to fabricate a microstructure with a convex corner on a Si{110} wafer, the extra structure must be added at the corner for its protection during the etching step. In the case of {100} wafers, the compensation structures provide well-defined corners if they are consumed by the lateral etching of vertical {100} planes. However, such kind of etching characteristics is not possible on {110} wafers. Furthermore, a {110} wafer is not suitable for the fabrication of a MEMS structure using surfactant-added TMAH as it has very low etch rate in this etchant, as presented in Fig. 6.10c [75, 83, 99, 103, 104, 109]. Figure 6.46 shows the different directions on a (110) plane in a cubic unit cell and the undercutting at the convex corners of a parallelogram formed by <112> directions. The undercut length and angle at acute and obtuse convex corners for a typical etch depth are l_a, α_a and l_o, α_o, respectively. To protect the corner on a {110} wafer, mainly three kinds of compensation geometry, namely triangular, rhombus, and parallelogramic beam, are reported, as illustrated in Figs. 6.23, 6.47, and 6.48 [56–59]. These designs are studied for the convex corners formed by the intersection of vertical {111} planes (i.e., <112> directions). The design methodology of these geometries is briefly described next.

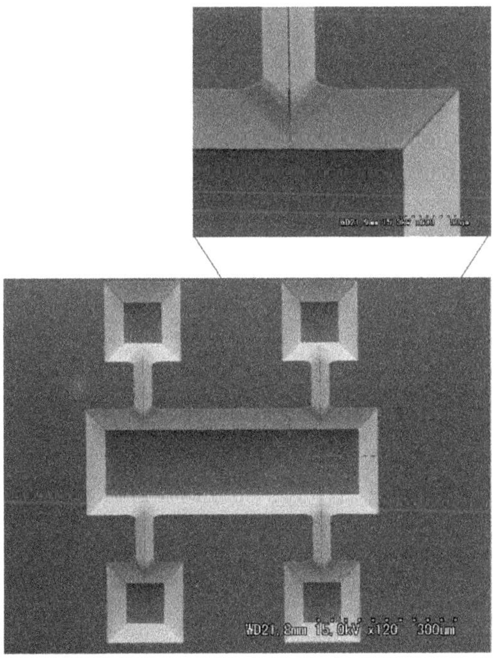

Figure 6.45 SEM picture of the pyramidal apertures connected to V-grooves.

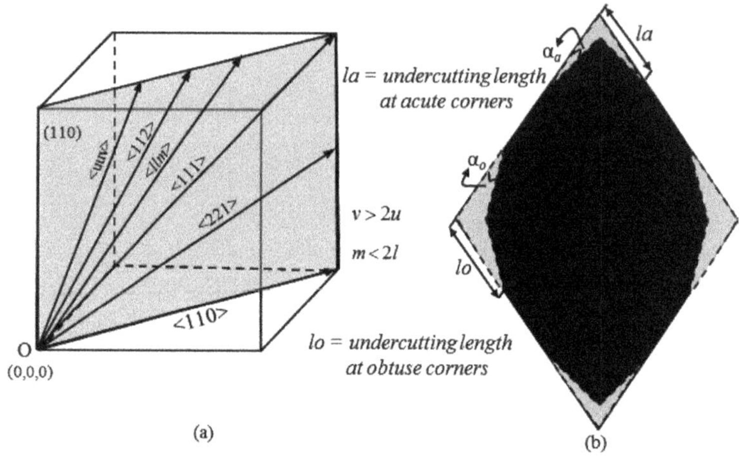

Figure 6.46 Schematic representation of (a) different directions on a (110) plane in a cubic unit cell and (b) the undercutting at the convex corners of a parallelogram formed by <112> directions.

- **Triangular:** As discussed in the case of a {100} wafer, the undercut shape of the corner (i.e., beveled angles) and the amount of undercutting (i.e., undercutting length l) versus etch depth (d) depend on the etching conditions and vary from etchant to etchant. Therefore these parameters must be known to investigate the design of the compensating geometry. The concept behind the design of this structure is the same as used for {100} wafers. However, the compensation structure cannot be identical for all corners as the included angles at convex corners are not same, as can be seen in Fig. 6.47. In this design, the directions of the sides of the triangle at obtuse corners are the same as the directions of undercutting lines at acute corners and vice versa, as shown in Fig. 6.47a. In the absence of a compensation pattern, the beveled shapes (or undercut shape) at obtuse and acute angles are shown by red- and yellow-colored lines, respectively. The successive etched profile of the compensating pattern during etching is represented by dotted lines. The sides JP' and KP' of the compensating triangle at acute corners are chosen to coincide with the <*uuv*> family of lines corresponding to the directions of the undercutting at obtuse corners, as indicated by red-colored lines. Similarly at obtuse corners the sides of the triangle, MO' and LO', should coincide with <*llm*> directions, which are the directions of undercutting at acute corners, as shown by yellow color. To calculate the dimensions of the triangle using the sine rule, the beveled angles and the lengths of undercutting along the <112> directions at obtuse (i.e., α_o and l_o) and acute (i.e., α_a and l_a) angles must be known. SEM photographs of acute and obtuse convex corners fabricated using triangular compensation patterns in 42.5 wt% KOH at 80°C KOH are shown in Figs. 6.49a and 6.49b, respectively [57]. It can be easily noticed that none of the corners is well shaped; elongated residues of different lengths are left at both types of corners.

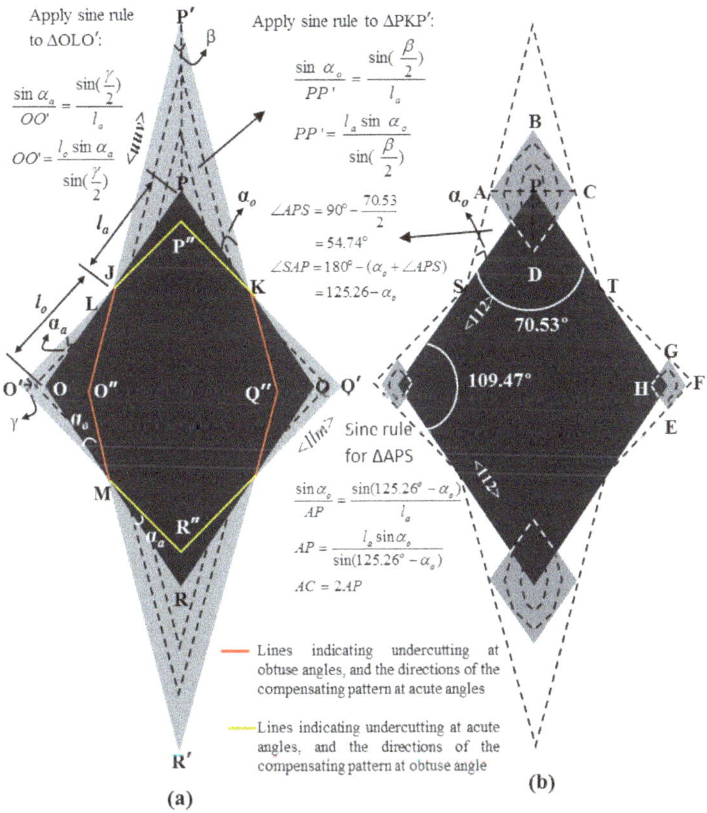

Figure 6.47 Schematic representation of (a) triangular and (b) rhombus (or square) corner compensation geometries for the protection of convex corners during fabrication on {110} silicon wafers. The dotted lines represent the shape of the structure as etching proceeds. The dimensions of the compensating geometries are calculated using the sine rule.

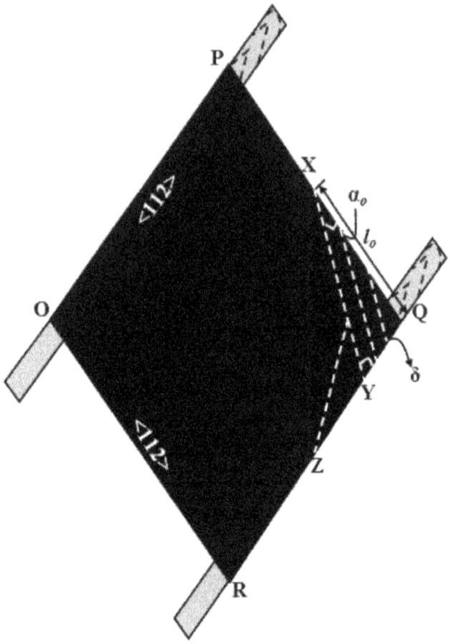

Length of the compensating beam for undercutting length QX at obtuse angle = QY

$$\frac{\sin \alpha_o}{QY} = \frac{\sin \delta}{l_o}$$

$$QY = \frac{l_o \sin \alpha_o}{\sin(180^\circ - 109.47^\circ - \alpha_o)} = \frac{l_o \sin \alpha_o}{\sin(70.53^\circ - \alpha_o)}$$

Figure 6.48 A <112> beam compensating pattern. The dimensions of the beam calculated for the obtuse corner using the sine rule are applicable to the acute corner as well.

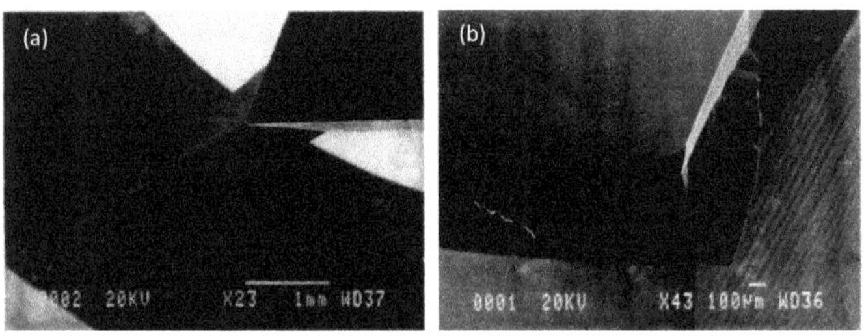

Figure 6.49 SEM photographs of (a) acute and (b) obtuse convex corners fabricated using triangular compensation patterns in 42.5 wt % KOH at 80°C. From Ref. [57]. Reproduced by permission of The Electrochemical Society.

- **Rhombus (or square)**: This design is a shortened version of the triangular-shaped pattern discussed in the previous section. The original idea of this kind of compensation structure is the same as the one reported by Puers et al. for the design of square-shaped compensation pattern for the formation of convex corners on Si{100} [26] . The sides of the compensating rhombus align along the <112> direction. The center of the rhombus coincides with the apex of the convex corners, and it is of maximum size, which can be inscribed in the triangle, as illustrated in Fig. 6.47b. In the etching process, the compensating rhombus is totally consumed by the undercutting that starts from the three convex corners (A, B, and C at acute corners; E, F, and G at obtuse corners), as illustrated by the dotted lines in Fig. 6.47b. The dimensions of the rhombic pattern can simply be calculated using the sine rule as done earlier for other structures. SEM images of the convex corners fabricated in pure and IPA-added 40 wt% KOH are shown in Figs. 6.50i and 6.50ii, respectively [56]. In all cases except the acute corner in IPA-added KOH (Fig. 6.50ii-c), an irregular-shaped residue can easily be observed.

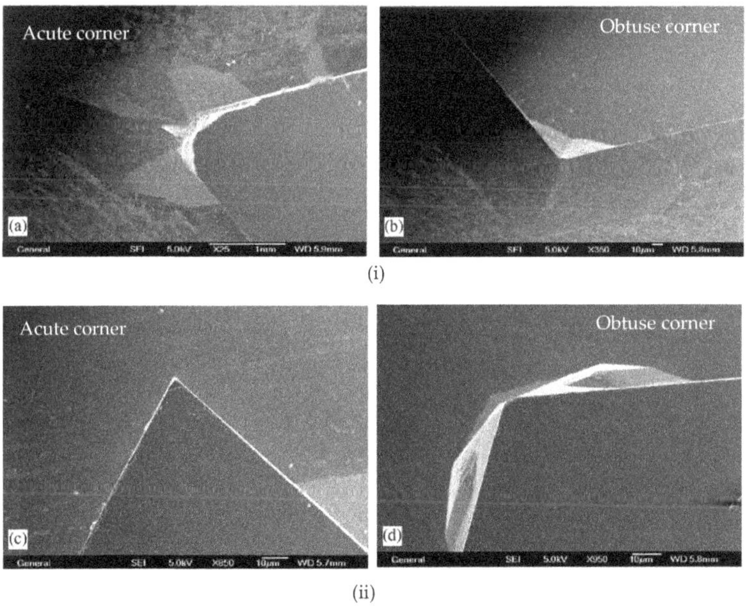

Figure 6.50 SEM images of convex corners fabricated in (i) pure and (ii) IPA-added 40% KOH using rhombus compensation geometry. Reprinted from Ref. [56], Copyright (2006), with permission from Elsevier.

- **Parallelogramic beam (or beam)**: In this method, a <112>-oriented beam (or parallelogramic beam) is fitted at the convex corner. To calculate the length of the beam to protect the corner for etch depth *d*, the undercutting versus etch depth data should be available. As illustrated in Fig. 6.48, the length of the compensating beam corresponding to undercutting length l_0 (i.e., QX) for etch depth *d* should be QY. In this design, the beam width does not play any role and it should be as small as possible, but it should not be consumed by the lateral undercutting of {111} planes which appear at its edges. The great feature of this structure is that the same-size beams are added at the acute and obtuse corners. SEM pictures of convex corners fabricated in 42.5 wt% KOH using beam-shaped compensation geometry are presented in Figs. 6.51a and 6.51b, respectively. An SEM image of a cantilever beam with vertical sidewalls in a {110} wafer realized in 25 wt% TMAH using this design is shown in Fig. 6.52.

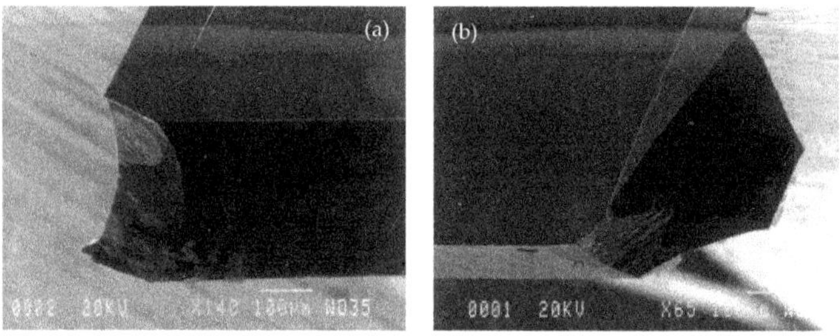

Figure 6.51 SEM photographs of (a) acute and (b) obtuse convex corners fabricated using parallelogramic beam compensation design in 42.5 wt% KOH at 80°C. From Ref. [57]. Reproduced by permission of The Electrochemical Society.

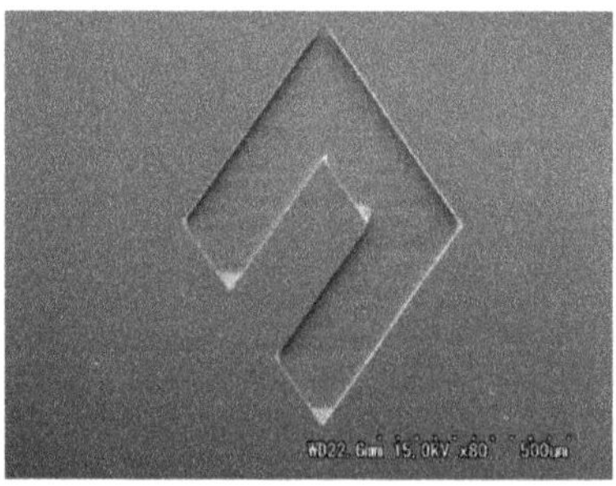

Figure 6.52 SEM picture of a cantilever beam with vertical sidewalls in a {110} wafer realized in 25 wt% TMAH using beam-shaped compensation pattern at its corners.

6.6.2 Perfect Convex Corners Using Two-Step Etching Techniques

The compensation technique relies on time-delayed etching of convex corners through some added extra geometry. The dimensions of the compensating structures depend on the etch depth up to which the convex corners have to be created. This technique, therefore, requires precise control of the etch rate and highly uniform wafer thickness. The convex corner results in a damaged corner if the etching is performed for more than the prescribed etching time. Moreover, in the corner compensation method, high spatial requirement is a big concern, especially when the spacing between adjacent structures is limited. Hence, achieving a very sharp corner without any damage at the convex edge (or any residual at the bottom surface) is a tedious task. To avoid these issues and to fabricate microstructures with ideally sharp corners, two techniques have been developed [62–65]. The convex corner is formed by the intersection of two {111} planes, and in these techniques, {111} planes are fabricated in two steps. After fabricating one of the two {111} planes of the convex corner, it is protected by a masking layer. Thereafter, in the next step of etching the second plane is realized. Hence both methods provide very clean and sharp convex corners, but lithography requires two masks and the wet etching is performed in two steps. The fabrication steps of these techniques are discussed in the following sections.

6.6.2.1 Sharp-edge convex corners using lithography on anisotropically etched patterns

To fabricate very-sharp-edge convex corners, Kwon and Kim have developed a method which employs two masks and two steps of anisotropic etching. It requires lithography after the first anisotropic etching step [62]. Processing steps used in this technique are briefly outlined in Fig. 6.53. After defining a pattern using mask #1, anisotropic etching is employed to form a cavity of desired depth (Fig. 6.53b). Thereafter a masking layer is deposited on exposed silicon. The masking layer can be a thermal oxide or chemical vapor deposition (CVD) nitride. If the oxide is used as a mask, KOH is not a suitable etchant as it exhibits a finite oxide etch rate. In that case, TMAH should be used as an etchant. Subsequent to the deposition of a masking layer, the second step of lithography is performed using mask #2 to pattern the masking layer on the sidewalls as well as the top surface. Due to uneven surface topography, it is very hard to coat the photoresist of uniform thickness using the spin-coating technique [111–113]. The use of a standard liquid photoresist, deposited by a spin-coating process, results in nonuniform thickness, especially on sloping sidewalls and convex edges/corners. At the convex edges, the film may be even discontinuous, as the photoresist reflows from these edges due to surface tension. A very thin or discontinuous photoresist coating at the convex edges fails to protect the masking layer at the sharp edges during the etching process. This, in turn, leads to etching of silicon during anisotropic etching. Moreover, due to reflow of the spin-coated photoresist, an excessive amount of photoresist is accumulated on the bottom of the cavities/trenches. To coat a constant-thickness photoresist, the spray-coating method is employed. The thickness, uniformity, and surface roughness of a spray-coated photoresist depend on several parameters such as viscosity of the photoresist, orifice of the spray nozzle, and air pressure applied to the nozzle [113–115]. After the patterning of the masking layer using the second step of lithography (Fig. 6.53c), the next step of anisotropic etching is then carried out (Fig. 6.53d). At last, the mask layer can be removed globally, if required. An SEM picture of a microstructure fabricated using this method is shown in Fig. 6.53f. It can be noticed that the convex corners are well protected.

Although this technique provides protected convex corners and the overetching does not distort the shape of the convex corner, the second step of lithography is onerous as it is performed on 3D surface topography. The 3D nature of the microstructure makes it very difficult to transfer the pattern [112, 113, 115–118]. There are three major issues in the lithography process on 3D topography: uniform photoresist coating, reflection of UV light from tilted sidewalls, and uneven gap between the mask and the bottom surface. The first concern is the uniform coating of the photoresist using the spin-coating method, which is most commonly used in microfabrication. Although electrodeposition can be employed to obtain a uniform-thickness photoresist on an uneven surface, it has its own limitations [116–118]. The electrodeposited photoresist needs a conductive base on which the photoresist electrodeposition takes place, and therefore it cannot be used at all stages of a process. Although the spray-coating method provides better results than spinning, it is also not an ideal solution to coat a uniform-thickness resist on a 3D topography surface. Variations in spray-coated resist thickness are observed if the wafer contains cavities with a large difference in size. The second problem in this method is the reflection of UV light from tilted sidewalls during UV exposure. This unwanted reflection exposes the photoresist of opposite sidewalls. Although it can be avoided using polarized UV light, it requires extra hardware in the exposure tool [116]. The third issue is the uneven gap between the mask and the wafer surface. It may cause a diffraction problem and can change the feature size transferred from the mask to the substrate.

6.6.2.2 Perfect convex corners using the LOCOS process

In this technique, two-step anisotropic etching and the local oxidation of silicon (LOCOS) process are used [59, 63, 64, 99, 119]. The LOCOS process is the same as that used in silicon semiconductor technology [120]. Although it requires two-step etching, it does not need any lithography after the first step of etching. Both steps of lithography using photomasks #1 and #2 are employed before the first step of etching. The sequence of process steps for the formation of a microstructure with perfect convex corners are illustrated in Fig. 6.54. The fabrication starts with the patterning of an oxide layer using mask #1. After this step, the wafer is coated with silicon nitride. Now lithography is performed using mask #2 for opening the windows for the first step of anisotropic etching (Fig. 6.54c). The nitride is etched out by the reactive ion etching (RIE) process and the oxide layer by RIE or buffered hydrofluoric acid (BHF). After this step, anisotropic etching is performed. This step of etching produces cavities with sloped sidewalls, as is shown in Fig. 6.54d. Now the thermal oxidation process is carried out. This step of oxidation is called LOCOS as only exposed silicon is oxidized and the remaining part is protected by the nitride layer [120]. After LOCOS, silicon nitride is globally striped out selectively (i.e., without damaging the oxide layer) by dry etching or wet chemical etching in hot phosphoric acid (H_3PO_4). If the nitride is etched out in hot phosphoric acid, the sample should be immersed in BHF for a short time (about 20 s) in order to remove any oxide layer formed over the nitride during the LOCOS process [121]. This step is attempted because the oxide etch rate in phosphoric acid is negligible in comparison to the nitride etch rate. Now, the second step of silicon etching is employed. This step of silicon etching should be performed in TMAH (or EDP) as the oxide layer is being used as the masking layer. EDP should be avoided due its toxic nature. At this step, undercutting will start from the corners marked *l*, *m*, *n*, and *o*. The undercutting at these corners proceeds and stops automatically where the {111} planes developed at edges intersect the adjacent {111} planes of cavities which are formed by the first step of silicon etching and protected by the oxide layer. The intersection of these planes forms the sharp convex edges, as is shown in Fig. 6.54f. In this method, the overlapped area (w_1) between the two masks is calculated using simple trigonometric relations. In this process, it should be equal to or greater than $\sqrt{2}T_{Si}$, where T_{Si} is the wafer thickness, as shown in Fig. 6.54c. The minimum width w_2 in mask #2 is decided by the consideration that a through-hole should be formed during the first anisotropic etching. The width (w_3) is not critical, except that the corners *l*, *m*, *n*, and *o* lie within the area defined by mask #1. It is taken according to the requirement of dimensions of the structure. The appearance of corners marked *l*, *m*, *n*, and *o* is essential for this process because these corners play a very critical role during the second anisotropic etching step for the realization of perfect convex corners, as mentioned earlier. The total time of the second anisotropic etching step is more than what is required for the etching of the total thickness of the wafer. This process is not time controlled, that is, extra etching (or overetching) does not affect the sharpness of the desired convex corners.

Figure 6.55 shows SEM photographs of microstructures with perfect convex corners/edges fabricated on a Si{100} wafer using this process. The structures have total symmetry and perfect convex corners/edges with a square-shaped top and bottom. The sharpness of a convex corner can be observed in its close-up view. No critically time-controlled etching of silicon is required, and the process gives perfect convex corners, even on wafers which have thickness variation. Although this technique provides perfect convex corners, it requires extra mask and processing steps such as silicon nitride deposition and its selective etching, LOCOS, etc. This technique is demonstrated only for freestanding structures. For the fabrication of non-freestanding structures, it requires an etch-stop layer, for example, silicon-on-insulator (SOI) wafers [99, 122].

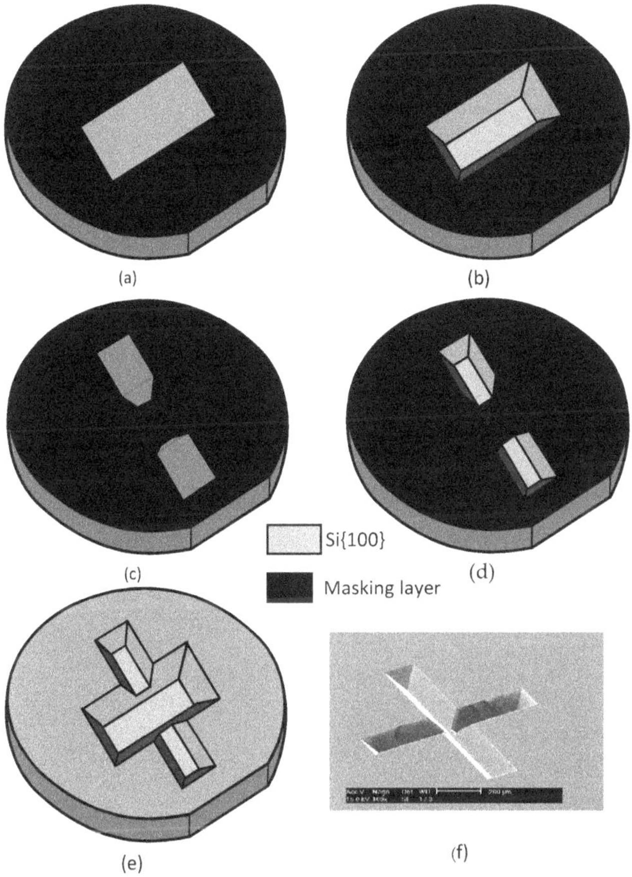

Figure 6.53 Fabrication sequence for the formation of a microstructure with protected convex corners using two-step etching. (a) Oxidation and patterning of oxide, (b) first step of silicon anisotropic etching, (c) deposition of a mask layer (e.g., thermal oxide or CVD nitride), followed by patterning, (d) second step of anisotropic etching, and (e) removal of the mask layer. (f) SEM picture of a structure fabricated using similar process steps. Reprinted from Ref. [62], Copyright (2002), with permission from Elsevier.

This technique is generic and can also be employed for Si{110} to form a microstructure with a perfect convex corner, as presented in Fig. 6.56 [59]. In this method, the overlapped area between the two masks is calculated using simple trigonometric relations. Accordingly the dimensions of the masks are determined. In the mask design, lateral undercutting and the angles of etched sidewalls should also be considered.

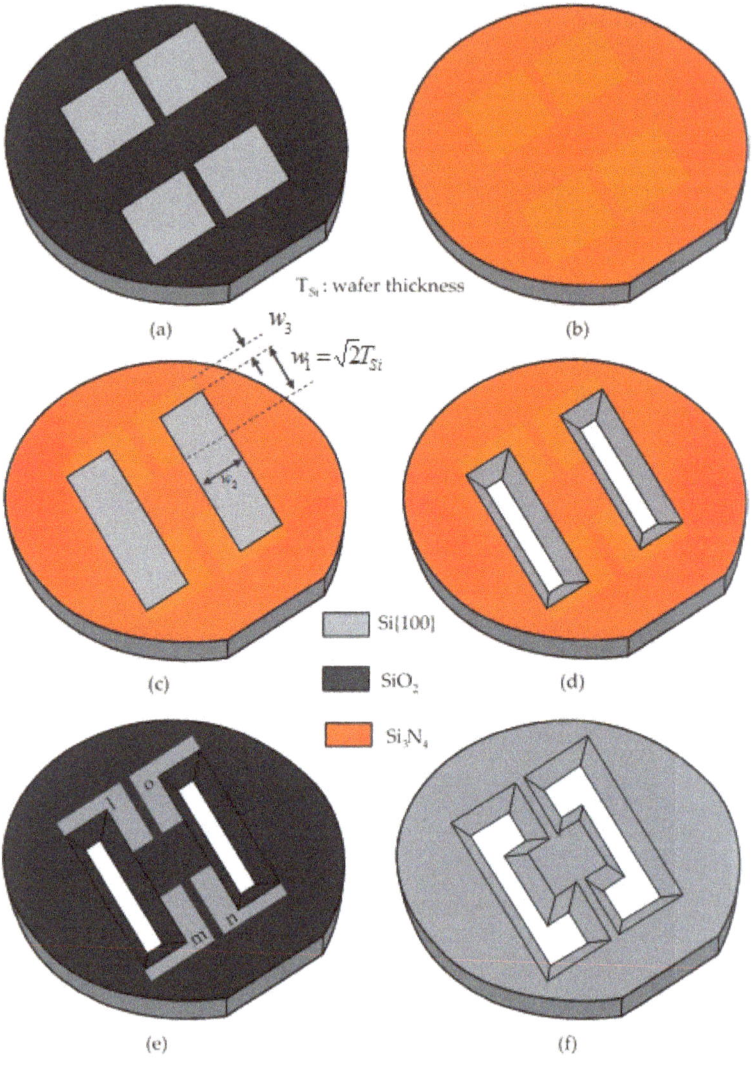

Figure 6.54 Fabrication sequence for the formation of a microstructure with protected convex corners using two-step etching: (a) oxidation and patterning of oxide, (b) deposition of Si_3N_4, (c) patterning of Si_3N_4 for first-step anisotropic etching, (d) anisotropic etching, (e) thermal oxidation of exposed silicon (i.e., LOCOS) followed by nitride etching, and (f) second step of anisotropic etching followed by removal of the oxide layer.

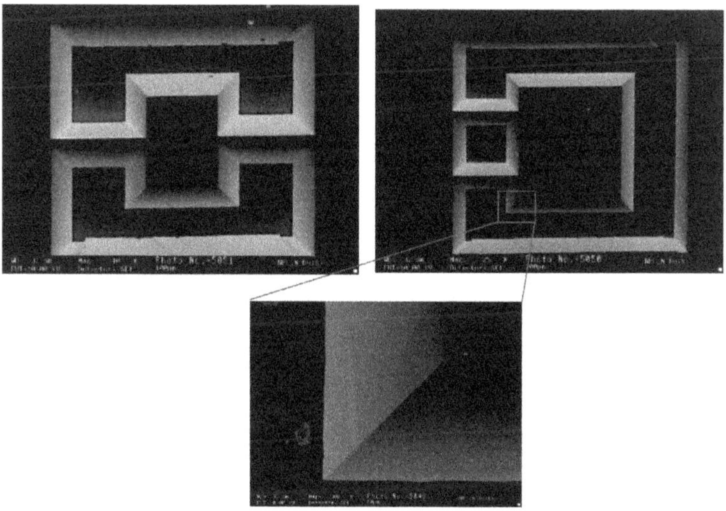

Figure 6.55 SEM photographs of microstructures with perfect convex corners/edges. From Ref. [64]. @ IOP Publishing. Reproduced with permission. All rights reserved.

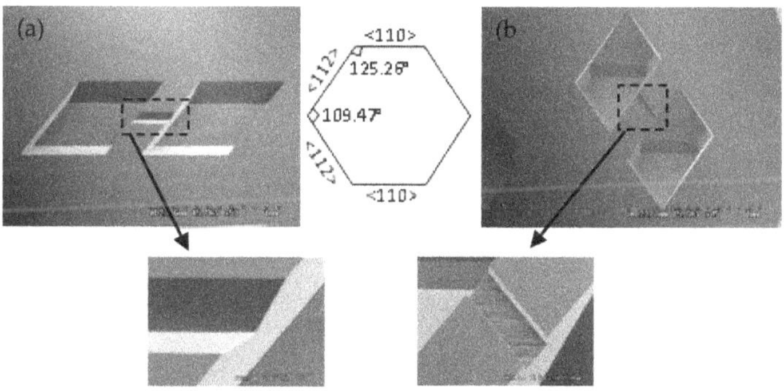

Figure 6.56 Different shapes of microfluidic channels with multiple reservoirs realized in a Si{110} wafer using two-step etching. Microstructures' edges are aligned along (a) <110> and <112> directions and (b) <112> directions. From Ref. [59]. @ IOP Publishing. Reproduced with permission. All rights reserved.

6.7 Summary

Various fabrication methods of convex corners in {100}- and {110}-oriented silicon wafers using anisotropic wet chemical etching are presented in this chapter. In the corner compensation method, the spatial requirement and the resultant shape of the etched convex corner are the major concerns. Although <100>-oriented compensation design provides a sharp-edge corner under certain conditions, its high spatial requirement is a main drawback, especially when the bent V-grooves are required to be realized. To reduce the spatial requirement and to get a sharp convex corner, a compensating geometry designed by superimposed squares is used, provided etchant characteristics fulfill certain requirements. A square-shaped design needs less space to fit at the corner, but it is not suitable to achieve sharp corners. In a surfactant-added TMAH solution, a triangular-shaped design is the best choice. In this etchant, the formation of bent V-grooves with sharp convex corners is easily achievable. In the corner compensation method, a <100> beam is an appropriate choice for the formation of a sharp convex corner in high-concentration TMAH (e.g., 20–25 wt%) and KOH, provided space around the convex corner is not a restriction. In the case of spatial restriction, such as bent V-grooves, surfactant-added high-concentration TMAH with triangular-shaped compensation geometry is a right choice. To protect the convex corner on a Si{110} surface, a <112>-oriented beam (or parallelogramic beam) is an optimal compensating geometry, owing to its simple design, and the same-size beams are used at the acute and obtuse corners.

Apart from the corner compensation method, two different methods which provide perfectly sharp-edge convex corners are discussed. Unlike the corner compensation method, the formation of a sharp convex corner in these methods does not require any time-controlled etching. However, these techniques require two-mask lithography and two-step etching, which makes them more expensive and complex.

References

1. Gad-el-Hak, M. (2002). *The MEMS Handbook* (CRC Press, Boca Raton).
2. Elwenspoek, M., and Jansen, H. (1998). *Silicon Micromachining* (Cambridge University Press, UK).
3. Takahata, K. (2013). *Advances in Micro/Nano Electromechanical Systems and Fabrication Technologies* (InTech).

4. Lindroos, V., Tilli, M., Lehto, A., and Motooka, T. (2010). *Handbook of Silicon Based MEMS Materials and Technologies* (William Andrew).
5. Marc, J. M. (2002). *Fundamentals of Microfabrication: The Science of Miniaturization*, 2nd ed. (CRC Press, Boca Raton).
6. Kovacs, G. T., Maluf, N. I., and Petersen, K. E. (1998). Bulk micromachining of silicon, *IEEE Proc.*, **86**, pp. 1536–1551.
7. Petersen, K. E. (1982). Silicon as a mechanical material, *IEEE Proc.*, **70**, pp. 420–457.
8. Lang, W. (1996). Silicon microstructuring technology, *Mater. Sci. Eng. R*, **17**, pp. 1–55.
9. Fruhauf, J. (2005). *Shape and Functional Elements of the Bulk Silicon Microtechnique: A Manual of Wet-Etched Silicon Structures* (Springer-Verlag, Berlin).
10. Pal, P., and Singh, S. S. (2013). A simple and robust model to explain convex corner undercutting in wet bulk micromachining, *Micro Nano Syst. Lett.*, **1**, pp. 1–6.
11. Pal, P., and Singh, S. S. (2013). A new model for the etching characteristics of corners formed by Si{111} planes on Si{110} wafer surface, *Engineering*, **5**, pp. 1–8.
12. Trieu, H. K., and Mokwa, W. (1998). A generalized model describing corner undercutting by the experimental analysis of TMAH/IPA, *J. Micromech. Microeng.*, **8**, pp. 80–83.
13. Chahoud, M., Wehmann, H. H., and Schlachetzki, A. (1998). Etching simulation of convex and mixed InP and Si structures, *Sens. Actuators A*, **69**, pp. 251–258.
14. Schroder, H., and Obermeier, E. (2000). A new model for Si{100} convex corner undercutting in anisotropic KOH etching, *J. Micromech. Microeng.*, **10**, pp. 163–170.
15. Shikida, M., Nanbara, K., Koizumi, T., Sasaki, H., Sato, K., Odagaki, M., Ando, M., Furuta, S., and Asaumi, K. (2000). A model explaining mask-corner undercut phenomena in anisotropic silicon etching: a saddle point in the etching-rate diagram, *Sens. Actuators A*, **97–98**, pp. 758–763.
16. Chang Chien, W. T., Chang C. O., Lo, Y. C., Li, Z. W., and Chou, C. S. (2005). On the Miller-indices determination of Si{100} convex corner undercut planes, *J. Micromech. Microeng.*, **15**, pp. 833–842.
17. Merlos, A., Acero, M. C., Bao, M. H., Bausells, J., and Esteve, J. (1992). A study of the undercutting characteristics in the TMAH:IPA system, *J. Micromech. Microeng.*, **2**, pp. 181–183.

18. Dong, W., Zhang, X., Liu, C., Li, M., Xu, B., and Chen, W. (2004). Mechanism for convex corner undercutting of (110) silicon in KOH, *Microelectron. J.*, **35**, pp. 417–419.

19. Schroder, H., Obermeier, E., Horn, A., and Wachutka, G. K. M. (2001). Convex corner undercutting of {100} silicon in anisotropic KOH etching: the new step-flow model of 3-D structuring and first simulation results, *J. Microelectromech. Syst.*, **10**, pp. 88–97.

20. Zubel, I. (2000). Silicon anisotropic etching in alkaline solutions III: on the possibility of spatial structures forming in the course of Si (100) anisotropic etching in KOH and KOH+IPA solutions, *Sens. Actuators A*, **84**, pp. 116–125.

21. Lee, D. B. (1969). Anisotropic etching of silicon, *J. Appl. Phys.*, **40**, pp. 4569–4575.

22. Bean, K. E. (1978). Anisotropic etching of silicon, *IEEE Trans. Electron Devices*, **ED-25**, pp. 1185–1193.

23. Bean, K. E., and Runyan, W. R. (1977). Dielectric isolation: comprehensive, current and future, *J. Electrochem. Soc.*, **124**, pp. 5C–12C.

24. Abu-Zeid, M. (1984). Corner undercutting in anisotropically etched isolation contours, *J. Electrochem. Soc.*, **131**, pp. 2138–2142.

25. Wu, X. P., and Ko, W. H. (1989). Compensating corner undercutting in anisotropic etching of (100) silicon, *Sens. Actuators A*, **18**, pp. 207–215.

26. Puers, B., and Sansen, W. (1990). Compensation structures for convex corner micromachining in silicon, *Sens. Actuators A*, **23**, pp. 1036–1041.

27. Mayer, G. K., Offereins, H. L., Sandmaier, H., and Kuhl, K. (1990). Fabrication of non-underetched convex corners in anisotropic etching of (100) silicon in aqueous KOH with respect to novel micromechanic elements, *J. Electrochem. Soc.*, **137**, pp. 3947–3951.

28. Offereins, H. L., Kuhl, K., and Sandmaier, H. (1991). Methods for the fabrication of convex corners in anisotropic etching of (100) silicon in aqueous KOH, *Sens. Actuators A*, **25**, pp. 9–13.

29. Sandmaier, H., Offereins, H. L., Kuhl, K., and Lang, W. (1991). Corner compensation techniques in anisotropic etching of (100)-silicon using aqueous KOH, *6th International Conf. on Solid State Sensors and Actuators* (Transducers 91, San Francisco), pp. 408–411.

30. Hui, W. C. (1991). Technique for protecting chip corners in wet chemical etching of silicon wafers, *Technical Report Lawrence Livermore National Lab.*, CA, United States.

31. Offereins, H. L., Sandmaier, H., Marusczyk, K., Kuhl, K., and Plettner, A. (1992). Compensating corner undercutting of (100) silicon in KOH, *Sens. Mater.*, **3**, pp. 127–144.

32. Bao, M., Burrer, C., Esteve, J., Bausells, J., and Marco, S. (1993). Etching front control of <110> strips for corner compensation, *Sens. Actuators A*, **37–38**, pp. 727–732.

33. Scheibe, C., and Obermeier, E. (1995). Compensating corner undercutting in anisotropic etching of (100) silicon for chip separation, *J. Micromech. Microeng.*, **5**, pp. 109–111.

34. Kampen, R. P., and Wolffenbuttel, R. F. (1995). Effects of <110>-oriented corner compensation structures on membrane quality and convex corner integrity in (100)-silicon using aqueous KOH, *J. Micromech. Microeng.*, **5**, pp. 91–94.

35. Zhang, Q., Liu, L., and Li, Z. (1996). A new approach to convex corner compensation for anisotropic etching of (100) Si in KOH, *Sens. Actuators A*, **56**, pp. 251–254.

36. Enoksson, P. (1997). New structure for corner compensation in anisotropic KOH etching, *J. Micromech. Microeng.*, **7**, pp. 141–144.

37. Chung, C. K., Lee, C. C., and Wu, C. Y. (1998). A novel approach to corner compensation of multistep Si(100) terraced structure for microlens, *IEEE/LEOS Summer Topical Meeting*, Monterey, CA.

38. Long, M. K., Burdick, J. W., and Antonsson, E. K. (1999). Design of compensation structures for anisotropic etching, *Proc. International Conference on Modeling and Simulation of Microsystems* (MSM 99).

39. Ma, L. (2001). *Robust Mask-Layout and Process Synthesis in Micro-Electro-Mechanical-Systems (MEMS) Using Genetic Algorithms*, PhD thesis, California Institute of Technology Pasadena, California.

40. Tellier, C. (2003). CAD design of mask compensation patterns, *Proc. IEEE Sens.*, **1**, pp. 517–522.

41. Wacogne, B., Sadani, Z., and Gharbi, T. (2004). Compensation structures for V-grooves connected to square apertures in KOH-etched (100) silicon: theory, simulation and experimentation, *Sens. Actuators A*, **112**, pp. 328–339.

42. Fan, W., and Zhang, D. (2006). A simple approach to convex corner compensation in anisotropic KOH etching on a (100) silicon wafer, *J. Micromech. Microeng.*, **16**, pp. 1951–1957.

43. Mukhiya, R., Bagolini, A., Margesin, B., Zen, M., and Kal, S. (2006). <100> bar corner compensation for CMOS compatible anisotropic TMAH etching, *J. Micromech. Microeng.*, **16**, pp. 2458–2462.

44. Biswas, K., Das, S., and Kal, S. (2006). Analysis and prevention of convex corner undercutting in bulk micromachined silicon microstructures, *Microelectron. J.*, **37**, pp. 765–769.

45. Pal, P., Sato, K., and Chandra, S. (2007). Fabrication techniques of convex corners in a (100)-silicon wafer using bulk micromachining: a review, *J. Micromech. Microeng.*, **17**, pp. R1–R23.

46. Wacogne, B., Zeggari, R., Sadani, Z., and Gharbi, T. (2006). A very simple compensation technique for bent V-grooves in KOH etched (100) silicon when thin structures or deep etching are required, *Sens. Actuators A*, **126**, pp. 264–269.

47. Kummamuru, R. K., Hu, L., Cook, L., Efremov, M. Y., Olson, E. A., and Allen, L. H. (2008). A close proximity self-aligned shadow mask for sputter deposition onto a membrane or cavity, *J. Micromech. Microeng.*, **18**, p. 095027.

48. Pal, P., Sato, K., Shikida, M., and Gosalvez, M. A. (2009). Study of corner compensating structures and fabrication of various shapes of MEMS structures in pure and surfactant added TMAH, *Sens. Actuators A*, **154**, pp. 192–203.

49. Pal, P., and Sato, K. (2015). A comprehensive review on convex and concave corners in silicon bulk micromachining based on anisotropic wet chemical etching, *Micro Nano Syst. Lett.*, **3**, pp. 1–42.

50. Yu, J. C. (2011). Convex comer compensation for a compact seismic mass with high aspect ratio using anisotropic wet etching of (100) silicon, *Symp. Design, Test, Integration and Packaging of MEMS and MOEMS (DTIP)*, Aix-en-Provence, France, pp. 197–199.

51. Mukhiya, R., Bagolini, A., Bhattacharya, T. K., Lorenzelli, L., and Zen, M. (2011). Experimental study and analysis of corner compensation structures for CMOS compatible bulk micromachining using 25wt% TMAH, *Microelectron. J.*, **42**, pp. 127–134.

52. Bagolini, A., Faes, A., and Decarli, M. (2010). Influence of etching potential on convex corner anisotropic etching in TMAH solution, *J. Microelectromech. Syst.*, **19**, pp. 1254–1259.

53. Shayan, M., Arezoo, B., Vanini, A. S., and Kotamjani, A. A. (2011). Silicon (100) undercutting modelling and corner compensation structures design in wet anisotropic etching process, *Proc. Inst. Mech. Eng. B*, **225**, pp. 1041–1049.

54. Apanius, M., Kaul, B., and Abramsona, A. R. (2007). Silicon shadow mask fabrication for patterned metal deposition with microscale dimensions using a novel corner compensation scheme, *Sens. Actuators A*, **140**, pp. 168–175.

55. Smiljanic, M. M., Jovic, V., and Lazic, Z. (2012). Maskless convex corner compensation technique on a (100) silicon substrate in a 25 wt% TMAH water solution, *J. Micromech. Microeng.*, **22**, p. 115011.

56. Jia, C., Dong, W., Liu, C., Zhang, X., Zhou, J., Zhong, Z., Xue, H., Zang, H., Xu, B., and Chen, W. (2006). Convex corners undercutting and rhombus compensation in KOH with and without IPA solution on (110) silicon, *Microelectron. J.*, **37**, pp. 1297–1301.

57. Kim, B., and Cho, D. D. (1998). Aqueous KOH etching of silicon (110) etch characteristics and compensation methods for convex corners, *J. Electrochem. Soc.*, **145**, pp. 2499–2508.

58. Ciarlo, D. R. (1987). Corner compensation structures for (110)-oriented silicon, *Proc. IEEE Micro Robots and Teleoperators Workshop*, Hyannis, MA, USA.

59. Pal, P., Gosalvez, M. A., Sato, K., Hida, H., and Xing, Y. (2014). Anisotropic etching on Si{110}: experiment and simulation for the formation of microstructures with convex corners, *J. Micromech. Microeng.*, **24**, p. 125001.

60. Giousouf, M., Assmus, F., and Kuck, H. (1999). Structuring of convex corners using a reoxidation process-application to a tuning fork resonator made from (110)-silicon, *Sens. Actuators A*, **76**, pp. 416–424.

61. Chu, H. Y., and Fang, W. (2004). A vertical convex corner compensation and non {111} crystal planes protection for wet anisotropic bulk micromachining process, *J. Micromech. Microeng.*, **14**, pp. 806–813.

62. Kwon, J. W., and Kim, E. S. (2002). Multi-level microfluidic channel routing with protected convex corners, *Sens. Actuators A*, **97–98**, pp. 729–733.

63. Pal, P., and Chandra, S. (2004). Recessed microstructures with perfect convex corners for accelerometers, *Sens. Lett.*, **2**, pp. 226–231.

64. Pal, P., and Chandra, S. (2004). A novel process for perfect convex corner realization in bulk micromachining, *J. Micromech. Microeng.*, **14**, pp. 1416–1420.

65. Kawakatsu, H., Saya, D., Kato, A., Fukushima, K., Toshiyoshi, H., and Fujita, H. (2002). Millions of cantilevers for atomic force microscopy, *Rev. Sci. Instrum.*, **73**, pp. 1188–1192.

66. Baryeka, I., and Zubel, I. (1995) Silicon anisotropic etching in KOH-isopropanol etchant, *Sens. Actuators A*, **48**, pp. 229–238.

67. Cho, W. J., Chin, W. K., and Kuo, C. T. (2004). Effects of alcoholic moderators on anisotropic etching of silicon in aqueous potassium hydroxide solutions, *Sens. Actuators A*, **116**, pp. 357–368.

68. Zubel, I., and Kramkowska, M. (2004). Etch rates and morphology of silicon (hkl) surfaces etched in KOH and KOH saturated with isopropanol solutions, *Sens. Actuators A*, **115**, pp. 549–556.

69. Philipsen, H. G. G., and Kelly, J. J. (2009). Influence of chemical additives on the surface reactivity of Si in KOH solution, *Electrochim. Acta*, **54**, pp. 3526–3531.

70. Zubel, I., and Kramkowska, M. (2001). The effect of isopropyl alcohol on etching rate and roughness of (100) Si surface etched in KOH and TMAH solutions, *Sens. Actuators A*, **93**, pp. 138–147.

71. Zubel, I., and Kramkowska, M. (2002). The effect of alcohol additives on etching characteristics in KOH solutions, *Sens. Actuators A*, **101**, pp. 255–261.

72. Pal, P., Gosalvez, M. A., and Sato, K. (2010). Silicon micromachining based on surfactant-added tetramethyl ammonium hydroxide: etching mechanism and advanced application, *Jpn. J. Appl. Phys.*, **49**, p. 056702.

73. Gosalvez, M. A., Pal, P., Tang, B., and Sato, K. (2010). Atomistic mechanism for the macroscopic effects induced by small additions of surfactants to alkaline etching solutions, *Sens. Actuators A*, **157**, pp. 91–95.

74. Tang, B., Pal, P., Gosalvez, M. A., Shikida, M., Sato, K., Amakawa, H., and Itoh, S. (2009). Ellipsometry study of the adsorbed surfactant thickness on Si{110} and Si{100} and the effect of pre-adsorbed surfactant layer on etching characteristics in TMAH, *Sens. Actuators A*, **156**, pp. 334–341.

75. Pal, P., Sato, K., Gosalvez, M. A., Kimura, Y., Ishibashi, K., Niwano, M., Hida, H., Tang, B., and Itoh, S. (2009). Surfactant adsorption on single crystal silicon surfaces in TMAH solution: orientation-dependent adsorption detected by in-situ infra-red spectroscopy, *J. Microelectromech. Syst.*, **18**, pp. 1345–1356.

76. Yang, C. R., Chen, P. Y., Yang, C. H., Chiou, Y. C., and Lee, R. T. (2005). Effects of various ion-typed surfactants on silicon anisotropic etching properties in KOH and TMAH solutions, *Sens. Actuators A*, **119**, pp. 271–281.

77. Yang, C. R., Yang, C. H., and Chen, P. Y. (2005). Study on anisotropic silicon etching characteristics in various surfactant-added tetramethyl ammonium hydroxide water solutions, *J. Micromech. Microeng.*, **15**, pp. 2028–2037.

78. Sato, K., Uchikawa, D., and Shikida, M. (2001). Change in orientation-dependent etching properties of single-crystal silicon caused by a surfactant added to TMAH solution, *Sens. Mater.*, **13b**, pp. 285–291.

79. Sarro, P. M., Brida, D., van der Vlist, W., and Brida, S. (2000). Effect of surfactant on surface quality of silicon microstructures etched in saturated TMAHW solutions, *Sens. Actuators A*, **85**, pp. 340–345.

80. Sekimura, M. (1999). Anisotropic etching of surfactant-added TMAH solution, *Proc. 21th IEEE Micro-Electro-Mech. Syst. Conf.*, Orlando, Florida, pp. 17–21.

81. Yang, C. R., Chen, P. Y., Chiou, Y. C., and Lee, R. T. (2005). Effects of mechanical agitation and surfactant additive on silicon anisotropic etching in alkaline KOH solution, *Sens. Actuators A*, **119**, pp. 263–270.

82. Resnik, D., Vrtacnik, D., Aljancic, U., Mozek, M., and Amon, S. (2005). The role of Triton surfactant in anisotropic etching of {110} reflective planes on (100) silicon, *J. Micromech. Microeng.*, **15** pp. 1174–1183.

83. Xu, Y. W., Michael, A., and Kwok, C. Y. (2011). Formation of ultra-smooth 45° micromirror on (100) silicon with low concentration TMAH and surfactant: techniques for enlarging the truly 45° portion, *Sens. Actuators A*, **166**, pp. 164–171.

84. Pal, P., Sato, K., Gosalvez, M. A., and Shikida, M. (2007). Study of rounded concave and sharp edge convex corners undercutting in CMOS compatible anisotropic etchants, *J. Micromech. Microeng.*, **17**, pp. 2299–2307.

85. Zubel, I., Kramkowska, M., and Rola, K. (2012). Silicon anisotropic etching in TMAH solutions containing alcohol and surfactant additives, *Sens. Actuators A*, **178**, pp. 126–135.

86. Rola, K. P., and Zubel, I. (2013). Triton surfactant as an additive to KOH silicon etchant, *J. Microelectromech. Syst.*, **22**, pp. 1373–1382.

87. Tanaka, H., Umeki, N., and Sato, K. (2013). Perfect adsorption of ppb-level surfactant in 5% KOH water solution on a silicon surface changing anisotropic etching properties, *Proc. Transducers 2013*, Barcelona, Spain, pp. 1978–1981.

88. Pal, P., Ashok, A., Haldar, S., Xing, Y., and Sato, K. (2015). Anisotropic etching in low concentration KOH: effects of surfactant concentration, *Micro Nano Lett.*, **10**, pp. 224–228.

89. James, T. D., Parish, G., Winchester, K. J., and Musca, C. A. (2006). A crystallographic alignment method in silicon for deep, long microchannel fabrication, *J. Micromech. Microeng.*, **16**, pp. 2177–2182.

90. Chang, W. H., and Huang, Y. C. (2005). A new pre-etching pattern to determine <110> crystallographic orientation on both (100) and (110) silicon wafers, *Microsyst. Technol.*, **11**, pp. 117–128.

91. Ensell, G. (1996). Alignment of mask pattern to crystal orientations, *Sens. Actuators A*, **53**, pp. 345–348.

92. Vangbo, M., and Backlund, Y. (1996). Precise mask alignment to the crystallographic orientation of the silicon wafers using wet anisotropic etching, *J. Micromech. Microeng.*, **6**, pp. 279–284.

93. Lai, J. M., Chieng, W. H., and Huang, Y. C. (1998). Precision alignment of mask etching with respect to crystal orientation, *J. Micromech. Microeng.*, **8**, pp. 327–329.

94. Tseng, F. G., and Chang, K. C. (2003). Precise [100] crystal orientations determinations on <110> -orientated silicon wafers, *J. Micromech. Microeng.*, **13**, pp. 47–52.

95. Singh, S. S., Veerla, S., Sharma, V., Pandey A. K., and Pal, P. (2016). Precise identification of <100> directions on Si{001} wafer using a novel self-aligning pre-etched technique, *J. Micromech. Microeng.*, **26**, p. 025012.

96. Abedinov, N., Grabiec, P., Gotszalk, T., Ivanov, T. Z., Voigt, J., and Rangelow, I. W. (2001). Micromachined piezoresistive cantilever array with integrated resistive microheater for calorimetry and mass detection, *J. Vac. Sci. Technol. A*, **19**, pp. 2884–2888.

97. Pal, P., and Chandra, S. (2004). Bulk-micromachined structures inside anisotropically etched cavities, *Smart Mater. Struct.*, **13**, pp. 1424–1429.

98. Chandra, S., Singh, J., and Chand, A. (1997). Design and development of microstructures for MEMS applications, *Proc. SPIE*, **3226**, pp. 22–30.

99. Pal, P., and Sato, K. (2009). Various shapes of silicon freestanding microfluidic channels and microstructures in one step lithography, *J. Micromech. Microeng.*, **19**, p. 055003.

100. Pal, P., and Sato, K. (2010). Fabrication methods based on wet etching process for the realization of silicon MEMS structures with new shapes, *Microsyst. Technol.*, **16**, pp. 1165–1174.

101. Yu, J. C., and Lan, C. B. (2001). System modelling of microaccelerometer using piezoelectric thin films, *Sens. Actuators A*, **88**, pp. 178–186.

102. Gosalvez, M. A. (2003). *Atomistic Modeling of Anisotropic Etching of Crystalline Silicon*, PhD thesis, Helsinki University of Technology.

103. Pal, P., Sato, K., Gosalvez, M. A., Tang, B., Hida, H., and Shikida, M. (2011). Fabrication of novel microstructures based on orientation dependent adsorption of surfactant molecules in TMAH solution, *J. Micromech. Microeng.*, **21**, p. 15008.

104. Gosalvez, M. A., Tang, B., Pal, P., Sato, K., Kimura, Y., and Ishibashi, K. (2009). Orientation and concentration dependent surfactant

adsorption on silicon in aqueous alkaline solutions: explaining the changes in the etch rate, roughness and undercutting for MEMS applications, *J. Micromech. Microeng.*, **19**, p. 125011.

105. Seidel, H., Csepregi, L., Heuberger, A., and Baumgartel, H. (1990). Anisotropic etching of crystalline silicon in alkaline solutions I: orientation dependence and behavior of passivation layers, *J. Electrochem. Soc.*, **137**, pp. 3612–3626.

106. Jaccodine, R. J. (1962). Use of modified free energy theorems to predict equilibrium growing and etching shapes, *J. Appl. Phys.*, **33**, pp. 2643–2647.

107. Shikida, M., Sato, K., Tokoro, K., and Uchikawa, D. (2000). Differences in anisotropic etching properties of KOH and TMAH solutions, *Sens. Actuators A*, **80**, pp. 179–188.

108. Sato, K., Shikida, M., Yamashiro, T., Asaumi, K., Iriye, Y., and Yamamoto, M. (1999). Anisotropic etching rates of single-crystal silicon for TMAH water solution as a function of crystallographic orientation, *Sens. Actuators A*, **73**, pp. 131–137.

109. Cheng, D., Gosalvez, M. A., Hori, T., Sato, K., and Shikida, M. (2006). Improvement in smoothness of anisotropically etched silicon surfaces: effects of surfactant and TMAH concentrations, *Sens. Actuators A*, **125**, pp. 415–421.

110. Pal, P., Haldar, S., Singh, S. S., Ashok, A., Xing, Y., and Sato, K. (2014). A detailed investigation and explanation to the appearance of different undercut profiles in KOH and TMAH, *J. Micromech. Microeng.*, **24**, p. 95026.

111. Kutchoukow, V. G., Mollinger, J. R., and Bossche, A. (1999). Novel method for spinning of photoresist on wafers with through-hole, *Proc. 13th European Conf. Solid-State Transducers (Eurosensors)*, pp. 256–272.

112. Craven, D. (1996). Photolithography challenges for micromachining industry, *BACUS Symp.*, pp. 1–10.

113. Venstra, W. J., Spronck, J. W., Sarro, P. M., and Eijk, J. V. (2009). Photolithography on bulk micromachined substrates, *J. Micromech. Microeng.*, **19**, p. 55005.

114. Luxbacher, T., and Mirza, A. (1999). Spray coating for MEMS, interconnects, and advanced packaging applications, *Sensors*, **16**, pp. 61–64.

115. Singh, V. K., Sasaki, M., Hane, K., and Esashi, M. (2004). Flow condition in resist spray coating and patterning performance for three-

dimensional photolithography over deep structures, *Jpn. J. Appl. Phys.*, **43**, pp. 2387–2391.

116. Linder, S., Baltes, H., Gnaedinger, F., and Doering, E. (1998). Photolithography in anisotropically etched grooves, *Proc. IEEE Micromachining Workshop III*, pp. 38–43.

117. Kersten, P., Bouwstra, S., and Petersen, J. W. (1995). Photolithography on micromachined 3-D surfaces using electrodeposited photoresist, *Sens. Actuators A*, **51**, pp. 51–54.

118. Heschel, M., and Bouwstra, S. (1998). Conformal coating by photoresist of sharp corners of anisotropically etched through-holes in silicon, *Sens. Actuators A*, **70**, pp. 75–80.

119. Pal, P., and Sato, K. (2009). Complex three dimensional structures in Si{100} using wet bulk micromachining, *J. Micromech. Microeng.*, **19**, p. 105008.

120. Appels, J. A., Kooi, E., Paffen, M. M., Schatorje, J. J. H., and Verkuylen, W. H. C. G. (1970). Local oxidation of silicon and its application in semiconductor devices Technology, *Philips Res. Rep.*, **25**, pp. 118–132.

121. Loewenstein, L. M., and Tipton, C. M. (1991). Chemical etching of thermally oxidized silicon nitride: comparison of wet and dry etching methods, *J. Electrochem. Soc.*, **138**, pp. 1389–1394.

122. Colinge, J. P. (2004). *Silicon-on-Insulator Technology: Materials to VLSI* (Springer).

Chapter 7*

Alignment of Mask Patterns to Crystallographic Directions

7.1 Introduction

Fabrication of microelectromechanical system (MEMS) structures using surface micromachining does not require the identification of precise crystallographic directions on silicon wafers. The wafer/substrate in this method of fabrication is mostly used as a support. Therefore, there is usually no issue of proper alignment of the mask edges/sides along crystallographic directions for the fabrication of micro-/nanostructures using surface micromachining. Similarly, dry etching- (anisotropic or isotropic) and wet isotropic etching-based silicon bulk micromachining, which, however, uses the bulk material for fabrication, is almost independent of the crystallographic directions; therefore, precise alignment of the mask edges with crystal direction does not have any significant importance. On the contrary, silicon bulk micromachining based on wet anisotropic etching depends strongly on the crystal directions, with the etching rate being different along different directions. Thus, it is of utmost importance to ensure that the mask edges are precisely aligned along the correct crystallographic directions to obtain structures with controlled dimensions. This makes the precise determination of crystallographic directions a vital step to avoid over sizing of the

*Chapter 7 is coauthored by Mr. Sajal Sagar Singh, IIT Hyderabad, India.

Silicon Wet Bulk Micromachining for MEMS
Prem Pal and Kazuo Sato

ISBN 978-981-4613-72-9 (Hardcover), 978-1-315-36492-6 (eBook)
www.panstanford.com

structure. This oversizing of the structures depends on the degree of misalignment as well as the length of the structures. In this chapter, the importance of precise mask alignment in wet anisotropic etching-based silicon bulk micromachining is discussed. Various methods, including the use of the orientation flat (OF) (or wafer flat [WF]) and the pre-etched patterns, to identify crystallographic directions on the wafer surface are described.

7.2 Role of Precise Alignment in Wet Bulk Micromachining

The role of precise alignment of mask edges along crystallographic directions during the lithography process is described by taking three examples of the most widely used structures in MEMS, namely microchannels, cantilever beams, and diaphragms.

7.2.1 Microchannels

Microchannels are extensively employed in bio-MEMS and microfluidics applications [1]. The typical length of a microchannel varies from several microns to a few millimeters. To fabricate a dimensionally accurate vertical microchannel in Si{110} wafers, the mask edges must be aligned precisely along the <112> crystallographic directions. Figure 7.1a shows the schematic diagram of etched microchannels when the edges of the mask patterns are perfectly aligned along the crystallographic direction. In this case, the channels are distinctly defined as per their dimensions, as shown in the zoomed image. However, misalignment in aligning the mask edges would lead to undercutting, as shown in Fig. 7.1b. This can also lead to merging of channels if they are patterned close to each other, as shown in the same figure. The effect of misalignment can also be seen from the cross-sectional view of the channels presented in Fig. 7.2.

Similarly, long cavities formed on a Si{100} wafer also becomes oversized due to misalignment by even a small angle. Figure 7.3 shows schematically the undercutting due to misalignment, which increases the dimensional inaccuracy of the structures. Although there would be some undercutting even in the case of precisely aligned mask owing to the finite etch rate of {111} plane, this undercutting is extremely less to cause any significant oversizing of the channels or cavities. The maximum undercutting (U) measured perpendicular to the mask edge due to misalignment by an angle δ for the channel/cavity of length l can be calculated using the relation given by

$$U = l\sin(\delta)\cos(\delta) \tag{7.1}$$

Thus, a microchannel of length 10 mm will result in an undercut of ~18 μm even for a misalignment of 0.1°.

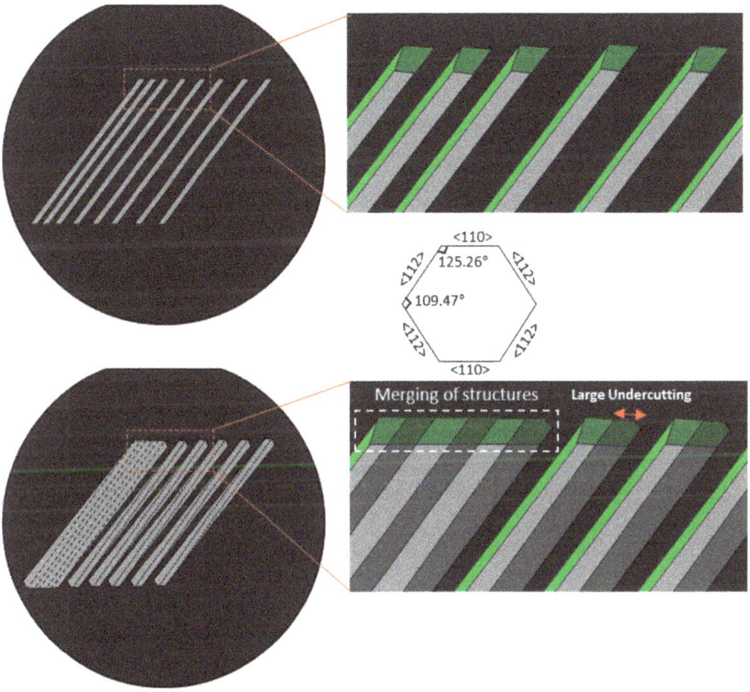

Figure 7.1 Effect of the misalignment of mask edges on a Si{110} wafer when the mask edges are (a) accurately and (b) inaccurately aligned along the crystallographic directions. The zoomed (and tilted) section shows that the microchannels are fabricated with exact dimensions when the alignment is accurate. However, in the case of misaligned edges the channels become oversized, leading to merging of channels.

Diodes and transistors are considered as basic building blocks of integrated circuits (ICs). In MEMS there are no elementary building blocks which can support all designs, as the structures are application dependent. However, cantilevers and diaphragms are extensively used for sensing and actuation purposes in biosensors, switches, temperature sensors, and many others applications [1–3]. Thus they can be called as the fundamental structures of MEMS devices.

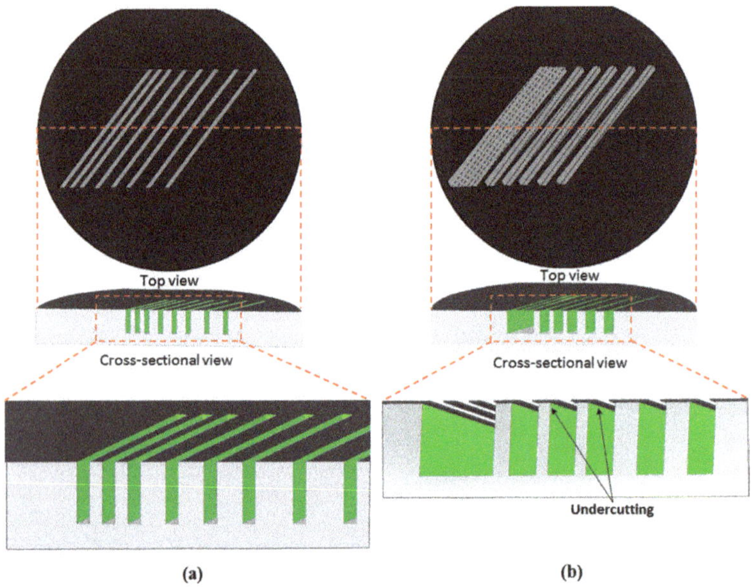

Figure 7.2 Schematic diagram of the cross-sectional view of the channels shown in Fig. 7.1, with (a) dimensionally accurate channels in the case of precise alignment of mask edges and (b) oversized channels in the case of inaccurately aligned channels. The channels experience large undercutting due to the misalignment of mask patterns along the <112> direction. At the same time, merging takes place for closely patterned channels.

7.2.2 Cantilever Beams

Arrays of cantilever beams are used in various sensors and actuators for developing high-performance devices [4]. These arrays can be fabricated either by front-side etching or back-side etching. To fabricate using front-side anisotropic wet etching, the phenomenon of convex-corner undercutting is used for releasing the beams, as discussed in Chapter 6 [5, 6]. Etching stops when it encounters the {111} planes.

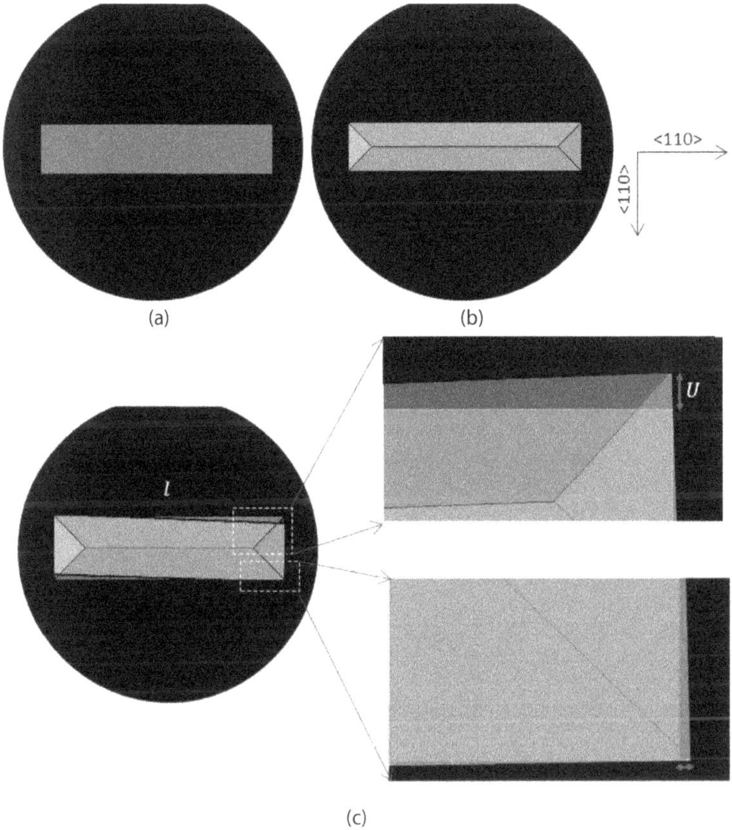

Figure 7.3 Effect of the misalignment of mask edges on the etched profile on a Si{100} wafer. (a) A channel (or rectangular) patterned using the wafer flat as the reference <110> direction. Etched channels when the mask edges are (b) accurately and (c) inaccurately oriented along the <110> directions. In the case of accurate alignment, the etched channel is dimensionally accurate without any undercutting, while undercutting occurs if the mask edges are slightly misaligned.

Thus, alignment of the mask edges along the precise crystallographic directions is necessary to obtain dimensionally accurate cantilever beams, as shown in Fig. 7.4a. Owing to the large number of beams in an array, even a slight misalignment would result in larger underetching at the anchor points, leading to additional overhang, as shown in Fig. 7.4b. The additional overhang results in coupling of the beams and affects their dynamics to a large scale, giving undesirable results. To fabricate high-performance devices, it is necessary to avoid such coupling and thus align the mask patterns precisely along the <110> direction on Si{100}.

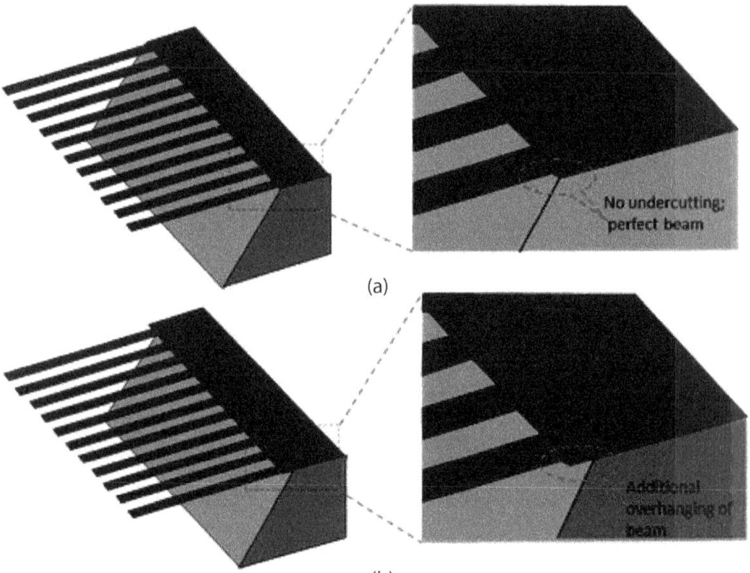

(b)

Figure 7.4 Schematic diagram of an array of cantilever beams fabricated when the mask edges are aligned along the <110> direction (a) accurately and (b) inaccurately. The zoomed view shows that there is no underetching and oversizing (or additional overhang) when the mask is accurately aligned with the <110> direction, while additional overhang occurs due to the misalignment of mask patterns. The additional overhanging leads to coupling between adjacent beams.

7.2.3 Diaphragms

Another important MEMS structure used in sensing devices is a suspended diaphragm. To fabricate the diaphragm, the silicon wafer is coated with the diaphragm material, as shown in Fig 7.5a. Patterning is then done on the back side of the wafer (Fig. 7.5b). The dimension of the patterned geometry depends on the dimensions of the required diaphragm, as well as the misalignment with the crystallographic direction, as will be explained later.

Finally etching is carried out throughout the wafer thickness *h* from the back side to obtain the suspended diaphragm, as illustrated by dotted lines in Fig. 7.5c. The cross section of the diaphragm is shown in Fig 7.5d. It is to be noted that the etchants used for etching the silicon must have high selectively to the diaphragm material. The details of fabrication issues are discussed in Chapter 8.

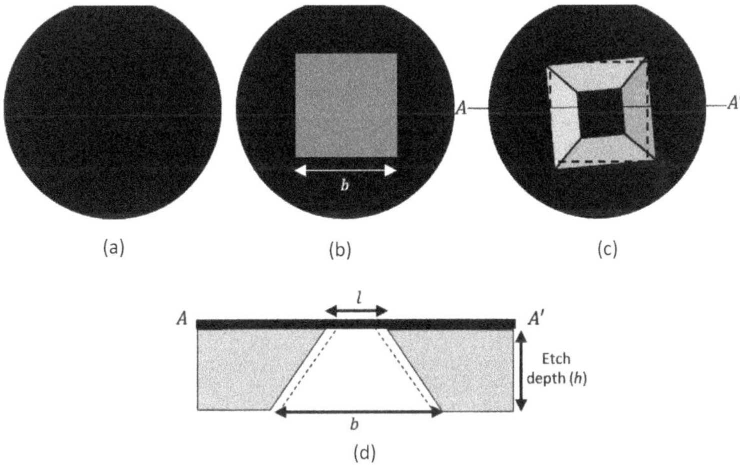

Figure 7.5 Schematic view showing the fabrication of a diaphragm on a Si{100} wafer: (a) deposition of a thin film of diaphragm material (e.g., SiO_2), (b) patterning of the back side, (c) wet anisotropic etching, and (d) cross-sectional view of the suspended diaphragm. The dotted lines show the geometry when the mask edges are aligned perfectly along the crystallographic direction, while the solid line indicates the overetched geometry due to misalignment.

As mentioned earlier, the dimensions of the diaphragm also depend on the precision with which the mask edges are aligned along the crystallographic directions. It can be seen from Fig. 7.5c,d that if the mask edges are aligned precisely along the <110> direction on a Si{100} wafer, the diaphragm would be of the required dimensions, as shown by the dotted line. However, any misalignment would lead to an oversized diaphragm, as shown by solid lines in the same figure. To fabricate a square diaphragm of side *l*, the dimension *b* of the mask pattern can be calculated using the following geometric relation when the mask is misaligned by an angle $0 \leq \delta < \frac{\pi}{2}$:

$$b = \frac{(l + 2h\cot(54.7))}{(\sin(\delta) + \cos(\delta))} \qquad (7.2)$$

It is evident from Eq. 7.2 that the size of the diaphragm for square mask patterns of side b depends on the misalignment angle δ. To control the dimensions of the diaphragm one needs to ensure that δ is negligible. In quantitative terms, fabricating a square diaphragm of side 50 μm through a 500 μm thick Si{100} wafer results in the increase of the diaphragm length by 13% if it is misaligned by even half a degree. This can alter the performance of the diaphragm-based sensor.

Owing to the importance of fabricating MEMS structures with controlled dimensions and smooth sidewalls, alignment of mask edges along precise crystallographic directions is in fact the first and one of the most important steps in wet anisotropic etching-based bulk micromachining. In the next section, several techniques for the alignment of mask edges with crystallographic directions are discussed.

7.3 Alignment Techniques

The basic method to align mask patterns along crystallographic directions is to use the WF as a reference direction. This method is accurate, provided the WF is precisely along the specific crystallographic direction. However, the determination of the WF is not done precisely by wafer manufacturing industries. Generally, the wafer manufacturing industries (e.g., Topsil) use an optical beam to focus at the etched surface and the reflected ray is used to determine the crystallographic direction (WF) of the whole ingot prior to slicing it to wafers [7]. The WF determined by this method has a crystallographic misalignment which varies from 1° to 5°, depending on the quality of the wafer [8]. An inaccurate WF would lead to inaccuracy in aligning the mask edges along the crystallographic directions. This inaccuracy is undesirable for the sophisticated and high-performance devices as they require extremely high precision control of dimensions. The effect of using an inaccurate WF to align the mask edges is demonstrated in Figs. 7.1–7.5. Thus, aligning the mask using the WF as a reference direction is not sufficient to achieve high precision control in dimensions. Another method is to use X-ray diffraction, which can precisely determine the crystallographic direction, but it is impracticable to mount the diffraction equipment onto the mask aligner. Due to the associated limitation of above-mentioned methods, other techniques to determine the crystallographic directions on Si{100} and Si{110} wafers with high accuracy have been developed. These include cleaved-edge method and wet anisotropically pre-etched patterns. These methods are described in the following sections.

7.3.1 Cleaved-Edge Alignment

A single-crystal material has a tendency to cleave along the weaker crystallographic planes caused by the alignment of weaker bonds between atoms in the crystal lattice. In single-crystalline silicon, a perfect cleavage takes places at the direction comprising the {111} and {110} planes as the bond density for these planes is lower than other planes such as {100}. Therefore, the slip lines and other defects at the edges of silicon wafers are usually responsible for wafer breakage. This property of the crystal lattice to break along the weaker direction is exploited to determine the crystallographic directions. In this method, a silicon wafer is cleaved along the scribed line, which is drawn along directions that comprise weaker bonds (e.g., <112> directions on the Si{110} surface). This cleaved edge then comprises the {111} planes. As a result, the cleaved edge is along the direction which comprises the {111} planes and is used as a reference for alignment. For instance, cleaved edges on a Si{110} wafer are aligned along the <110> and <112> directions, as shown in Fig. 7.6, for the wafer whose flat is oriented along <110> direction. To fabricate the microstructures, for example, long and deep channels on a {110} wafer, the edges of the microchannel patterns are then aligned with cleaved-edges oriented along the <112> directions [9].

Although this method is very simple to identify the crystallographic directions, the wafer is separated into pieces. At the same time, the cleaved surface may not be formed by a single {111} planes. Therefore, this method of alignment is detrimental owing to the wafer disintegration as well as uncertainty in obtaining the cleaving along the precise crystal direction, leading to poorly fabricated structures. Besides, reproducibility of this method is questionable. In addition to that, this method also requires expertise in wafer handling to scribe lines on the wafer surface.

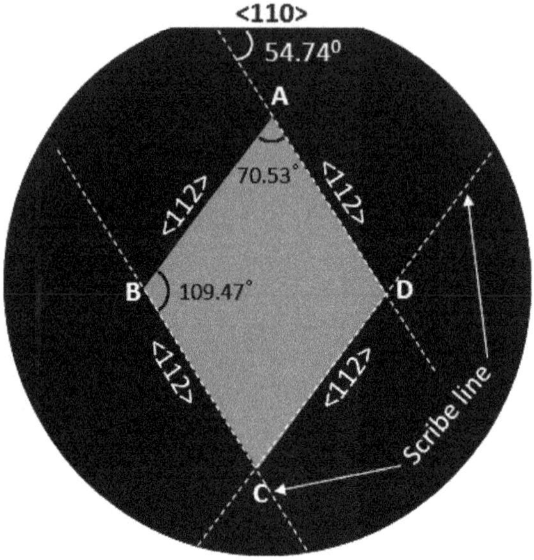

Figure 7.6 Cleaved-edge method to find the crystallographic direction on a Si{110} wafer surface. The scribe lines are drawn with respect to the wafer flat and the chip is cleaved along the scribed lines.

7.3.2 Pre-etched Patterns

The cleaved-edge method relies on the accuracy of the WF as well as the handler's accuracy to cleave (or dice) the wafer along the scribed line. However, it is not a robust and accurate method to ensure precise determination of the crystallographic directions. To find crystallographic directions with high accuracy, a more accurate method called *pre-etched patterns* was developed. In this method, different kinds of mask patterns are etched by wet anisotropic etching on a silicon wafer to identify the crystallographic direction (e.g., <110>, <100>) with high precision [7–16]. These patterns are etched on the periphery of the wafer and thus the usable wafer space is not significantly occupied. These pre-etched patterns act as aids for the subsequent aligning of mask edges. This method of alignment has proven to be accurate in determining the crystallographic directions with an accuracy as high as 0.01°. It is widely used to assist in the fabrication of dimensionally accurate microstructures for various applications [1–3]. Several types of pre-etched patterns have been proposed to date which can determine the principal crystallographic directions with high accuracy on both Si{100} and Si{110} wafers. These techniques with their pros and cons are explained in subsequent sections. The techniques are categorized to determine the principal directions (e.g., <110>, <100>) on Si{100} and Si{110} wafers.

7.3.2.1 Identifying crystallographic directions on a Si{100} wafer

For the fabrication of microstructures on a Si{100} wafer using wet anisotropic etching, identification of the <110> direction is of great interest due to the appearance of stable {111} planes at these directions. Nevertheless, <100> directions are also important due to the appearance of stable {110} planes when etching is done in surfactant-added tetramethylammonium hydroxide (TMAH). These {110} planes are used as 45° micromirrors in optical MEMS applications. To determine the <110> direction on a Si{100} wafer, Ensell proposed a set of circular window features, as shown in Fig. 7.7 [10]. After etching, these circular windows take the shape of inverted pyramids enclosed by <110> directions, as shown in Fig. 7.7. The number of circles in the pre-etched pattern depends on the accuracy of the WF (e.g., <110>). If the <110> direction lies within ±2° of the WF, the circular patterns should cover an arc of 2° on both sides of the reference direction given by the WF. At the same time, the diameter of the circles of the pre-etched patterns also depends on the accuracy with which the crystallographic direction needs to be identified. Although smaller circles will give better accuracy, it is difficult to investigate structures of small sizes under the mask aligner. Ensell fabricated the pre-etched pattern consisting of a series of 41 circular openings of diameter 75 μm and spaced 78.5 μm with an angular pitch of 0.1° on an arc of radius 45 mm. The 41 circles require 20 circles on each side of the reference direction, essentially the WF, and 20 circles with an angular pitch of 0.1° correspond to 2° on either side of the direction as given by the flat. After patterning the geometry on diametrically opposite ends, as shown in Fig. 7.7, etching is carried out in an anisotropic etchant.

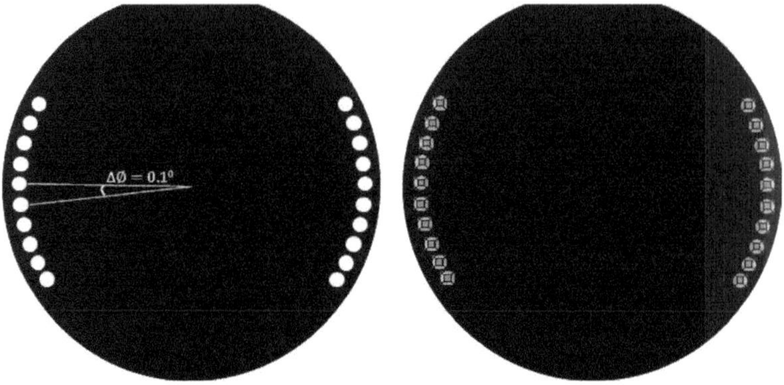

Figure 7.7 Schematic view of the pre-etched pattern, before and after etching, proposed by Ensell [10]. The etched pattern consisting of inverted pyramids formed at each circular opening.

Upon prolonged etching the circular openings take the shape of inverted pyramids, as presented in Fig. 7.8. The masking layer can be removed for better visualization. The misalignment of the edges of the neighboring structures is used as a measure to determine the precise <110> directions. As shown in Fig 7.8, as one moves toward the center from any end, the base of the pyramid shifts to one direction. Moving down from top (structure A), one sees that the structure B is shifted (offset) by Δx. The structures in B, however, are approximately aligned in a straight line. Moving further down to structure C, it can be seen that the base shift changes its direction. This implies that the precise <110> direction lies between structures A and C. If at all the shift Δx is the same between A to B and B to C, then the <110> direction passes exactly through the central pyramid of structure B. This extent of offset is a measure of the misalignment with the <110> direction. The author proposed to inspect the etched patterns to find the three pyramids (structures B) with the closest alignment of edges to determine the crystal direction. After selecting the three pyramids whose edges are almost exactly aligned to each other, the central pyramid is considered to be aligned along the precise <110> direction. For an angular interval of $\Delta\varphi = 0.1°$, the minimum value of Δx for the given dimension of the pre-etched patterns is 0.07 μm and with every increase of 0.1° in $\Delta\varphi$, Δx increases by 0.14 μm. After finding the cavities across which the offset is minimum and changes its direction, the precise <110>

direction can be identified on the basis of the value of Δx of the neighboring cavity, as explained earlier. To increase the accuracy, the dimensions of the circles can be reduced, in which case there would be fewer number of pyramids whose edges align precisely.

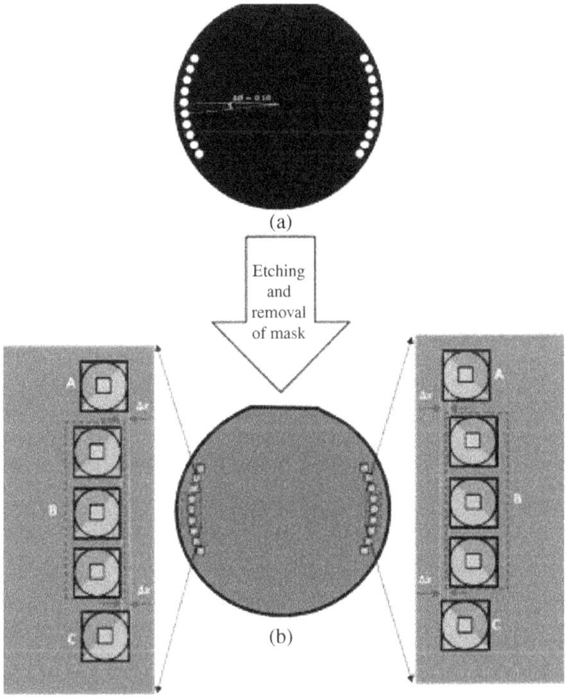

Figure 7.8 Schematic representation of the technique proposed by Ensell [10]: (a) circular openings patterned on an arc and (b) etched profile of the pre-etched patterns. Neighboring etched pyramids are misaligned, and the degree of misalignment depends on the alignment of the line between their centers and the <110> direction. Moreover, the offset Δx changes its direction across the perfectly aligned direction.

This is a very simple technique both in terms of design and in terms of its implementation in finding the crystal direction. The author visually inspected the structures under a microscope to determine the pyramids with almost exactly aligned edges. However, this is not the most accurate methodology as the submicron offset is not clearly visible under the microscope. For the most accurate results, the offset Δx should be measured for each structure to determine the one with the minimum offset which would correspond to the precise

<110> direction. This minimum offset length is of the order of the submicron range, which is difficult to measure on a normal mask aligner and thus it needs sophisticated equipment.

Vangbo and Backlund proposed an alignment fork as the pre-etched pattern to determine the <110> directions [7]. The property of symmetric undercutting across the <110> direction is exploited to determine the <110> directions. The pattern comprises a set of three ridges tapered by small angles (α, β), as shown schematically in Fig. 7.9. This set of structure is then repeated at an angular interval of 0.1°. Using this pre-etched pattern, the <110> direction can be determined using two effects. The first is the tapered ridge effect. Upon etching, the dimension of the ridge will start to increase where it is narrowest and finally will lead to an oversized opening. A smaller angle of tapering ensures faster etching time to form the final undercut structure. If the ridges happen to be lying symmetrically across the <110> direction on the patterned wafer, then the undercut lengths of the two ridges will be same. To determine the <110> direction, the etched patterns are carefully inspected to find the pair of ridge which is symmetrically etched. This is one method to determine the precise <110> direction. Another method is by using the tapered groove effect. In this method, the mask pattern comprises of ridges which widen with a small angle. As etching progresses, the ridges take the shape of grooves and the bottom plane continues to diminish and the final etched profile comprises of intersecting {111} planes. If the undercutting rate is low (i.e., near the <110> direction), the grooves will be narrower and more complete (i.e., the base will have a longer intersection). However, if undercutting rate is more (i.e., away from <110> direction) the groove will be wider with a longer length of incomplete portion of the base for the same etching time. To determine the correct <110> direction, the grooves can be inspected under a microscope to determine and compare the lengths of the incomplete portion of the formed grooves. For the alignment of subsequent mask edges, a square window is added corresponding to each alignment fork. The entire set is now repeated at an angular interval of 0.1° with the center point of rotation at the opposite side of the wafer. Additionally, a similar square is also patterned at the center of rotation itself. Now, the subsequent mask is also added with two similar squares on each side. During subsequent alignment, one is overlapped with the square at the center of rotation and the other is overlapped with the square corresponding to the fork where symmetric undercutting occurs. This is a simple method as the precise direction can be determined by a visual investigation of set of structures to find one with symmetric undercut. However, this can be a tedious task as the structures near the <110> direction can have a similar undercut which may not be distinguishable. The source of error can be removed by measuring the undercut lengths to determine the symmetric undercut structures but it would require sophisticated equipment. At the same time, the varying undercut rate at different places on the wafer may also trick the human eye and give erroneous results. Similarly, determining the incomplete length of the base of the groove under a microscope will depend on the judgement of the user. This can lead to errors in determining the crystal direction.

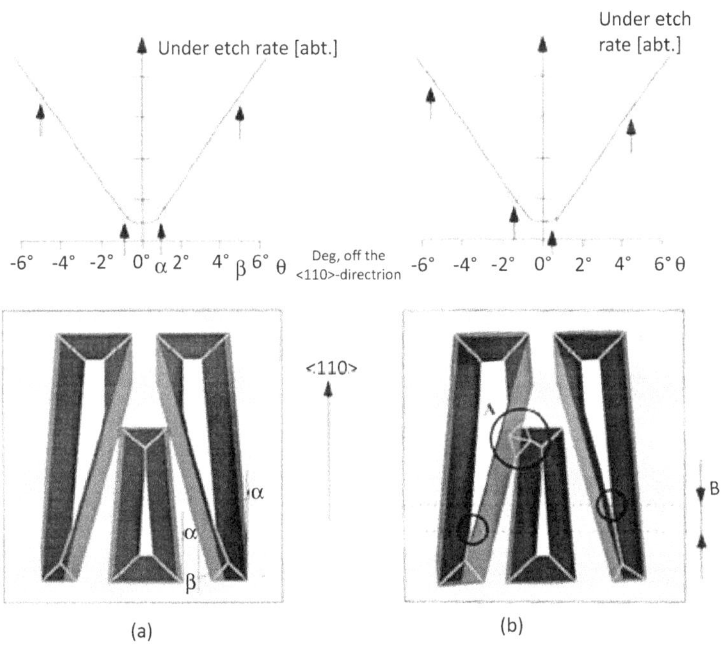

Figure 7.9 Technique to determine the <110> direction on the Si{100} surface proposed by Vangbo and Backlund exploiting the symmetric etch rate behavior across the <110> direction: (a) symmetric underetch due to accurate alignment of fork-shaped structure along the <110> direction and (b) asymmetric underetch owing to misalignment. The set of ridges with symmetric undercutting constitutes the precise <110> direction. From Ref. [7].

Another method to determine the <110> direction on a Si{100} wafer was proposed by Lai et al., who used a pre-etched pattern comprising rectangular and square openings stacked on an arc of radius *R* which is greater than the diameter of the wafer [11]. These patterns were fabricated on one side of the wafer and a series of only square windows called an auxiliary set on the diametrically opposite sides. These two sets of patterns fabricated on opposite ends lie on an arc of concentric circles. The schematic diagram of the pre-etched patterns is shown in Fig. 7.10.

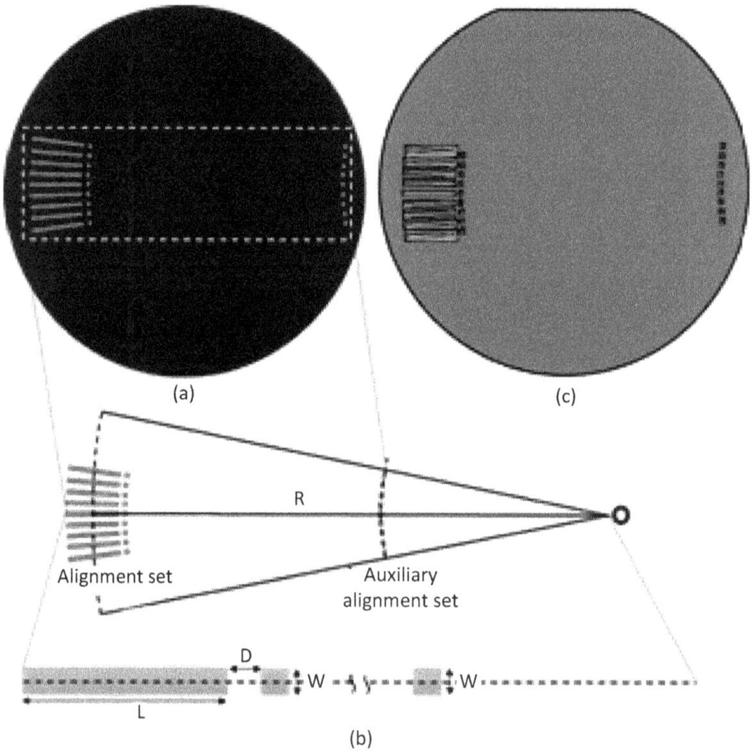

Figure 7.10 Schematic diagram of the technique used by Lai et al. [11]: (a) patterning of pre-etched patterns, (b) dimension of each pattern of the set on one side and an auxiliary set consisting of only square windows on the other side, and (c) etched profile after anisotropic etching. From Ref. [11].

The dimensions of the pre-etched pattern used by the authors are L = 1000 μm, D = 16 μm, and W = 15 μm. The width (W) of the square geometry as well as the rectangular geometry is kept the same because the vertical distance between the edges of the square and rectangular patterns after etching is a measure of the extent of misalignment. On prolonged etching, the rectangular and square openings took the shape of inverted pyramids. Depending on the degree of misalignment of the pattern edges with the <110> direction, the structures became oversized. As a result the vertical distance between the edges of the rectangular and square openings increased. Figures 7.10c and 7.11 show the etched profile of the patterns on prolonged etching. It is clear from the geometry that the vertical distance Δy is given by

$$\Delta y = (L + D) \sin\delta. \tag{7.3}$$

At the precise <110> direction, δ is zero, and thus the vertical distance between the edge of the square and rectangular geometry (Δy) is zero. However, as we move away from the <110> direction, the vertical distance keeps on increasing and finally merges in the case of a large deviation from the <110> direction, as shown in Fig. 7.11. The pattern with minimum Δy is selected and aligned with dimensionally similar pattern on the subsequent mask. The advantage of this method is that the structures far away from the precise <110> direction merge with each other, while structures near the <110> direction remain un-merged and distinctly defined. This reduces the domain of investigation to find out the precise <110> direction. However, from Eq. 7.3 we see that for a misalignment of 0.01°, the value of Δy is 0.18 mm when the rectangular opening is 1000 mm long and the distance (D) between rectangle and square is 16 mm. Using an optical microscope, the structure with this minimum undercut can be located to be corresponding to the <110> direction within the tolerance. The next minimum undercut length for a misalignment of 0.02° is 0.35 μm. As a result, it should be ensured that the microscope is capable of distinguishing between these submicron distances.

For precise identification, these undercut lengths can be measured to determine the structure with minimum undercutting. However, measurement of such small distances firstly requires sophisticated equipment and secondly makes the task tedious. The benefit of these patterns is that they provide a longer length for aligning the mask pattern at one side of the wafer; however, at the other side of the wafer the aligning length is only 16 μm. This makes the technique prone to theta error. We can, however, reduce the theta error by patterning the pre-etched feature on the auxiliary set as well. This would provide a longer length on diametrically opposite sides of the wafer for subsequent mask alignment and thus would reduce the error. Although the etching time will remain the same, the space used on the wafer will increase. To distinguish the structure with the minimum undercut, the undercut length can be increased by increasing either L or D (or both), but this is generally not preferred because first it would significantly increase the etching time and second it would occupy more wafer space.

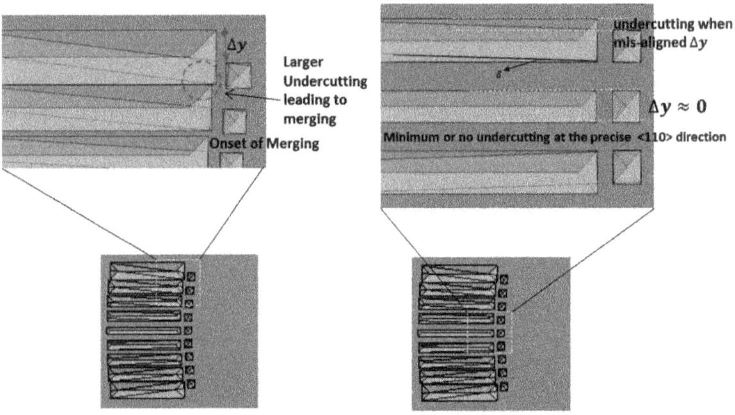

Figure 7.11 Schematic view of the etched pattern of the geometry shown in Fig. 7.10. An etched pattern consisting of an inverted V-groove is formed at each opening. The etched pattern near the perfectly aligned direction has minimum undercutting (minimum offset). The undercutting increases at directions away from the <110> direction and the structure starts to merge at a larger distance.

In another study by Chang and Huang, a pre-etched pattern is utilized to determine the <110> directions on both Si{100} and Si{110} wafers [12, 16]. Additionally, they carried out the statistical analysis to calculate the precision of determination of crystal directions. Their analysis suggested that the precision with which the crystal orientation can be determined does not solely depend on the resolution of the pre-etching mask patterns but is limited by the finite step resolution of the mask generator. As a result, they optimized the dimensions and the tilt angle so as to limit the error due to finite step of the pattern generator within the design's tolerance. However, here we discuss only the technique for precise determination of crystal direction. The proposed pre-etched pattern comprises a series of trapezoids/ridge with the shorter sides parallel to each other and the longer sides inclined at different angles. The trapezoid with a title angle of 0° is designed to be parallel to the reference direction essentially the WF (<110> direction). The dimensions of the height and the smaller edges of the parallel sides of each trapezoid is fixed as H = 500 μm and W = 5 μm. The distance between consecutive trapezoids is also fixed as D = 17.5 μm. A symmetric set is patterned across the reference line, as shown schematically in Fig. 7.12.

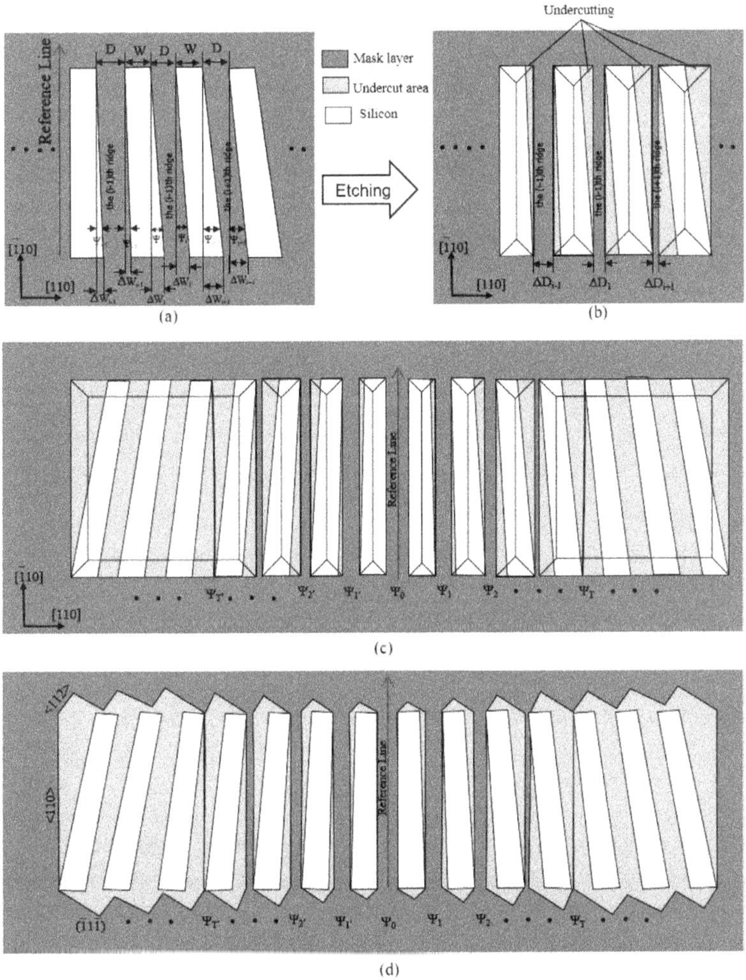

Figure 7.12 Schematic representation of the technique using a pre-etched pattern with tapered ridges, as proposed by Chang and Huang [12]: (a) part of the patterned structures, (b) etched profile on Si{100}, (c) the entire etched patterns on Si{100}, and (d) etched profile on Si{110}. The index of ridge with ΔD_i = 0 (indexed by T and T') on both sides of the reference line is used to determine the misalignment. From Ref. [12]. With permission of Springer.

Upon etching, the trapezoidal opening takes the shape of rectangular grooves (Fig. 7.12b). The undercutting occurs due to misalignment of mask edges and the final structure is defined by {111} planes. Due to the undercutting, the distance ΔD_i reduces and finally becomes zero at a certain index T, as shown in Fig. 7.12c. Similarly, the other side of the reference direction also consists of a structure with index T' where ΔD_i becomes zero. To determine the misaligned angle, the following relation is used:

$$\psi_{\mathrm{M}} = \Delta\psi(T - T')/2$$

where $\Delta\psi$ is defined by $\psi_i - \psi_{i-1}$, which is 0.0125°. A positive value of ψ_{M} implies that the crystal direction lies in the region of a positive tilt angle ψ_i, while a negative value of ψ_{M} implies that the crystal direction lies in the region of a negative tilt angle ψ_i'. To align the subsequent mask precisely along the crystal orientation, an alignment accessory pattern consisting of a series of circular opening is fabricated on an arc of radius R with the center on the opposite side of the wafer where another circle is fabricated, as shown in Fig. 7.13. The angular pitch of the circles on the alignment accessory pattern is the same as that of the pre-etched pattern. Additional structures called dial structures are also fabricated beside each ridge to indicate the index of the structure. After etching, the pre-etched pattern takes the shape of rectangular groove, and the circles on the alignment accessory pattern as well as the circle at the other end of the wafer take the shape of square cavity. Then, the index of the ridge corresponding to T and T′, as discussed earlier, are obtained using the dial structure. Subsequently, the misorientation angle (ψ_{M}) is calculated. Now, the rectangle corresponding to this misorientation angle is selected from the alignment accessory pattern for subsequent alignment of the mask. The subsequent mask which comprises two squares—one for aligning the fabricated square on the other end of the wafer and the other for aligning the corresponding square from the alignment accessory pattern—is then aligned, as shown in Fig. 7.13. In the cases where the WF is perfectly aligned along the <110> direction, the index of T and T′ will be same. This will make the misorientation angle ψ_{M} equal to zero. The rectangle corresponding to this misorientation angle on the alignment accessory pattern is the one beside the zeroth ridge, which is essentially parallel to the WF. This method can also be used to determine the <110> directions on a Si{110} wafer. The etched geometry in the case of a Si{110} wafer is shown in Fig. 7.12d. This methodology does not require the measurement of small undercut lengths for precise determination of crystal directions. However, distinguishing the target ridge from the set of etched ridges is a difficult task. At the same time the etching time increases due to long structures. Additionally, it takes a lot of usable wafer space due to the long length of the patterns, which is itself complex in designing. Overall this methodology to determine the crystal direction is complicated as compared to several other simple methodologies which have been proposed.

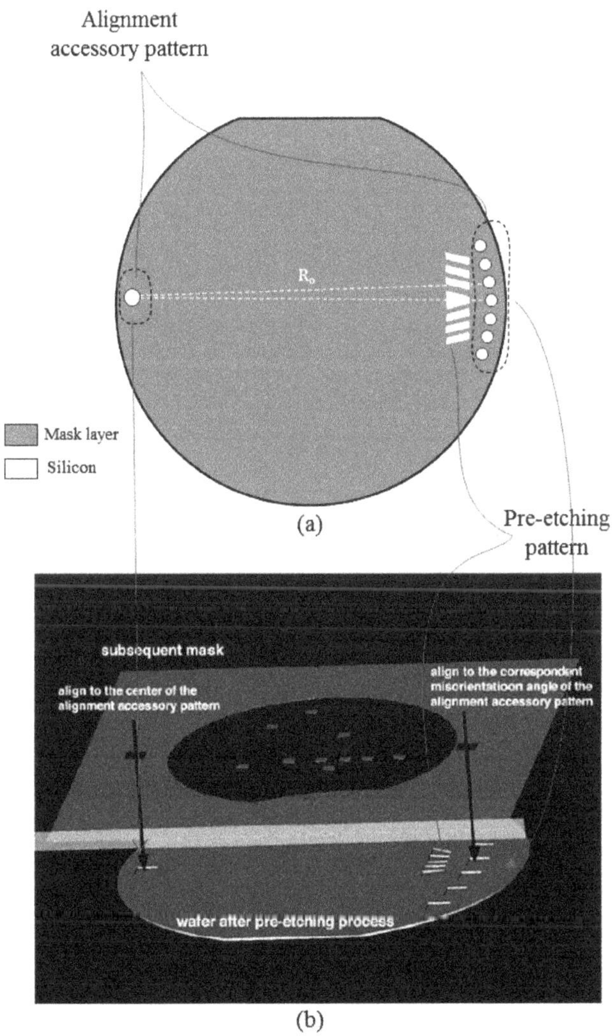

Figure 7.13 Schematic diagrams presenting (a) pre-etching and alignment accessory pattern on a wafer and (b) the methodology of subsequent mask alignment using the technique proposed by Chang and Huang [12]. From Ref. [12]. With permission of Springer.

As discussed in Chapter 5, the relative etch rates of silicon planes such as {100} and {110} can be altered significantly by the addition of a small amount of surfactant in high-concentration TMAH (e.g., 20–25 wt%). In surfactant-added 25 wt% TMAH, the etch rate of {110} planes is reduced to a considerably low level, while that of the {100} plane remains almost unaltered. This property of surfactant-added TMAH is used to fabricate 45° micromirrors, which are employed in optical MEMS to provide 90° out-of-plane reflection, as illustrated in Fig. 7.14 [17–19].

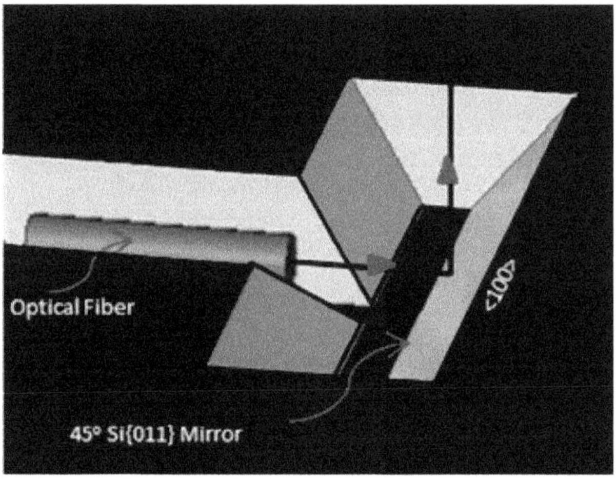

Figure 7.14 Schematic diagram showing the use of Si{110} planes as 45° micromirrors [14]. These planes appear at the <100> direction on a Si{100} surface and reflect the incoming beam at an angle of 90°. From Ref. [14]. © IOP Publishing. Reproduced with permission. All rights reserved.

To ensure that the reflected beam is exactly perpendicular to the incoming beam, it is vital to ensure that the mirror walls are precisely at an angle of 45°. On a Si{001} wafer, {011} planes oriented at an angle of 45° to the wafer surface appear at the <100> directions. To achieve a smooth and exactly 45° sidewall, the mask edges must be precisely aligned along the <100> direction on the Si{001} surface. Earlier attempts at determining the <100> direction on a Si{100} wafer were done by Chen et al. [13]. They exploited the phenomenon of undercutting at the contact line of the masking layer and the silicon substrate as a tool to determine the crystallographic directions. The pre-etched pattern comprises 21 circles of 50 mm diameter fabricated on an arc of 43 mm. The angular pitch of the circles is 0.1°. Figure 7.15 shows the schematic diagram of the proposed patterns.

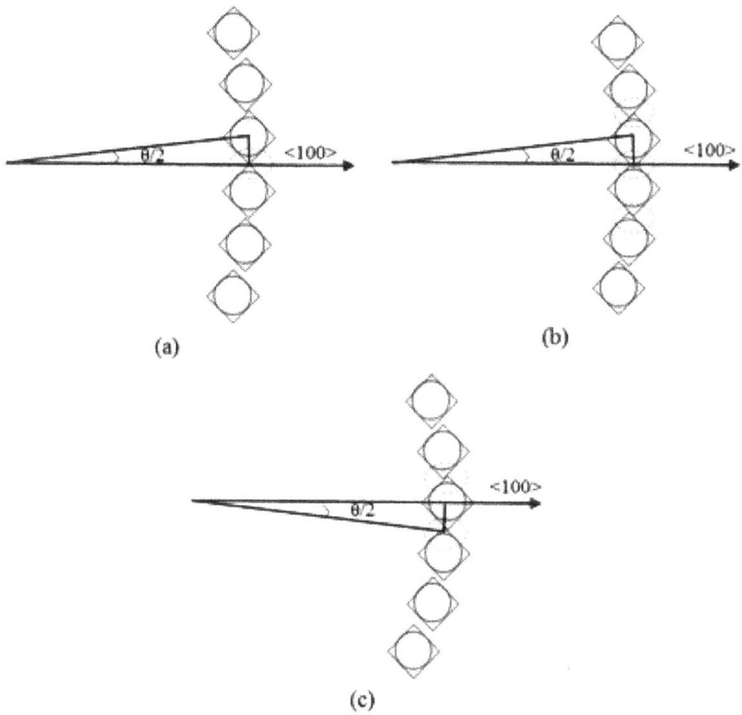

Figure 7.15 Schematic representation of the technique comprising of circular openings suggested by Chen et al. [13]. After etching, due to underetching at the mask edges, the corners of the square merge with each other with (a) just one corner merging, (b) an odd number of corners merging, and (c) an even number of corners merging. In the first case, the <100> direction lies between the two merged corners. In the case of an odd number of corners merging, the precise <100> direction passes through the middle corner. In the cases where there is even number of merges of corners, the <100> direction passes through the radial diagonal of the central square. From Ref. [13]. With permission of SPIE.

Upon etching, the circular opening starts to distort. As the etching continues further the corners of adjacent squares tend to merge with each other. The authors observed the number of such mergings to determine the precise direction. In the cases where there is only one merging point, the precise <100> direction lies through the merged point. If the number of merged corners is even, then the precise direction passes through the center of the central square, and if the number of merged structures is odd, then the <100> direction passes through the central merged point. In this technique, the number of merged points may depend on the etching time. This technique does not require any measurement to determine the undercut lengths, and the location of the precise direction can be done by simply observing the number of merging of corners. However, this method is correct under the assumption that the undercutting is uniform for all the structures on the wafer. In practical applications, the undercutting varies for structure to structure, even on the same wafer [20, 21]. As a result, the reproducibility and correctness of this methodology is questionable as the number of mergings can vary depending on the etch rate and can lead to error in the determination of precise crystallographic directions. At the same time, for subsequent alignment, one has to orient the wafer so as to align the mask edge to either small corner points or diagonal of the formed square. Aligning along both is a tough task, as well as being prone to error owing to their small size.

Singh et al. proposed a measurement-free technique to precisely identify the <100> directions on a Si{100} wafer [14, 16]. The pre-etched pattern comprises a set of four circles of 100 μm diameter each, separated by a distance of 43.5, 44.5, and 45.5 μm from one another along the radial direction. This differential spacing is incorporated in order to overcome the possibility of merging of notching due to overetching. The center of all the four circles lies on a straight line passing through the center of the wafer. This set of four circles is then repeated at an angle of 0.17° on both sides of the reference line, which is at an angle of 45° to the WF. The schematic and optical image depicting the geometry and dimensions of the pattern are shown in Fig. 7.16. The similar set is also patterned on diametrically opposite end of the wafer to reduce the theta error during alignment of the subsequent mask. The pattern is etched in an anisotropic etchant and the circular opening takes the shape of a square V-groove with {111} sidewalls, as explained in the previous and next chapters. The optical image of the etched pattern is shown in Fig. 7.17. Near the perfectly aligned <100> direction, the radial diagonal of all the four squares lies on a straight line and the notches of the squares align perfectly to each other. As we move away from the precise <100> direction, the notches start to misalign and are no longer aligned to each other in a straight line, as shown in Fig.

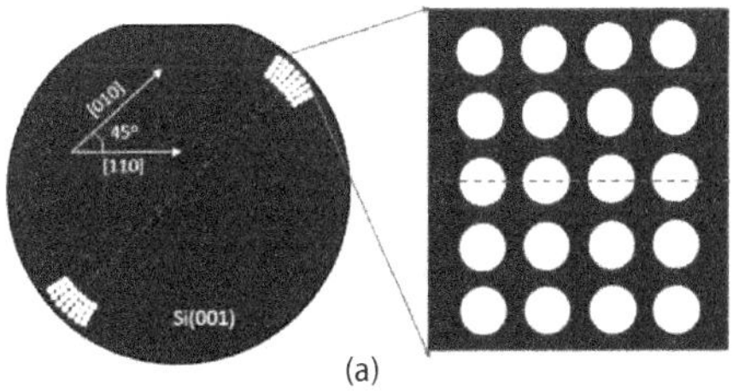

(a)

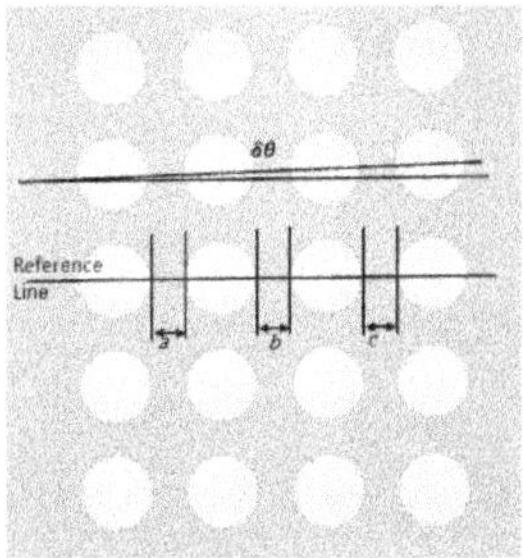

(b)

Quantity	Value
Number of circles on each side of wafer	4*49=196
Diameter of each circle	100 μm
(a,b,c)	(43.5,44.5,45.5) μm
$\delta\theta$	0.17^0

(c)

Figure 7.16 Patterns comprising circular opening proposed by Singh et al. [14] for identifying the <100> direction on the Si{100} surface: (a) schematic diagram and (b) optical images of the arrangement of the pre-etched pattern on diametrically opposite ends of the wafer. The table lists the details of the patterns. From Ref. [14]. © IOP Publishing. Reproduced with permission. All rights reserved.

7.17. The extent of misalignment depends on the actual deviation from the precise <100> direction. The misalignment of the notches also changes its direction across the <100> direction. As a result, the inspection to find the structure whose notches align to each other is to be done in the vicinity of the structure where the alignment of notches changes its direction, as can be seen in Fig. 7.17. This reduces the domain of inspection to determine the precise direction. As a result, with a simple visual inspection under an optical microscope, the precise direction can be located without any need of measuring the undercut length, which requires sophisticated equipment and is a tedious job. This technique requires the identification of structures whose notches align to each other and is therefore simpler than other techniques where one needs to make a careful judgment of structures with minimum undercut. At the same time, there is no need of any measurement as the precise crystal direction appears obvious. The aligning of the subsequent mask can be done by prefabricating a thin line at 45° to the <110> direction on diametrically opposite ends of the subsequent mask. This line can then be aligned along the self-aligned patterns which constitute the <100> direction on both sides of the wafer. The set of four circles provides a longer length for the subsequent alignment of mask edges, ensuring the minimization of the theta error.

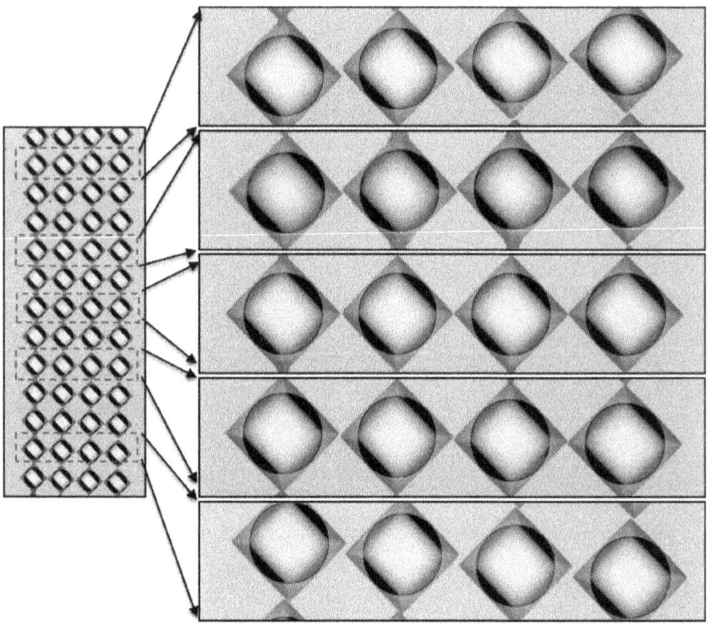

Figure 7.17 Optical images of the etched profile of the pattern shown in Fig. 7.16 at two different magnifications. At the precise <100> direction (center pattern), the notches self-align to each other, making them easily distinguishable with visual inspection. Moving away from the precise direction leads to an increase in the misalignment of the notches. This self-aligning feature makes the precise <100> direction obvious and inhibits the need for measurement of any kind. From Ref. [14].

7.3.2.2 Identifying crystallographic directions on a Si{110} wafer

On a Si{110} wafer there are primarily two principal directions, namely the <110> and <112> directions, which are useful owing to the appearance of slanted and vertical {111} planes at these two directions, respectively. The appearance of vertical {111} sidewalls makes the Si{110} wafer indispensable to fabricate high-aspect-ratio microchannels and gratings using wet anisotropic etching for different applications. There have been a few attempts to identify the above-mentioned principal directions on a Si{110} wafer.

In earlier attempts to precisely identify the <110> direction, Ciarlo fabricated a fan-like structure with rectangular openings of sides 8 μm × 3 mm with the longer length parallel to the WF (<110> direction) [22]. This structure is then repeated at an angular interval of 0.1°. The obtained pattern is similar to that as shown in Fig. 7.10a, although with dimensions of L = 3 mm and W = 8 μm. At the same time, the square adjacent to each rectangle is not present. Prolonged etching leads to undercutting at the mask edges due to misalignment of the edges with the <110> direction. The structure with the minimum undercut would be the least misaligned with the precise <110> direction. This method again deals with the disadvantage of selecting structures with small undercut lengths and increased etching time and using more wafer space, etc.

In another method, James et al. used just one circle of diameter 1 mm as the pre-etched pattern, which formed a hexahedron on prolonged etching [9]. The subsequent mask comprising the structures to be fabricated also comprised a dimensionally similar hexagon (alignment feature), which was formed after etching. The structures on the subsequent mask were fabricated in accordance with the edges of the alignment feature, as shown in Fig. 7.18. The edges of the hexagon on the mask were then aligned along the edges of the hexahedron formed by wet etching on the wafer, as shown in Fig. 7.18. This way all the other structures to be fabricated could get aligned to the correct crystallographic direction. While this method is a simple method to determine the <110> and <112> directions (given by the sides of the hexahedron) and requires less etching and processing time, it, however, again has the disadvantage of relying on aligning only along one side of the wafer, making it prone to theta error.

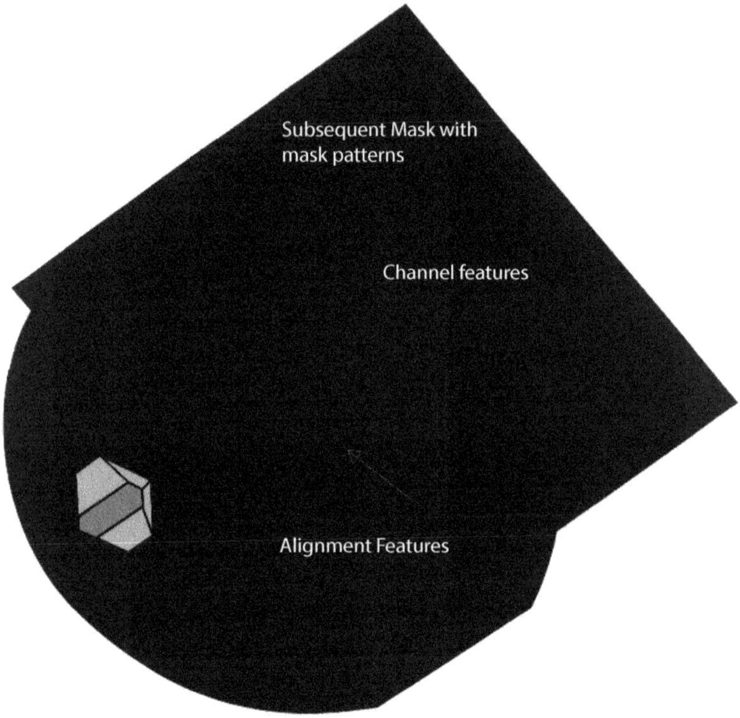

Figure 7.18 Schematic diagram of the etched pattern recommended by James et al. [9]. The subsequent mask comprises a dimensionally similar hexagon which is aligned with the hexagon fabricated after wet etching. From Ref. [9].

Tseng and Chang attempted to determine the <100> directions on a Si{110} wafer employing the hexagonal shape of a circular openings obtained after etching, as illustrated in Fig. 7.19 [8]. In their technique, they fabricated circular openings of diameters 153, 74, and 35 μm with a center-to-center distance of 192, 96, and 48 μm on an arc of radius 48.9 mm at 55° and –125° to the <111> directions. After etching, all the circular openings take the shape of a hexagonal groove, as explained in the Introduction section.

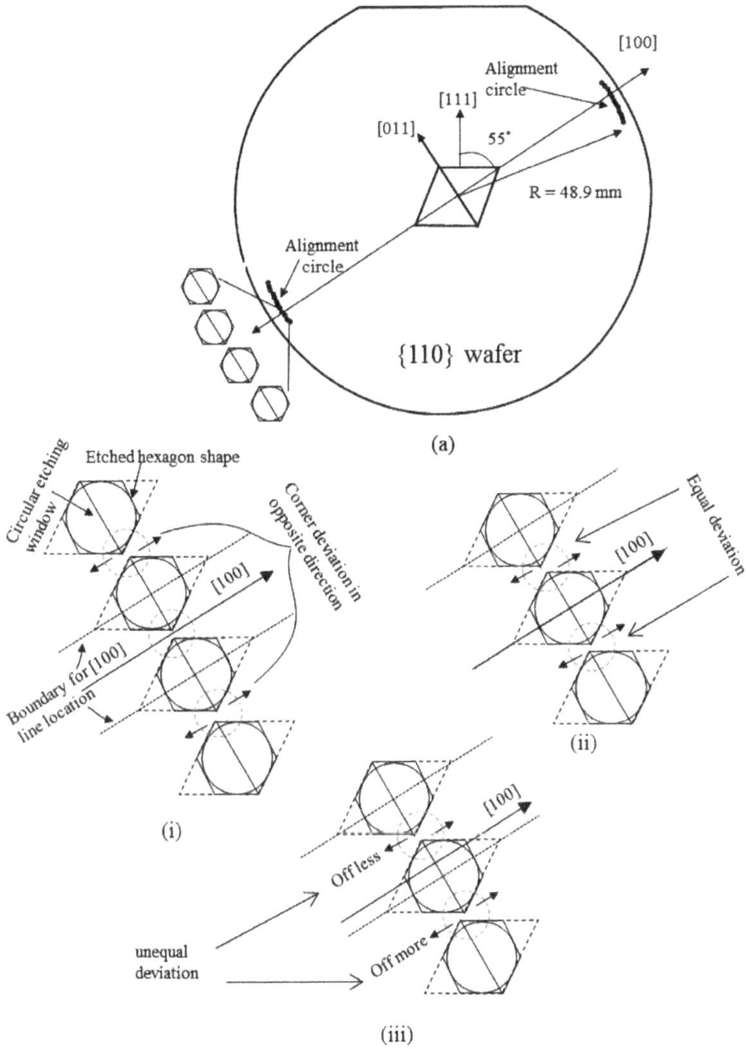

Figure 7.19 Pre-etched pattern proposed by Tseng and Chang [8] showing the (a) location of the patterns and the deviation of the adjacent corners of the etched profile when there is (i) no deviation, (ii) equal deviation in opposite directions, and (iii) unequal deviation in opposite directions. In the case of no deviation, the <100> directions pass through the middle of the corners of the hexagons. In the case of equal deviation, the <100> direction passes through the center of the hexagon. In the case of unequal deviation, the <100> direction passes in between the two corners depending on the magnitude of the deviation. From Ref. [8].

Figures 7.20 and 7.21 show the scanning electron microscopy (SEM) images of the fabricated structures. From the fabricated structures, that particular hexagonal structure is selected across which the direction of lateral deviation of corners changes its direction. The relative positions of the corners of the neighboring hexagons to this hexagon are used to determine the precise <100> direction. Figure 7.19 depicts schematically the top view of the etched profile of hexagonal grooves. It can be seen that there are

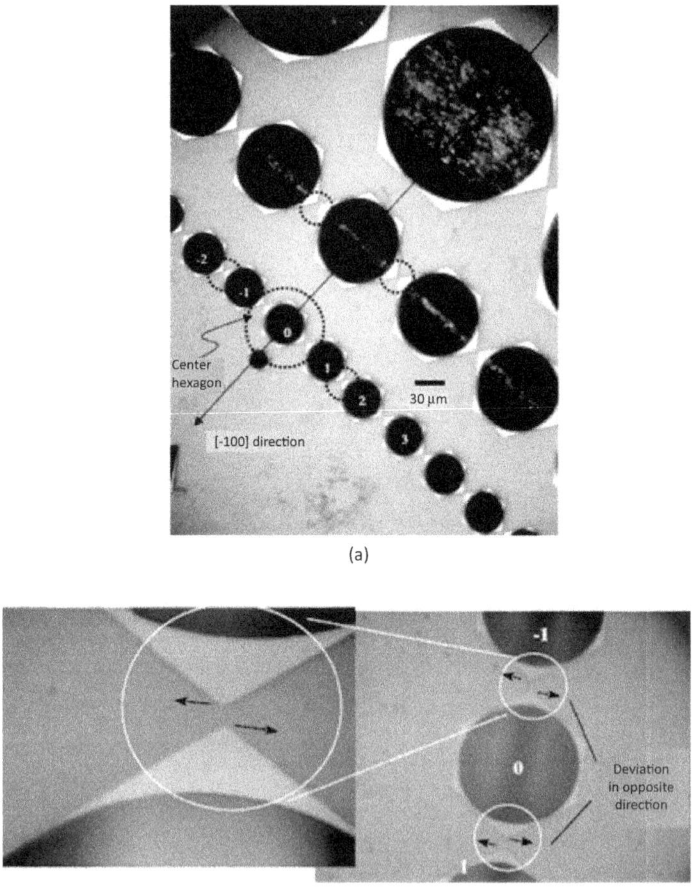

Figure 7.20 SEM image of the etched profile corresponding to the patterns shown in Fig. 7.19: (a) circles takes the form of hexagonal grooves and (b) zoomed image showing the deviation of corners of adjacent hexagons. From Ref. [8].

Figure 7.21 SEM images of the etched profile showing the lateral deviation of the etched hexagonal grooves. From Ref. [8]. © IOP Publishing. Reproduced with permission. All rights reserved.

three possible cases of the orientation of the corners of the adjacent hexagons. One of the possible orientations is that the corners of adjacent hexagons are precisely aligned to each other and the neighboring hexagons have corners deviated in opposite directions. In this case the precise <100> direction will lie perfectly in between the corners of the two aligning hexagons, as presented in Fig. 7.19i. The second case is that the corners of hexagons on either side are deviated in opposite directions but with equal magnitude. In this case the precise <100> direction will pass exactly through the center of the hexagon, as illustrated in Fig. 7.19ii. The third case is the intermediate case where the precise <100> direction will pass through a point in between the two cases mentioned above, depending on the magnitude of lateral deviation, as shown in Fig. 7.19iii. Using this technique, the <100> directions on a Si{110} wafer can be determined. This technique has the advantage of being measurement free. However, for subsequent alignment, one has to orient the wafer so as to align the mask edge with respect to two corner points, unlike long edges in other cases. This task becomes even more difficult when the 35 μm circles are used as reference for better accuracy where a high-magnification microscope is required to see the deviation. Other circles of larger diameter help in reducing the domain of investigation. However, it increases the etching time and occupies more usable space.

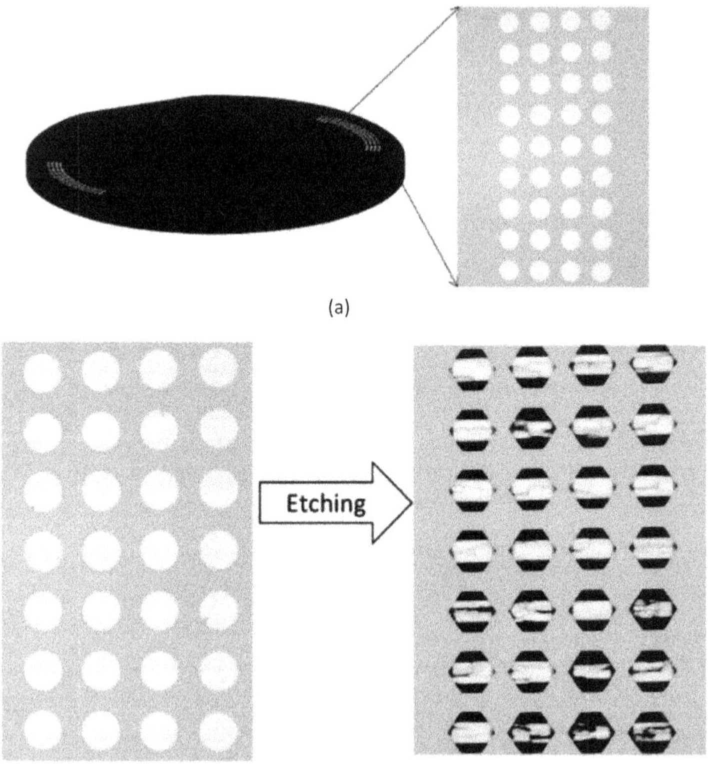

(b)

Figure 7.22 Pre-etched pattern consisting of circular openings proposed by Singh et al. [15]: (a) schematic view showing the location and the optical image of the pattern and (b) after etching the circular opening takes the shape of hexagonal grooves. From Ref. [15]. With permission of Springer.

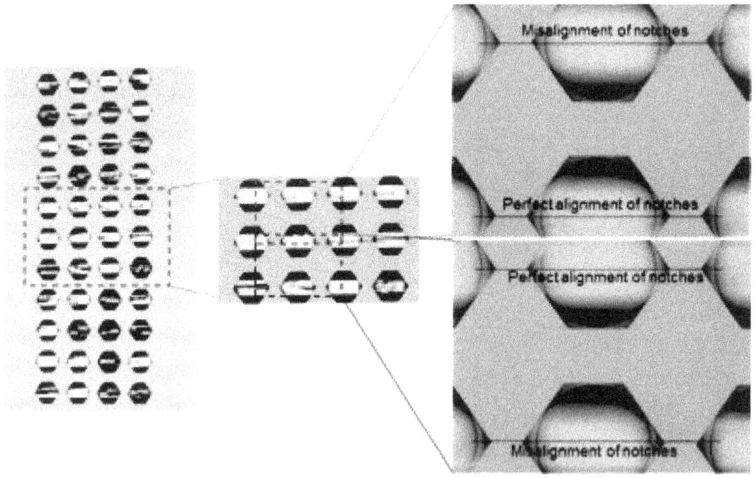

Figure 7.23 Optical images of the etched patterns of a technique presented in Fig. 7.22 showing the self-aligning nature of the structures zoomed at different magnifications. At the precise <110> direction, the notches of the four hexagons aligns exactly to each other, while getting misaligned at directions away from <110>. The central pattern exhibits the precise alignment of notches. At the same time, the immediately adjacent patterns (both top and bottom) can be seen misaligned. As a result, the <110> direction appears obvious. From Ref. [15]. With permission of Springer.

Recently, Singh et al. presented an improved measurement-free technique to determine the <110> direction on a Si{110} wafer [15]. A series of circular openings of diameter 100 mm were fabricated on four concentric arcs. Upon etching the circular openings took the shape of hexagonal grooves, as shown in the optical image in Fig. 7.22. Near the precise <110> direction, the notches of all the four radial hexagons align to each other in a straight line, while getting misaligned to each other at directions away from <110>. The zoomed optical images presented in Fig. 7.23 clearly describe the self-alignment nature of the hexagons near the precise <110> directions. This method is completely based on visual inspection and does not require any measurement and therefore can be employed using only an optical microscope. The large misalignment at directions away from <110> direction reduces the domain over which a closer inspection would be required to determine the precise <110> direction. This technique is simple, robust, and accurate. Additionally, it provides a longer length for aligning the edges of subsequent mask patterns. This method overcomes the shortcomings of the previous method where the mask edges needed to be aligned with respect to two corners. In this technique, the prefabricated reference line at the wafer end on the mask can be aligned with the notches of four circles along their diagonals. This not only reduces the efforts in alignment of subsequent mask but also reduces the associated errors.

7.4 Summary

The use of simple pre-etched patterns which can be used to precisely determine the principle directions in a Si{100} and a Si{110} wafer have been summarized in this chapter. Since the WF is not precisely aligned with the exact crystallographic directions, the use of pre-etched patterns aids in aligning the mask precisely along the crystallographic directions in order to obtain dimensionally accurate structures. Although the pre-etched method, in general, is employed for the determination of the <110> direction, it can be utilized for the identification of other directions as well. At the same time the <110> direction can be used as a reference to find out the other directions using the angular relations for subsequent mask alignment.

References

1. Wang, W., and Soper, S. A. (2006). *Bio-MEMS: Technologies and Applications* (CRC Press, Boca Raton).
2. Gad-el-Hak, M. (2005). *MEMS: Applications* (CRC Press, Boca Raton).
3. Madou, M. (2002). *Fundamentals of Microfabrication: The Science of Miniaturization*, 2nd ed. (CRC Press, Boca Raton).
4. Baller, M. K., Lang, H. P., Fritz, J., Gerber, Ch., Gimzewski, J. K., Drechsler, U., Rothuizen, H., Despont, M., Vettiger, P., Battiston, F. M., Ramseyer, J. P., Fornaro, P., Meyer, E., and Güntherodt, H.-J. (2000). A cantilever array-based artificial nose, *Ultramicroscopy*, **82**(1), pp. 1–9.
5. Pal, P., and Singh, S. S. (2013). A simple and robust model to explain convex corner undercutting in wet bulk micromachining, *Micro Nano Syst. Lett.*, **1**(1), pp. 1–6.
6. Pal, P., and Sato, K. (2015). A comprehensive review on convex and concave corners in silicon bulk micromachining based on anisotropic wet chemical etching, *Micro Nano Syst. Lett.*, **3**, pp. 1–42.

7. Vangbo, M., and Baecklund, Y. (1996). Precise mask alignment to the crystallographic orientation of silicon wafers using wet anisotropic etching, *J. Micromech. Microeng.*, **6**(2), pp. 279–284.

8. Tseng, F. G., and Chang, K. C. (2003). Precise [100] crystal orientation determination on <110> oriented silicon wafers, *J. Micromech. Microeng.*, **13**(1), pp. 47–52.

9. James, T. D., Parish, G., Winchester, K. J., and Musca, C. A. (2006). A crystallographic alignment method in silicon for deep, long microchannel fabrication, *J. Micromech. Microeng.*, **16**, pp. 2177–2182.

10. Ensell, G. (1996). Alignment of mask patterns to crystal orientation, *Sens. Actuators A*, **53**, pp. 345–348.

11. Lai, J. M., Chieng, W. H., and Huang, Y. C. (1998). Precision alignment of mask etching with respect to crystal orientation, *J. Micromech. Microeng.*, **8**, pp. 327–329.

12. Chang, W. H., and Huang, Y. C. (2005). A new pre-etching pattern to determine <110> crystallographic orientation on both (100) and (110) silicon wafers, *Microsyst. Technol.*, **11**(2–3), pp. 117–128.

13. Chen, P. H., Hsieh, C. M., Peng, H. Y., and Chyu, M. K. (2000). Precise mask alignment design to crystal orientation of (100) silicon wafer using wet anisotropic etching. *Proc. SPIE*, **4174**, *Micromachining and Microfabrication Process Technology VI*, pp. 462–466.

14. Singh, S. S., Veerla, S., Sharma, V., Pandey, A. K., and Pal, P. (2016). Precise identification of <100> directions on Si{001} wafer using a novel self-aligning pre-etched technique, *J. Micromech. Microeng.*, p. 025012.

15. Singh, S. S., Avvuru, N. V., Veerla, S, Pandey, A. K., and Pal, P. (2016). A measurement free pre-etched pattern to identify the <110> directions on Si{110} wafer, *Microsys. Technol.*, pp. 1–7.

16. Singh, S. S., Pal, P., Pandey, A. K., Xing, Y., and Sato, K. (2016). Determination of precise crystallographic directions for mask alignment in wet bulk micromachining for MEMS: a review, *Micro Nano Syst. Lett.*, **4**(5), pp.1–29.

17. Xu, Y. W., Michael, A., and Kwok, C. Y. (2011). Formation of ultra-smooth 45° micromirror on (100) silicon with low concentration TMAH and surfactant: techniques for enlarging the truly 45° portion, *Sens. Actuator A*, **166**, pp. 164–171.

18. Rola, K. P., Ptasiski, K., Zakrzewski, A., and Zubel, I. (2014). Silicon Typesetter: 45° micromirrors fabricated by etching in alkaline solutions with organic additives, *Microsyst. Technol.*, **20**, pp. 221–226.

19. Yagyu, H., Yamaji, T., Nishimura, M., and Sato, K. (2010). Forty-five degree micromirror fabrication using silicon anisotropic etching with surfactant-added tetramethylammonium hydroxide solution, *Jpn. J. Appl. Phys.*, **49**, p. 096503.

20. Pal, P., Haldar, S., Singh, S. S., Ashok, A., Xing, Y., and Sato, K. (2014). A detailed investigation and explanation to the appearance of different undercut profiles in KOH and TMAH, *J. Micromech. Microeng.*, **24**, p. 095026.

21. Gosalvez, M. A., Li, Y., Ferrando, N., Pal, P., Sato, K., and Xing, Y. (2016). Fluctuations during anisotropic etching: local recalibration and application to Si{110}, *J. Microelectromech. Syst.*, **25**, pp. 788–798.

22. Ciarlo, D. R. (1992). A latching accelerometer fabricated by the anisotropic etching of (110) oriented silicon wafers, *J. Micromech. Microeng.*, **2**, pp. 10–13.

Chapter 8

Simple to Complex Structures Using Wet Bulk Micromachining

8.1 Introduction

Wet bulk micromachining is employed for the fabrication of a wide range of microelectromechanical system (MEMS) structures, from simple cavities to complex structures. To select an etchant for bulk micromachining, various factors such as undercutting at mask edges and corners, sidewall profile, complementary metal-oxide-semiconductor (CMOS) process compatibility, etch rate, selectivity to mask material and its availability, selectivity to metals, anisotropy (crystal plane selectivity), etched surface morphology, control of etch parameters, safety of reactants and products, etch-stop characteristics, etc., are considered. The etching characteristics of alkaline solutions are primarily sensitive to two things, concentration of the etchant and incorporation of additives. The relative etch rates of crystallographic planes and the etched morphology vary significantly if an additive is added into the etchant or the concentration of the solution is changed. For instance, the addition of a surfactant in high-concentration tetramethylammonium hydroxide (TMAH) reduces the etch rates of {110} and its vicinal planes dramatically. However, the etch rates of {100} and its vicinal planes remain almost unaffected. Considering

Silicon Wet Bulk Micromachining for MEMS
Prem Pal and Kazuo Sato

ISBN 978-981-4613-72-9 (Hardcover), 978-1-315-36492-6 (eBook)
www.panstanford.com

the effect of the different etching parameters, the selection of the etchant (i.e., without or with an additive, low or high concentration) depends on the type of structure to be fabricated.

As discussed in Chapter 5, the following points must be kept in mind while selecting a TMAH solution to fabricate microstructures:

- If the etched surface morphology is not an issue, low-concentration TMAH should be selected as it provides a high etch rate of the {100} orientation.
- To achieve smooth surface morphology, high-concentration TMAH (20–25 wt%) should be preferred.
- In the fabrication of suspended structures using one side of the wafer, high undercutting is desirable for fast release of the structure. Therefore high-concentration TMAH (20–25 wt%) should be preferred for the etching process as it provides significant undercutting at convex corners.
- To achieve minimum undercutting with smooth surface morphology on a Si{100} wafer, 25 wt% TMAH with a surfactant should be used.
- The etch rates of the Si{110} surface and its vicinal orientations are supressed to a significantly low level in surfactant-added TMAH. Therefore pure TMAH should be employed for the formation of microstructures in Si{110} wafers.
- The effect of a surfactant is highly influenced by mechanical agitation and open-circuit potential (OCP) during the etching process. Therefore the following points must be taken into account to control the etched profile:

    a. No agitation should be used in the solution and/or samples if minimum undercutting at the mask edges is needed.
    b. Due to galvanic protection of the developed {111} facets, the undercutting rate at the mask edges of straight and/or bent V-grooves increases significantly after the grooves are formed. To minimize the undercutting at the mask edges, one should use a large enough sacrificial {100} area so that a large fraction of the etch front remains {100} oriented after the grooves are formed and/or the etching process must be stopped as soon as the grooves are formed.

- As described in Chapter 6, to fabricate a protected convex corner either the corner compensation method or the two-step etching technique is employed. In the corner compensation method, selection of the compensation design depends on the amount of space available around the corner and the type of etchant (i.e., pure or with additives, low or high concentration). Thus following points must be taken into consideration while selecting a compensation geometry for the protection of convex corners:
  a. In the case of high-concentration TMAH (i.e., 20–25 wt%), <100> beam shape compensation geometry is a better choice if space around the corner is not a concern.
  b. In low-concentration TMAH, a triangular (or square) shape design is an appropriate compensation structure.
  c. To protect the convex corner on a Si{110} surface, a <112>-oriented beam (or parallelogramic beam) is an optimal compensating geometry owing to its simple design, and the same-size beams are used at the acute and obtuse corners.
- For the fabrication of MEMS structures with perfectly sharp edges and corners, the local oxidation of silicon (LOCOS)-assisted two-/three-step etching technique should be adopted.

Similarly in the case of KOH-based etching, the following points should be taken into account while selecting the KOH solution:

- To achieve a highly smooth etched surface, high-concentration KOH should be employed.
- If the undercutting at the rounded/extruded corners/edges is undesirable, the etching should be performed in isopropyl alcohol (IPA)-added KOH solution.

Three types of silicon wafers, namely {100}, {110}, and {111}, are most widely produced commercially and hence are easily available. Among these orientations, Si{100} wafers are frequently employed for the fabrication of semiconductor devices (e.g., p-n diode, transistors, CMOS and BiCMOS devices, etc.) and the micromachined structures applied in MEMS/MOEMS. Other orientations are used for specific applications. Wet-bulk-micromachined structures can be divided into two categories: fixed (e.g., mesa, cavity, etc.) and freestanding (e.g., cantilever, diaphragm, etc.). In this chapter, a wide range of fabrication processes are described for the realization of these types of structures.

8.2 Cavities, Channels, and Mesa Structures

Cavities, channels, and mesa structures are fabricated using a one-mask process. To fabricate these types of structures, firstly a masking layer is patterned on the silicon surface by lithography and etching steps. After patterning of the masking layer, anisotropic etching is carried out. Figure 8.1 presents the schematic cross-sectional views of square-/rectangular-shaped grooves/cavities in single, silicon-on-insulator (SOI) and p–n structure {100} wafers. Figure 8.2 shows the scanning electron microscopy (SEM) pictures of a V-groove (or inverted pyramid) and square-shaped cavity corresponding to schematic diagrams shown in Figs. 8.1a and 8.1b, respectively. In the case of the microstructures shown Figs. 8.1a, 8.1c, and 8.1d, surface morphology is not an issue as these structures are formed by {111) planes and an etch-stop layer. The intersection of {111} planes (e.g., Fig. 8.1a) is called geometrical etch stop (or self-limiting etch) and provides V-shaped grooves/channels or inverted pyramids. In the fabrication of truncated-shape cavities, as shown in Fig. 8.1b, an etchant should be selected according to the requirement of etched surface morphology, as discussed in Chapters 4 and 5.

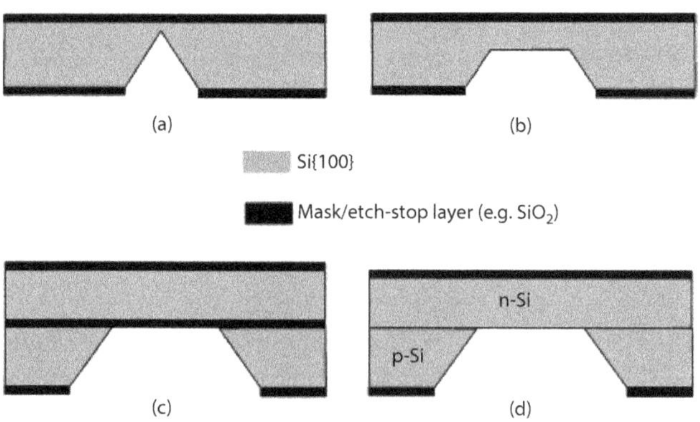

Figure 8.1 Schematic cross-sectional views of the square-/rectangular-shaped cavities/grooves formed by wet etching in single and bonded wafers. The etch depth is controlled by (a) {111} planes, (b) etching time, (c) etch-stop layer (e.g., SiO_2), and (d) electrochemical etch stop.

Figure 8.2 SEM pictures of etched cavities in Si{100} bounded by (a) {111} planes and (b) {111} and {100} planes corresponding to schematic diagrams shown in Figs. 8.1a and 8.1b, respectively.

In pure and surfactant-added high-concentration TMAH (20–25 wt%), the etch rates of Si{100} and its vicinal planes are almost the same, while the etch rates of Si{110} and its vicinal planes, undercutting at convex corners/curved edges or the edges not comprising {111} planes, are significantly low in surfactant-added TMAH in comparison to pure TMAH. In other words, pure TMAH exhibits severe undercutting at convex/rounded corners and the edges not comprising {111} planes, while surfactant-added TMAH provides minimum undercutting at all kinds of corners and edges. Moreover, the high-concentration TMAH without and with the addition of surfactant provides smooth etched surface morphology on the {100} surface. The similar etching characteristics are provided by pure and IPA-added KOH, as discussed in Chapter 4. The availability of two choices (low and high undercutting) promotes the application of wet anisotropic etching in MEMS in order to extend the range of microstructures. The type of the etchant (i.e., with or without surfactant/IPA) can be selected according to the requirement. For instance, to release the structures (e.g., cantilever) from the substrate by removing the silicon underneath the structural layer, high undercutting is desirable. Therefore high-concentration TMAH (20–25 wt%) (or pure KOH) is an optimal choice for the fast release of the structures. For the formation of an etched profile similar to mask patterns with smooth etched surface morphology, surfactant-added 25 wt% TMAH (or IPA-added KOH) should be used. Figure 8.3 exhibits various types of MEMS structures on {100} wafers with highly smooth etched surface morphology fabricated using a single lithography step and one-step etching up to the desired etch depth in surfactant-added 25 wt% TMAH.

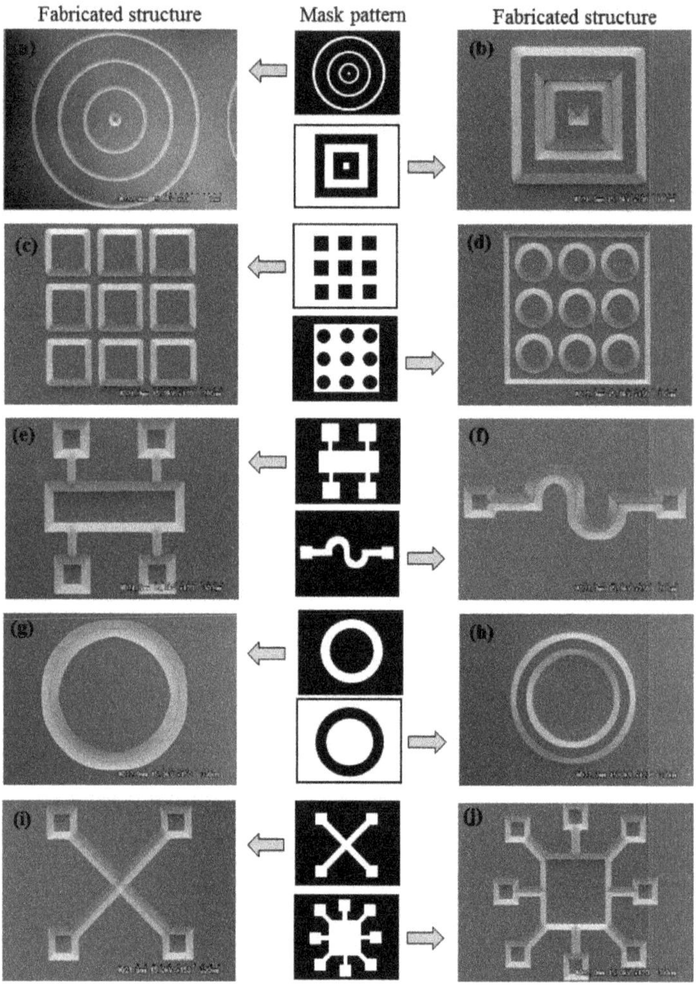

Figure 8.3 SEM pictures of different-shaped structures fabricated in a Si{100} wafer using wet anisotropic etching in surfactant-added 25 wt% TMAH.

8.3 Microstructures with 45° Slanted Sidewalls

The mask edges aligned along the <100> directions on a {100} surface comprise {110} planes, which make an angle of 45° with the {100} planes [1–3]. The etch rate of {110} planes is significantly low in surfactant-added 25 wt% TMAH (or IPA-added KOH). Thus, to fabricate a microstructure with 45° slanted sidewalls on a {100} wafer, the mask edges must be aligned along the <100> directions on the {100} wafer surface, as illustrated in Fig. 8.4. After patterning of the mask layer, etching should be carried out in surfactant-added 25 wt% TMAH. Figures 8.4c and 8.4d present SEM photographs of microstructures with 45° sidewalls.

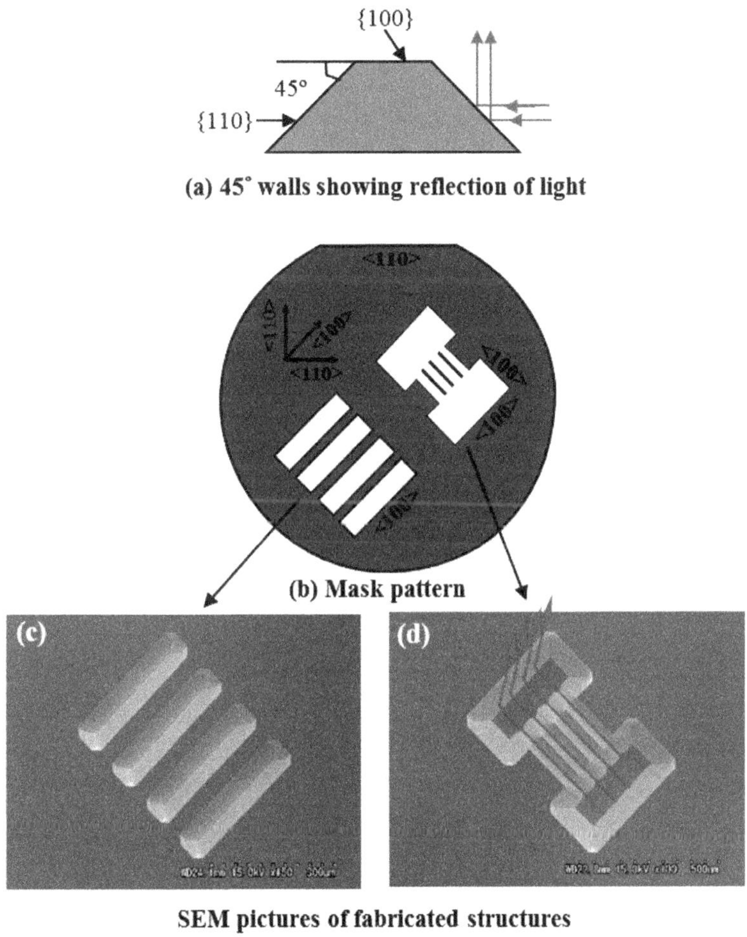

Figure 8.4 Microstructures with 45° sidewalls on a {100} wafer fabricated by aligning the mask pattern along the <100> direction and etching in surfactant-added 25 wt% TMAH.

8.4 Silicon Gratings

Wet anisotropic etching is a simple and effective method to fabricate high-quality diffraction gratings which have a wide range of applications in the infrared (IR) region [4–8]. They are used for high-resolution IR spectrographs, monochromators, and laser tuning. A diffraction grating comprises blazed grooves and therefore is also called an echelle grating. To fabricate a diffraction grating in a silicon

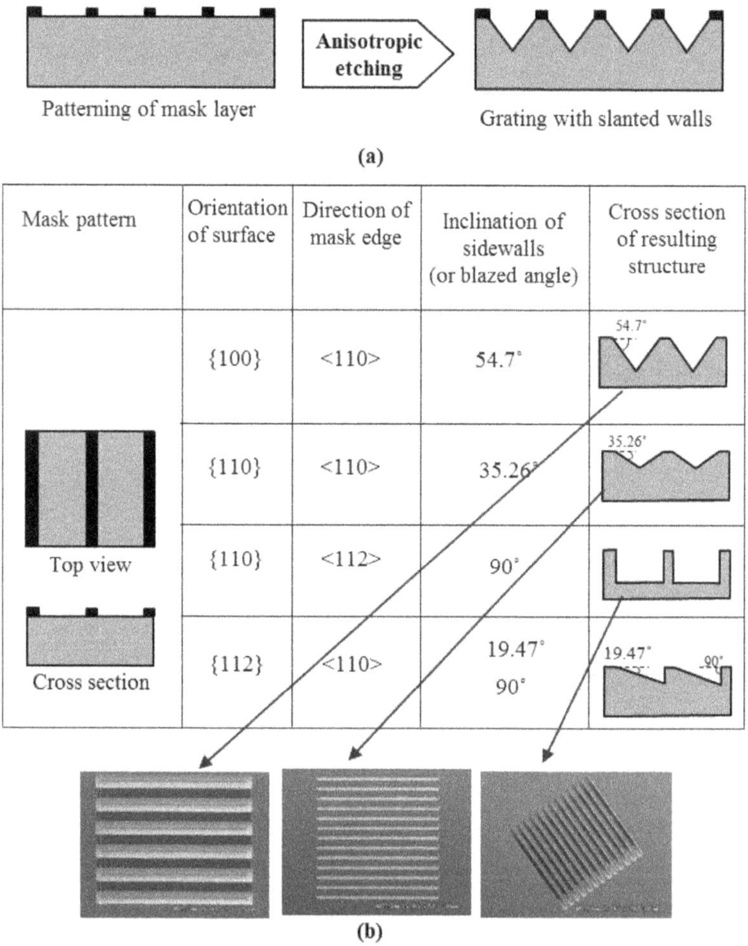

Figure 8.5 Linear gratings with different profiles. The inclination angle of {111}-sidewalls (or blazed angle) depends on the wafer orientation and the direction of the mask edges on the wafer surface.

wafer, a mask layer on a polished surface is patterned with parallel equidistant stripes using the photolithography process, as presented in Fig. 8.5a. Thereafter anisotropic etching is carried out to form blazed grooves. The groove profile (or blazed angle) of the etched diffraction grating depends on the wafer orientation, the mask edge direction on the wafer surface, and the etchant type (pure or surfactant added). Figure 8.5b shows linear gratings with different blazed angles. For example, the alignment of mask edges along the <110> direction on a {100} wafer provides an echelle pattern with a blaze angle of 54.7° (i.e., angle between the {100} and {111} planes). The apex angle of the grooves is fixed by the angle between opposite pairs of {111} planes. Different blaze angles can be achieved by use of a wafer cut at a nonzero angle relative to the {100} planes. After a groove pattern is completed, the mask layer is removed from the wafer. The blazed grooves are separated by narrow unblazed stripes (i.e., flat-topped grooves). The width of these flat tops can be decreased by using narrower mask stripes and/or by overetching after the formation of grooves. A combination of two-step etching and LOCOS, which is discussed later, is used to fabricate a silicon grating with a triangular profile.

8.5 Suspended Microstructures

Diaphragms and cantilever beams are two basic types of suspended structures which are extensively used in MEMS-based sensing devices and therefore considered as the backbone of MEMS. Freestanding structures are fabricated using either silicon or nonsilicon materials. In the case of nonsilicon materials, a material exhibiting high etch selectivity with silicon (e.g., SiO_2, Si_3N_4, etc.) should be used. The fabrication of square/rectangular freestanding diaphragms requires only one lithography step. Schematic cross-sectional views of suspended diaphragms fabricated in a Si{100} wafer using a material exhibiting high etch selectivity with Si{100} (e.g., SiO_2, Si_3N_4, P^+-Si, etc.), time-controlled etching from the back side, etch-stop layer in a SOI wafer, and electrochemical etch stop are presented in Figs. 8.6–8.9, respectively. In the case of the diaphragm shown in Fig. 8.7, the thickness of the membrane is controlled by etching time. To achieve a smooth etched surface morphology, etching should be carried out in pure or surfactant-/IPA-added high-concentration TMAH/KOH. The thickness of a P$^+$-Si-suspended diaphragm is tailored by prediffusion and drive-in cycles [9]. As the fabrication of diaphragms shown in Figs. 8.8 and 8.9 utilizes the etch-stop method, a TMAH/KOH solution of any concentration can be employed. The size of the back-side mask openings depends upon the thickness of the wafer to be etched out. For example, the dimensions of the back-side mask opening *B* for the diaphragm of side *F* (Fig. 8.9) are calculated as follows:

$$B = F + 2\frac{t}{\tan 54.7^\circ} \tag{8.1}$$

where *t* is the silicon thickness to be etched out from the back side.

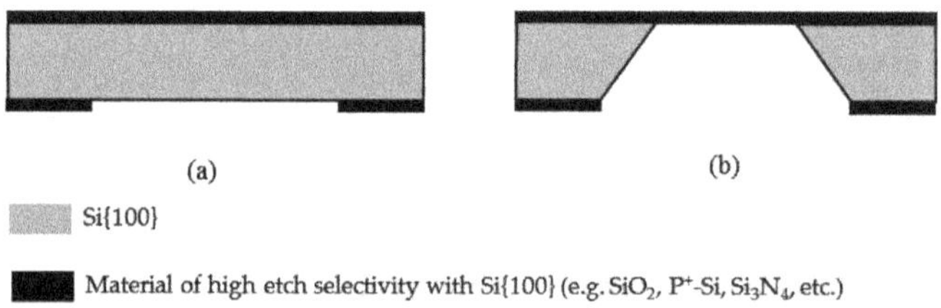

Figure 8.6 Process steps for the fabrication of a diaphragm using a material having high etch selectivity with Si{100} (e.g., SiO_2, Si_3N_4, P$^+$-Si, etc.): (a) deposition and patterning of a mask layer and (b) anisotropic etching.

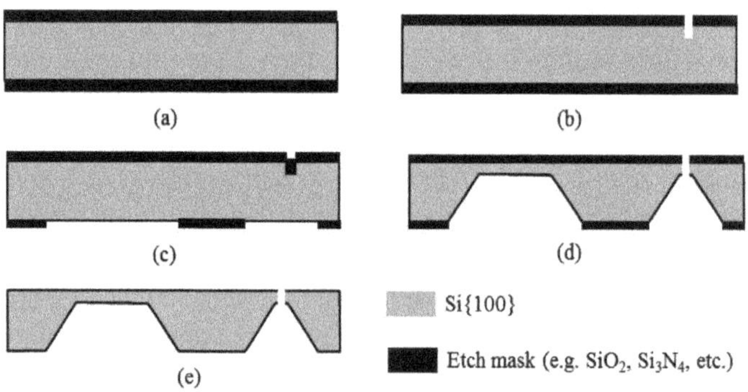

Figure 8.7 Process steps for the realization of a suspended diaphragm in Si{100} wafers using time-controlled etching: (a) deposition of an etch mask layer, (b) patterning of a front-side mask layer, followed by silicon etching to define the thickness of the diaphragm, (c) deposition of a mask layer on the front side and patterning of a back-side mask, (d) silicon anisotropic etching, and (e) removal of the mask layer.

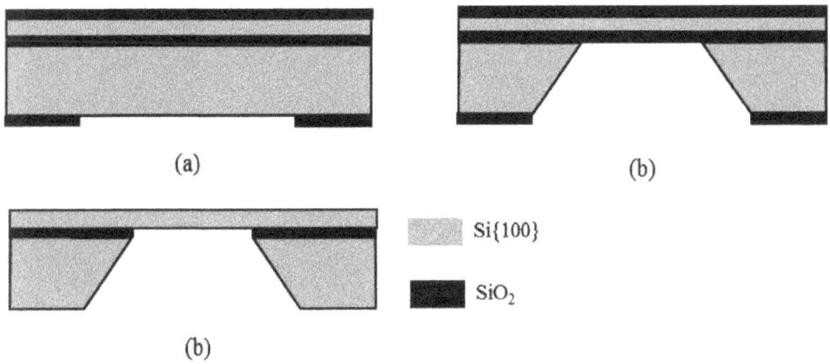

Figure 8.8 Fabrication of a Si diaphragm using a SOI wafer in which the buried oxide acts as the etch-stop layer: (a) patterning of a mask layer on the back side, (b) wet anisotropic etching, and (c) removal of the mask layer.

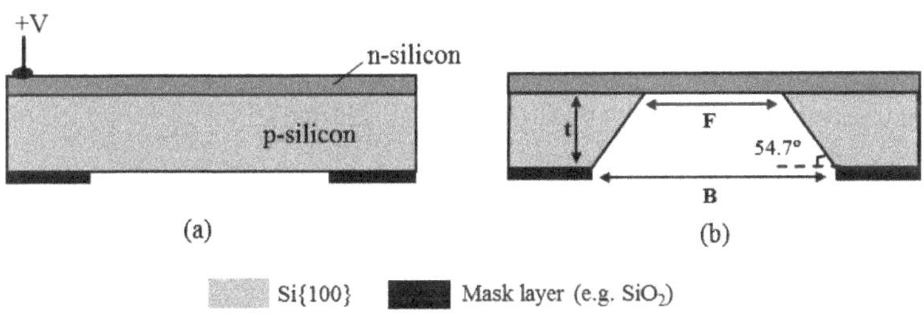

Figure 8.9 Process steps for the fabrication of a Si diaphragm using the electrochemical etch-stop method: (a) patterning of a mask layer on the back side and formation of metal contacts, and (b) electrochemical etching.

Freestanding cantilever beams on silicon wafers are formed either processing one side of the wafer or processing both sides of the wafers. The main process steps for the fabrication of a cantilever using only one side of the wafer are shown in Fig. 8.10 [10]. Firstly a structural layer is patterned, as presented in Fig. 8.10a. Thereafter, anisotropic etching is carried out in an etchant, providing high undercutting at convex corners and non-<110> edges, for instance, high-concentration 25 wt% TMAH or a KOH solution of any concentration. High undercutting is desirable for fast release of the structure. Figure 8.10c presents SEM images of various shapes overhanging SiO_2 MEMS structures.

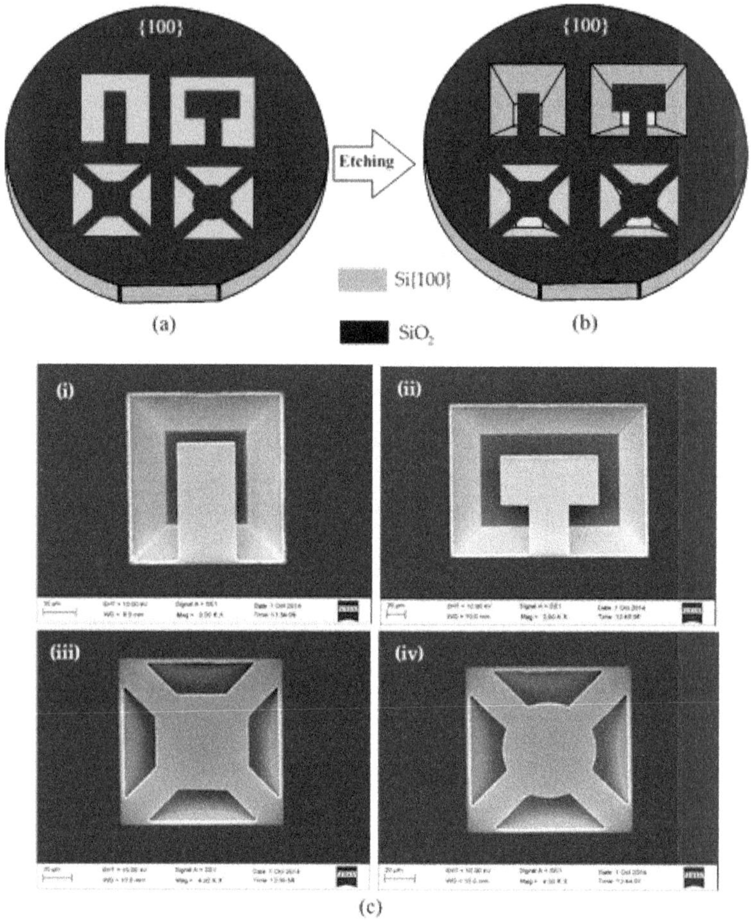

Figure 8.10 Different shapes of SiO_2 freestanding structures fabricated using 25 wt% TMAH by etching from one side of the wafer: (a) patterned oxide layer on a {100} wafer, (b) released structures after anisotropic etching, and (c) SEM pictures.

Fabrication steps for the realization of suspended cantilever beams using both sides of the wafer are illustrated in Fig. 8.11. To control the thickness of freestanding silicon microstructures using wet bulk micromachining, several methods have been employed. These are briefly summarized below, taking a conventional simple cantilever beam, as presented in Fig. 8.12, as an example.

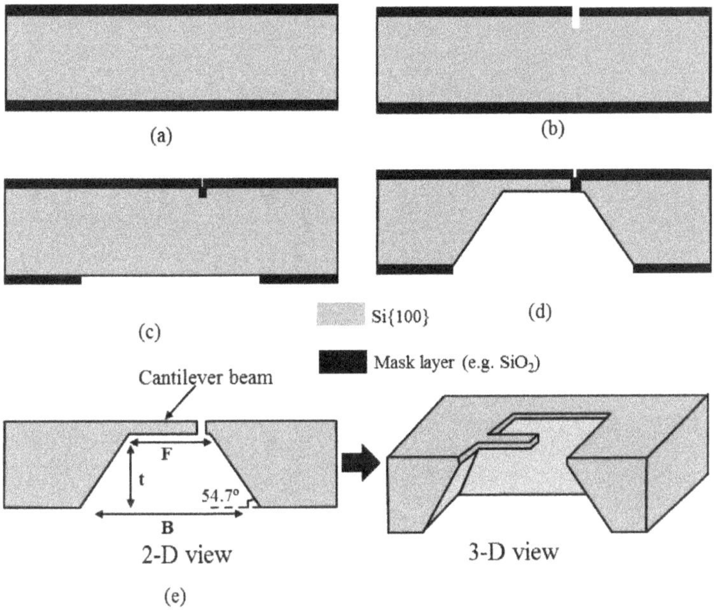

Figure 8.11 Schematic representation of the fabrication sequence of a freestanding cantilever beam in Si{100} wafers using time-controlled etching: (a) deposition of an etch mask layer, (b) patterning of a front-side mask layer, followed by silicon etching to define the thickness of the cantilever beam, (c) deposition of a mask layer on the front side and patterning of a back-side mask, (d) silicon anisotropic etching, (e) removal of the mask layer.

8.5.1 Time-Controlled Back-Side Etching

In this method, suspended silicon MEMS structures of desired thickness on silicon wafers are fabricated by time-controlled back-side etching after defining the shapes and thickness at the front side, as shown in Fig. 8.11 [11–16]. To fabricate the structures, silicon dioxide (or any other masking material) is deposited on the wafer. The process begins with the patterning of a mask layer on the front side, followed by silicon etching to define the thickness of the structures, as illustrated in Fig. 8.11b. The wafer is again coated by a masking layer and patterned on the back side with respect to the front-side patterns using a front-to-back alignment process, as shown in Fig. 8.11c [17, 18]. The back-side anisotropic etching is then carried out. The etching is stopped as soon as a through-hole is formed with the etch patterns formed by the first etching (Fig. 8.11b). The dimensions of the back-side mask opening *B* for the front-side square opening of side *F* (Fig. 8.11b) are calculated in the same way as done for the suspended diaphragm presented in Fig. 8.9. The schematic view and SEM picture of a suspended proof mass for an accelerometer, which is fabricated by processing both sides of the wafer and time-controlled etching, are shown in Fig. 8.13 [19]. If the thickness of the structure is controlled by etching time, the etch rate must be stable throughout the etching process, which is difficult in practice as a slight variation in etching temperature and the dissolution of silicon during etching may cause a minor variation in the etch rate. Therefore it is advisable to measure the etch depth after a certain etch time (or regular interval) to control the thickness (or etch depth) of the structure.

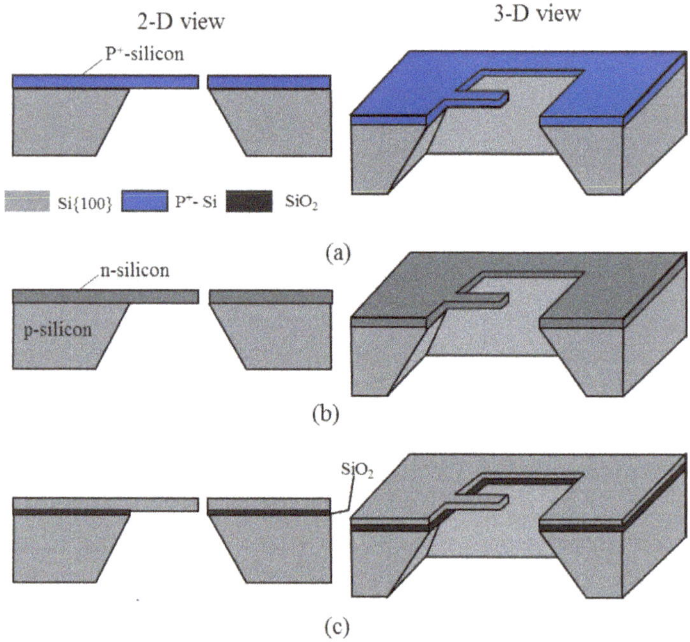

Figure 8.12 Schematic views of suspended silicon cantilever beams realized in a Si{100} wafer using (a) P$^+$-Si; (b) electrochemical etch stop; (c) etch-stop layer in a SOI wafer. Fabrication of these structures requires two-step etching: one step is used for the front side to define the shape of the cantilever, while the second one is employed from the back side to release the structure as discussed in Fig. 8.11.

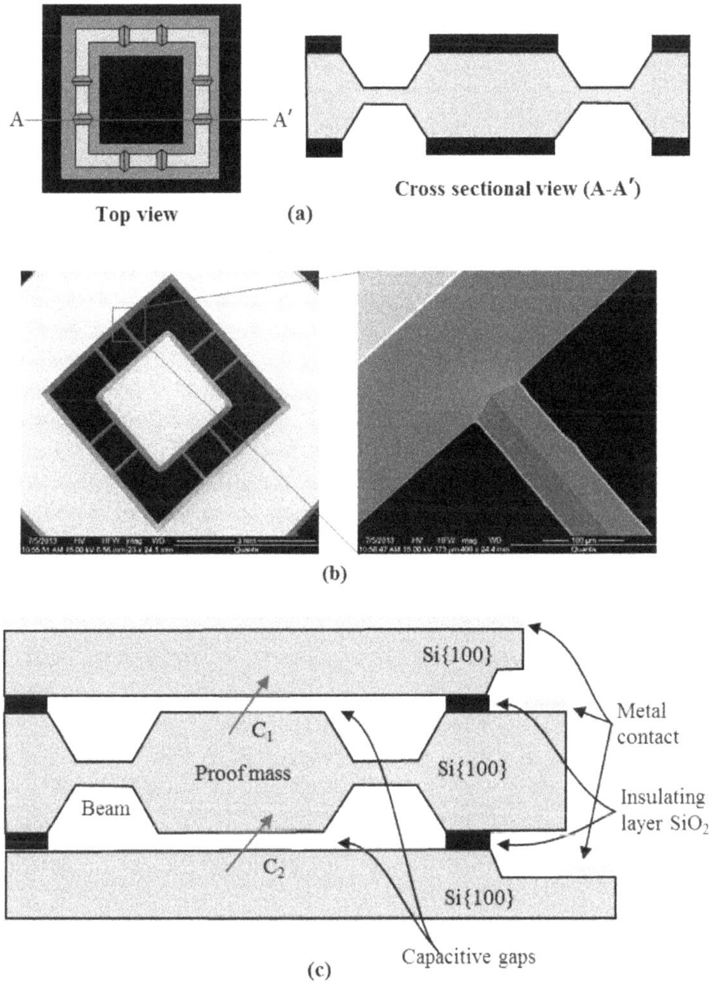

Figure 8.13 Accelerometer structure fabricated by processing both sides of a wafer: (a) top and cross-sectional views of the suspended proof mass, (b) SEM of the fabricated structure, and (c) accelerometer structure with a counterelectrode. To avoid undercutting at convex corners, etching is done in surfactant-added 25 wt% TMAH [19].

8.5.2 P⁺ Silicon

In the time-controlled etching method, a suspended structure can be fabricated using moderately doped silicon, but it requires a stable etch rate during etching. As discussed in Chapters 4 and 5, highly doped p-type silicon (i.e., P^+-Si) exhibits a negligible etch rate in wet anisotropic etchants. Therefore it can be used as a structural (or etch-stop) layer for the fabrication of MEMS components [20, 21]. The process steps for the fabrication of a P^+-Si cantilever beam, as presented in Fig. 8.12a, are the same as illustrated in Figs. 8.10 and 8.11. Although this method provides very-uniform-thickness structures, the thickness is limited by the diffusion depth of p-type impurities. Additionally, the mechanical properties of the material are affected by the large impurity concentration, increasing the internal stress.

8.5.3 Electrochemical Etch Stop

This method is adopted on p-type wafers containing n-type doped silicon (or n-type epitaxial layer) on the top surface (i.e., a p–n junction), as depicted in Fig. 8.12b [22–26]. In this case, the thickness of the suspended structure can be controlled precisely as discussed in Chapter 5. However, it requires a power supply, a wafer holder, and electrodes to achieve an etch stop at n-type silicon. Moreover, it is not suitable for batch processing as only one wafer can be processed at a time.

8.5.4 Silicon-on-Insulator Wafers

P^+-Si and electrochemical etch-stop methods can precisely control the thickness of the microstructures (e.g., cantilever, membrane, etc.). The former technique may cause contamination due to the high-temperature process and the residual stress owing to a heavily doped region, while the latter involves difficulties in the batch process of wet etching. To realize microstructures with uniform thickness throughout the wafer surface, SiO_2-based SOI wafers are used [11, 14, 27, 28]. The thickness of the device (or structural) layer of the SOI wafer defines the thickness of the released structures. Analogous to the previous method, after defining the shape of the structure on the front side, back-side etching is performed for release, as presented in Fig. 8.12c. However, in this method, time-controlled back-side etching is not needed as the etching is automatically stopped when it encounters the oxide layer. As can be seen in Fig. 8.12c, this method also needs precise front-to-back alignment and an additional mask for back-side processing.

It can be noted from Fig. 8.12c that the size of back-side opening (i.e., *B*) depends on the thickness of the wafer. Consequently, the same mask cannot be adopted for back-side lithography if the wafer-to-wafer thickness varies. To achieve microstructures of uniform thickness using time-controlled etching (Fig. 8.11) on all wafers using batch processing of wet etching, all wafers must be of the same thickness, which may be difficult in practice. Due to the large-size back-side mask opening, the methods illustrated in Figs. 8.11 and 8.12 are not suitable for the fabrication of densely placed structures.

8.5.5 Micromachined Cavities Sealed by Wafer Bonding

In this technique, the device (or structural) layer of the required thickness is transferred over micromachined cavities by the bonded and etched-back method [29–33]. Process steps for the fabrication of a cantilever beam using this method are briefly described in Fig. 8.14. It can be noted from the fabrication steps that the gap between the released structure and the bottom of the cavity in the supporting wafer can be tailored easily, which is desirable to control the actuation voltage or capacitance. SEM pictures of suspended cantilever beams fabricated using this technique are shown in Fig. 8.15. In this method, suspended structures of any shape can be formed as the removal of the underneath material for release is not needed. However, it requires an extra mask for the formation of micromachined cavities and a specific environment to control the pressure inside the sealed cavities [30, 33].

8.5.6 Silicon Nitride–Based SOI Wafers

The structures fabricated by back-side etching, as shown in Figs. 8.11 and 8.12, cannot be placed very closely due to the large-size mask opening on the back side of the wafer. To fabricate densely arrayed suspended silicon microstructures of the desired thickness on silicon wafers, the microstructure should be released by front-side etching. In this fabrication technique, suspended silicon microstructures of

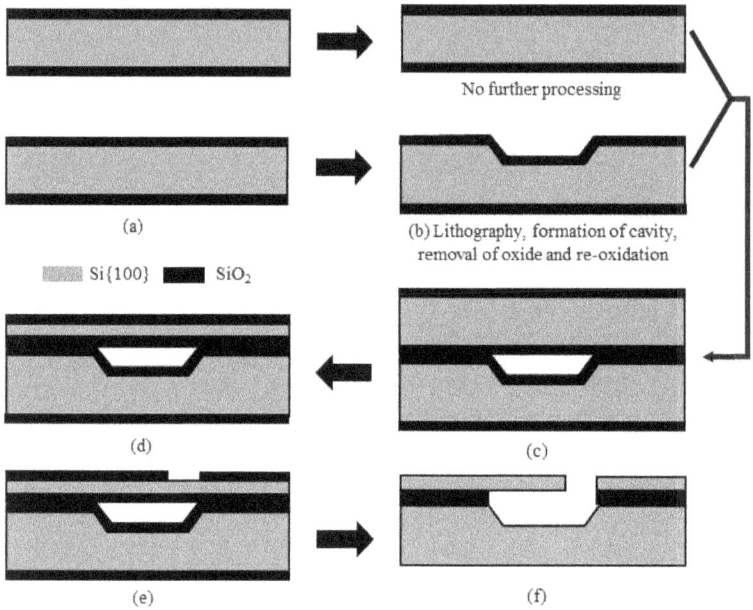

Figure 8.14 Fabrication of silicon freestanding structure of desired thickness over required depth cavity: (a) Wafers are thermally oxidized and split into two batches. (b) One set of wafers is patterned for etching to form the cavities of the required depth, while another set is kept unprocessed. (c) Direct wafer bonding, (d) thinning of the top wafer using pure TMAH and reoxidation, (e) lithography and etching of oxide, and (f) etching in surfactant-added TMAH, followed by removal of oxide to release the structure.

the desired thickness on Si{100} wafers are formed by processing only one side of the wafer [11]. This method involves direct wafer bonding with silicon nitride (Si_3N_4) as an intermediate layer, LOCOS, and wet anisotropic etching in pure and surfactant-added 25 wt% TMAH. Figure 8.16 illustrates the process steps for the realization of simple silicon cantilever beams of the required thickness with one photolithography step. The process steps are generic and can be used for other types of structures. Either nitride-based SOI wafers can be prepared in the laboratory using direct wafer bonding, etching, and chemical-mechanical polishing, or commercially available ones can be employed. Firstly, the SOI wafers are oxidized, followed by patterning using photolithography and oxide etching, as shown in

Fig. 8.16a. After this step, anisotropic etching is performed in surfactant-added 25 wt% TMAH to define the shape of the structure (Fig. 8.16b) as this etchant exhibits a reasonable Si{100} etch rate with minimum undercutting at all kinds of edges and corners. Thereafter, the oxide is removed and LOCOS is carried out (Fig. 8.16c). The silicon nitride is then removed. It can be removed by either dry etching or wet etching in hot phosphoric acid. Since the structure is now to be released, high undercutting at sharp convex corners and non-<110> mask edges is desirable. Thus, the structures are released in pure 25 wt% TMAH solution, followed by oxide and nitride etching, as shown in Fig. 8.16e. Simple and T-shaped cantilever beams fabricated using this method are shown in Fig. 8.17. This method provides densely arrayed structures of uniform thickness throughout the wafer and does not require any time-controlled etching during the etch release step. Furthermore, the process utilizes only one photomask and a single photolithography step.

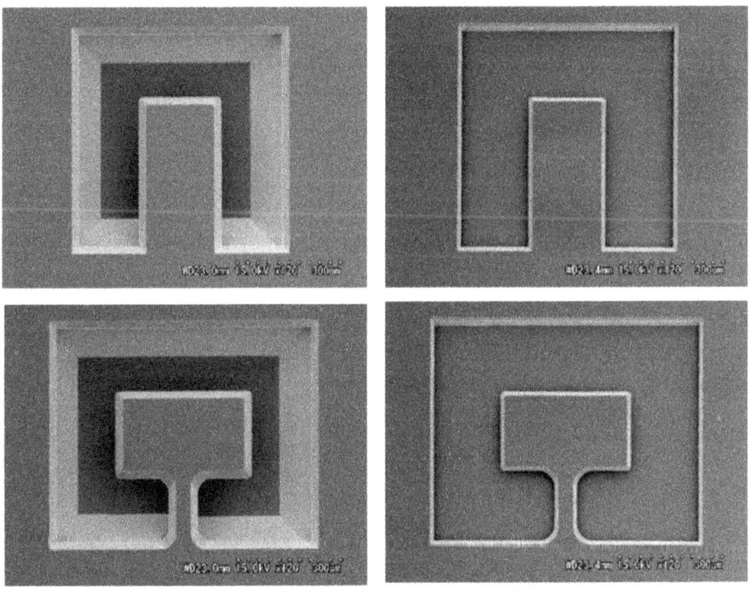

Beam thickness = 27 μm,
Cavity depth = 135 μm

Beam thickness = 10 μm,
Cavity depth = 17 μm

Figure 8.15 SEM pictures of cantilever beams of different shapes fabricated using the processed steps described in Fig. 8.14.

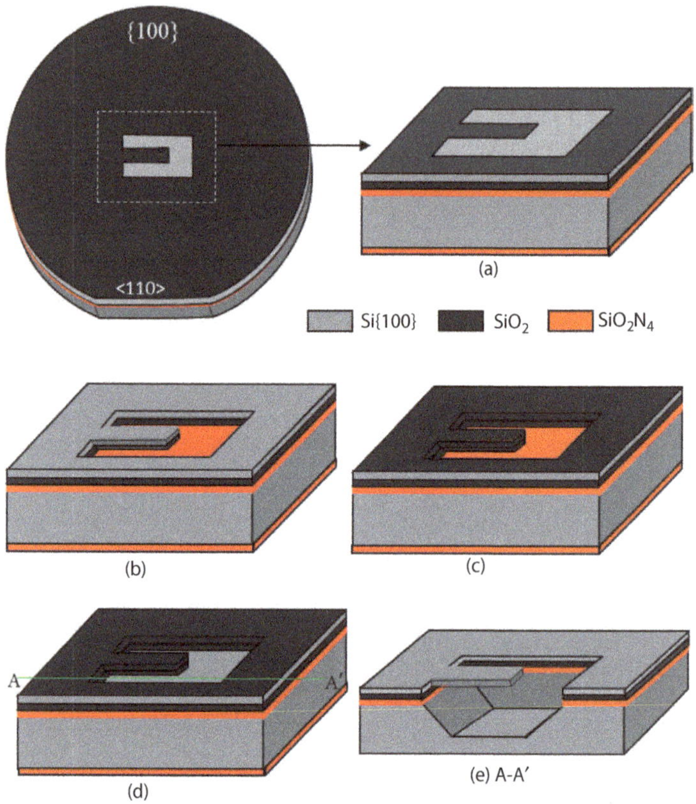

Figure 8.16 Process steps for the fabrication of silicon cantilever beams using one lithography step and two-step wet anisotropic etching on a silicon nitride–based SOI wafer: (a) patterning of the oxide layer, (b) silicon etching in 25 wt% TMAH+surfactant to define the shapes of the structures, followed by oxide etching, (c) LOCOS, (d) silicon nitride etching, while protecting the back side, and (e) silicon anisotropic etching in pure 25 wt% TMAH for releasing the structures, followed by removal of nitride and oxide layers.

8.6 Microstructures with Perfectly Sharp Edges and Corners

Wet bulk micromachining is a promising technique to fabricate microstructures with ideally sharp edges and corners in Si{100} and Si{110} wafers using two-step etching and LOCOS [34–38]. After the LOCOS process, a thermally grown oxide is used as an etch mask.

Therefore TMAH-based etching should be preferred as it exhibits a minimum etch rate for the oxide. The selection of pure or surfactant-added TMAH for silicon etching depends on the requirement of undercutting or the type of etched profile to be realized. Figure 8.18 shows the process steps for the fabrication of adjacent square grooves with sharp edges in a Si{100} wafer [38]. Firstly, nitride-deposited wafers are patterned using photolithography (Fig. 8.18a), followed by thermal oxidation and patterning of the oxide layer (Fig. 8.18b). In this step of oxidation, the oxide layer grows only at the exposed silicon, thus corresponding to LOCOS. Thereafter, the first step of etching is carried out in TMAH+surfactant as the undercutting at the convex corners of the mask pattern is undesirable. The LOCOS process is then carried out to grow an oxide layer on the exposed silicon (Fig. 8.18d). After this step, the nitride layer is etched out globally. This step is called mask inversion from Si_3N_4 to SiO_2. Thereafter, a second etching is performed in TMAH+surfactant (or pure TMAH) as the undercutting is not in picture at this step. Finally, the oxide layer is removed globally in buffered hydrofluoric acid (BHF), as shown in Fig. 8.18f. SEM images of silicon microstructures of different shapes with sharp edges formed using this method are shown in Fig. 8.19. The concentric rectangular and circular corrugation patterns presented in Figs. 8.19c and 8.19d, respectively, are promising structures for the formation of corrugated diaphragms using P^+-Si (and other materials) on a Si{100} wafer, as presented in Fig. 8.20, for the realization of MEMS-based devices such as capacitive pressure sensors, microphones, etc. [39–41].

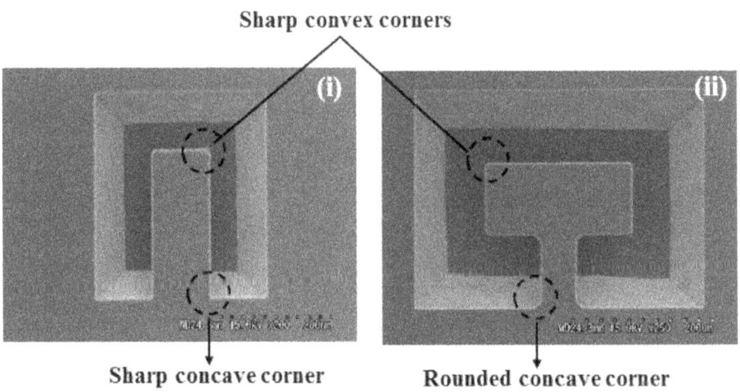

Figure 8.17 SEM pictures of freestanding (i) simple and (ii) T-shaped silicon cantilevers (thickness of the structures = 4.5 μm) fabricated in silicon nitride–based SOI wafers using the process steps illustrated in Fig. 8.16.

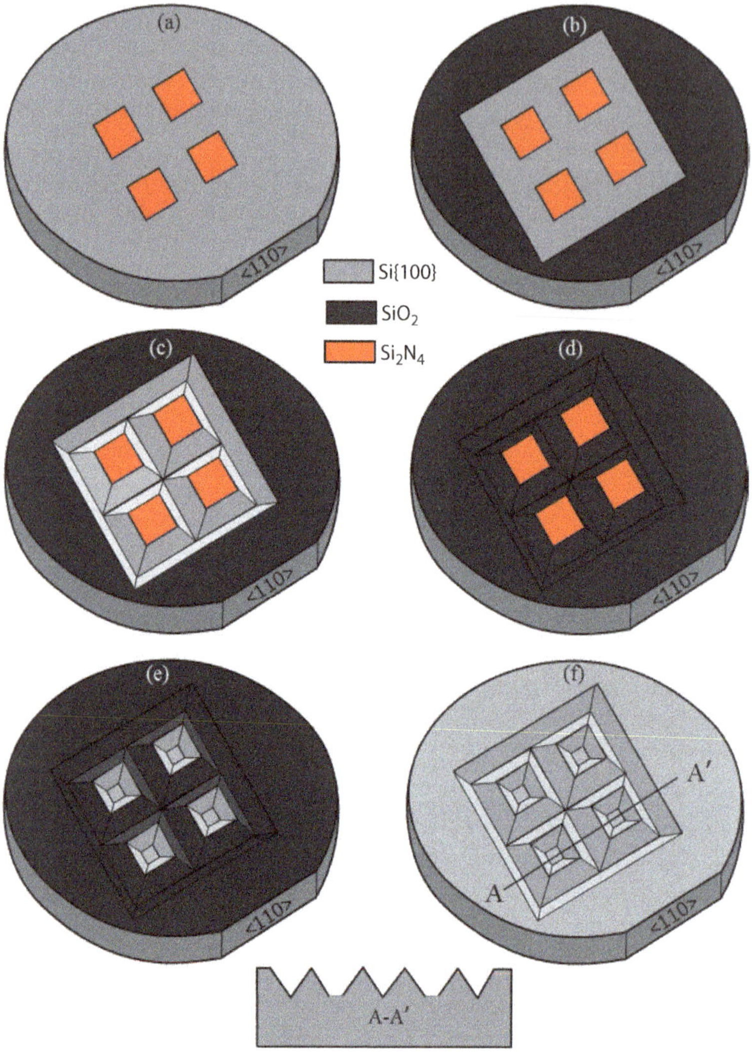

Figure 8.18 Process steps for the fabrication of microstructures with perfectly sharp edges on the top profile: (a) deposition and patterning of the nitride layer, (b) thermal oxidation and patterning of the oxide, (c) Si etching in 25 wt% TMAH+surfactant, (d) LOCOS, (e) removal of Si_3N_4, followed by silicon etching in TMAH+surfactant (or pure TMAH), and (f) final shape of the structure after oxide etching.

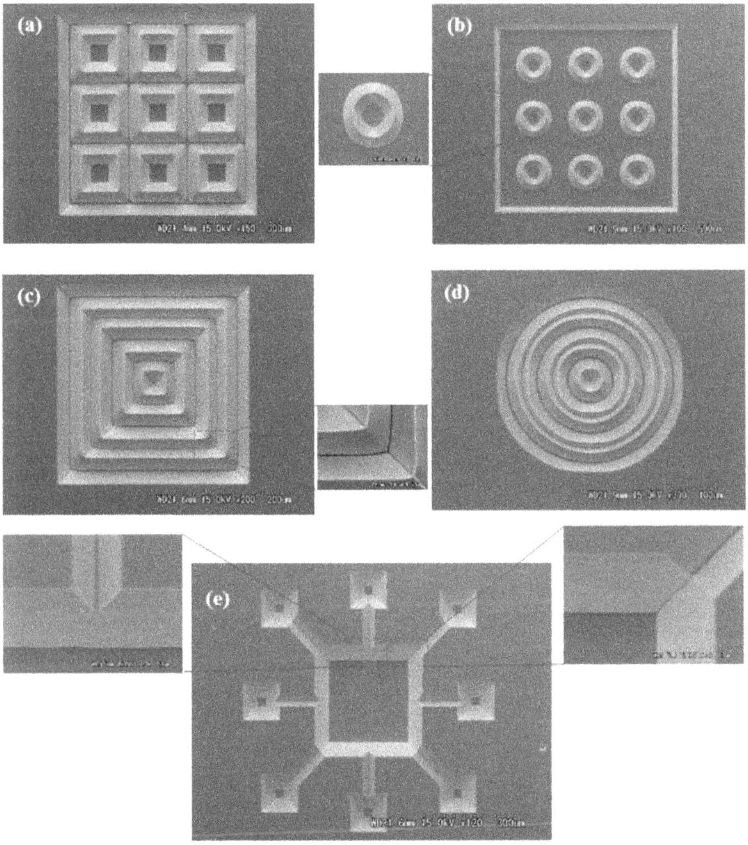

Figure 8.19 SEM pictures of various structures with protected edges and corners fabricated in {100} silicon wafers: (a) adjacent square grooves, (b) donut-shaped structure, (c) square- and (d) circular-shaped concentric corrugations, and (e) microfluidic channels with multiple reservoirs.

Figure 8.20 Schematic view of a corrugated P$^+$-Si diaphragm for capacitive pressure sensor and microphone applications.

The same fabrication methodology can be employed for the formation of silicon gratings with triangular profiles on silicon wafers with different orientations, as illustrated in Fig. 8.21 [4, 38]. Depending on the wafer orientation, the groove (or mask edge) direction on the wafer, and etchant type (pure or surfactant added), the triangular profiles with different blazed angles can be produced. For instance, to fabricate a grating with a 45° blaze angle, mask edges are aligned along the <100> direction on a {100} wafer and the etching is carried out in surfactant-added 25 wt% TMAH [38]. Figure 8.22 presents SEM images of silicon gratings with triangular profiles on {100} and {110} silicon wafers.

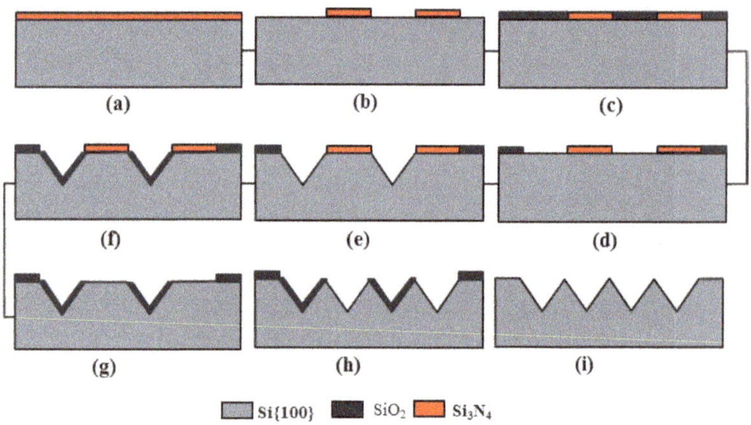

Figure 8.21 Fabrication steps for the formation of a triangular grating with ideally sharp edges using two-step etching and LOCOS: (a) deposition of nitride, (b) patterning of the nitride layer, (c) LOCOS, (d) patterning of oxide, (e) silicon etching, (f) LOCOS, (g) removal of Si_3N_4, (h) silicon etching, and (i) removal of the oxide layer.

To demonstrate the process for the fabrication of microstructures with perfectly sharp edges on a {110} wafer, microfluidic channels with two (or more) reservoirs are realized [36]. This method requires two mask patterns, employed at different stages of the photolithography process, which is fully completed before wet etching. The first mask defines the reservoirs (Mask #1) and the second one the channels (Mask #2). The alignment and overlap of the channel mask with respect to the edges of the reservoirs depend on the choice of reservoir geometry and channel direction (<110> or <112>), as illustrated in Fig. 8.23. The detailed fabrication steps are described in Fig. 8.24. Fabrication starts with the deposition and patterning of a silicon nitride layer (thickness = ~150 nm), as shown in Fig. 8.24a. Thereafter, the wafers are selectively oxidized using LOCOS to grow an oxide layer (Fig. 8.24b). After this step, the oxide is patterned for the first step of silicon anisotropic etching (Fig. 8.24c). Now, a second LOCOS is carried out to grow an oxide layer on the exposed silicon, followed by removal of the nitride layer and the underneath buffer oxide layer (Fig. 8.24d). Finally, the second step of anisotropic etching is performed and the oxide layer is removed in BHF (Fig. 8.24e). Different shapes of microfluidic channels with multiple reservoirs realized in a Si{110} wafer using the process sequence illustrated in Fig. 8.24 are presented in Fig. 8.25.

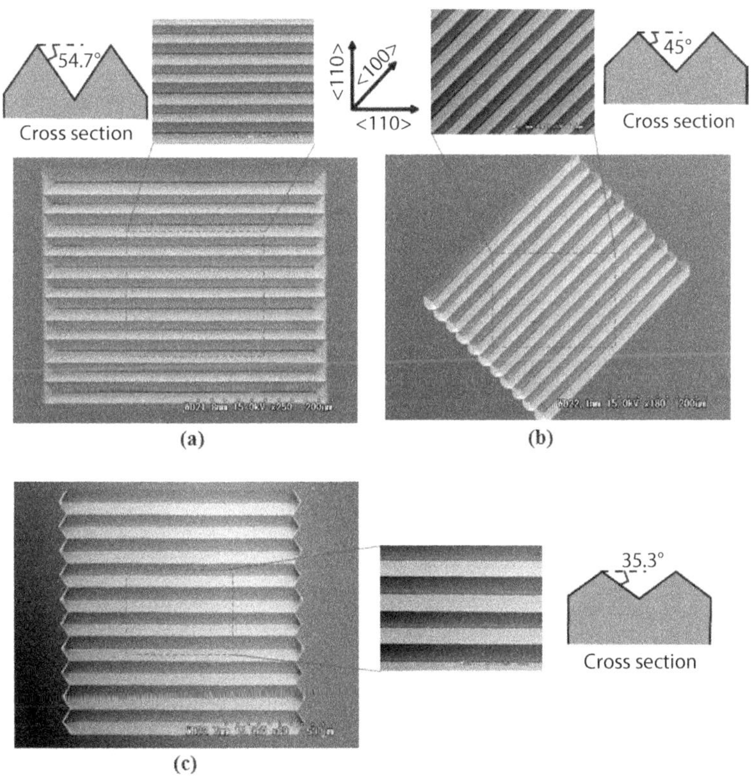

Figure 8.22 SEM images of silicon gratings with triangular profiles fabricated on (a and b) Si{100} and (c) Si{110} wafers.

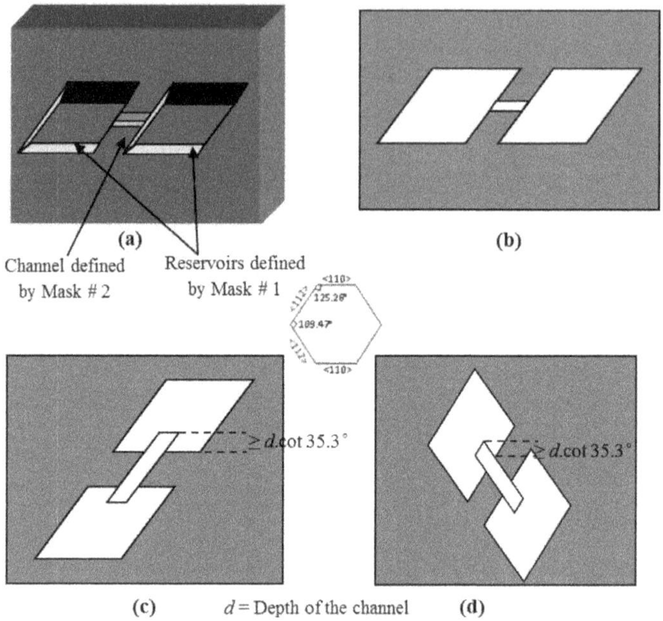

Figure 8.23 Design methodology for the fabrication of microstructures with ideally sharp convex corners on a {110} silicon wafer using two masks and the two-step etching process: (a) <110>-oriented microfluidic channel between two reservoirs, (b) no overlap between masks #1 and #2 for case (a), and (c and d) require overlap to obtain a <112>-oriented channel between reservoirs.

8.7 Suspended Microfluidic Channels

Apart from the freestanding MEMS structures (e.g., cantilever beams, diaphragms, etc.), microfluidic channels are also very useful structures for developing a system (or a device) to carry and deliver liquids [42]. Figure 8.26 presents the process steps for the fabrication of suspended channels (or grooved structures) using a thermally grown oxide as a structural layer in single {100} wafers [43]. The process starts with the patterning of a silicon nitride layer (Si_3N_4). After this step, nitride-patterned wafers are thermally oxidized to grow a thin oxide layer (Fig. 8.26b), followed by photolithography and oxide etching (Fig. 8.26c). Thereafter, anisotropic etching is carried out in TMAH+surfactant to define the shapes of the channels. The oxide layer is then etched out globally and a fresh thick oxide (1–2 μm) is deposited by thermal oxidation. The etched profile formed during the first etching (Fig. 8.26d) works as a mold for the oxide layer (Fig. 8.26e). The oxide layer at this step works as a structural layer. The nitride layer is then selectively etched and silicon etching is performed in pure 25 wt% TMAH to release the structure from the substrate, as shown in Fig. 8.26f. Various shapes of SiO_2 freestanding microfluidic channels, tray-shaped cantilevers, and side-released circular and square cups using the process steps illustrated in Fig. 8.26 are shown in Fig. 8.27.

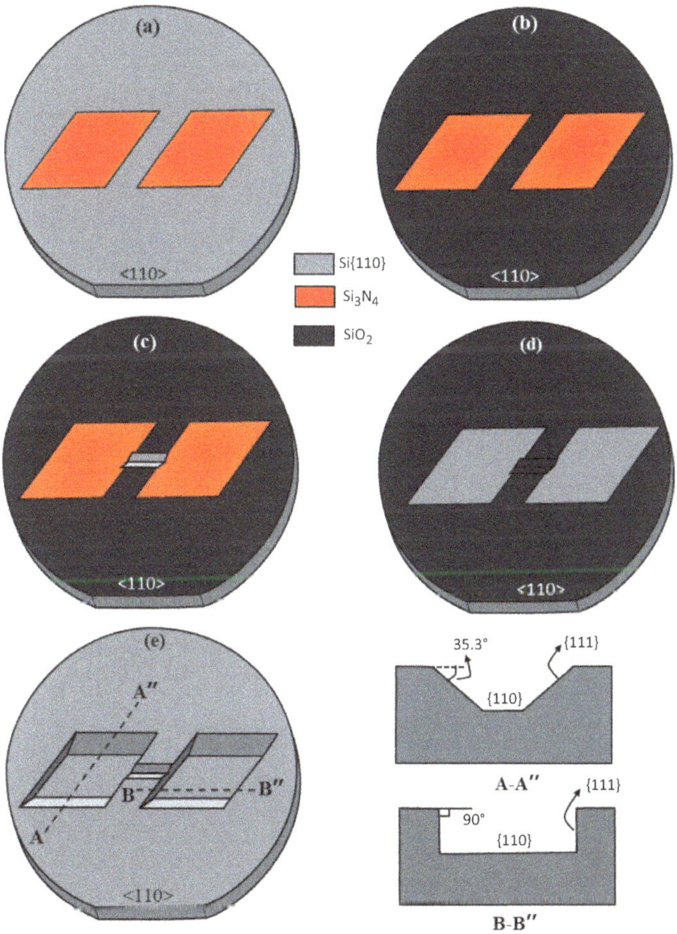

Figure 8.24 Fabrication steps for the realization of a MEMS microfluidic channel with protected edges and corners in a Si{110} wafer: (a) deposition and patterning of nitride, (b) LOCOS, (c) patterning of oxide, followed by anisotropic etching, (d) reoxidation and removal of nitride (i.e., mask inversion), and (e) again anisotropic etching and global removal of oxide in BHF.

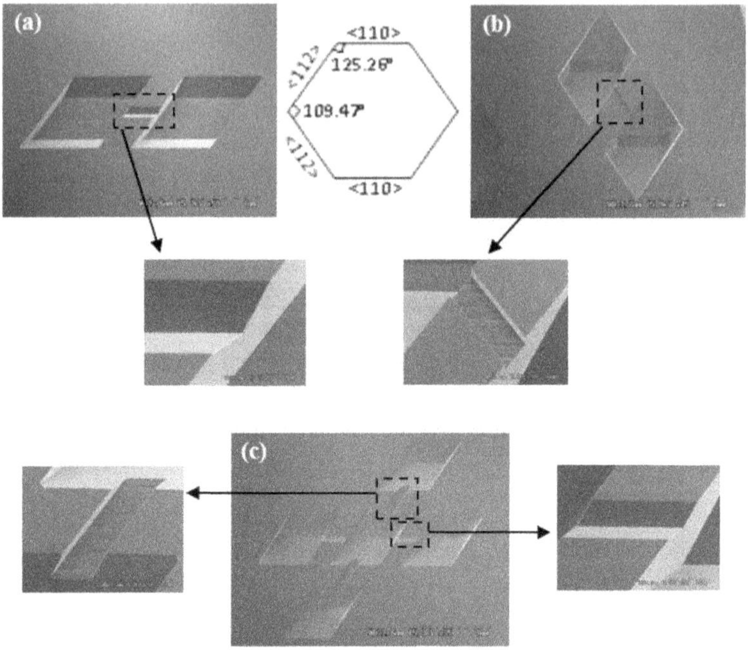

Figure 8.25 Different shapes of microfluidic channels with multiple reservoirs realized in a Si{110} wafer using the process steps illustrated in Fig. 8.24. The edges of microstructures are aligned along (a) <110> and <112>, (b) <112> directions, and (c) <110> and <112> directions.

As discussed in Section 8.5.6 in Fig. 8.16, suspended silicon cantilever beams are fabricated using nitride-based SOI wafers. The same process steps can be employed for the fabrication of overhanging microfluidic channels. Figures 8.28a and 8.28b illustrate the schematic views of a diagonally supported silicon microfluidic channel after the first and second anisotropic etching steps, respectively, corresponding to process steps shown in Figs. 8.16b and 8.16e, respectively [11]. Analogous to the process steps presented in Fig. 8.16, the first step of etching is carried out in surfactant-added TMAH, while the second step of etching is performed in pure 25 wt% TMAH. After the first anisotropic etching in surfactant-added TMAH, {110} planes emerge at <100> mask edges and {111} surfaces develop at <110> sides/edges. Various dimensions illustrated in Fig. 8.28a after the first etching step can be formulated using simple trigonometry:

$$X_2 = X_1 - \sqrt{2}\frac{2 \times T}{\tan 54.7^\circ} = X_1 - 2T \tag{8.2}$$

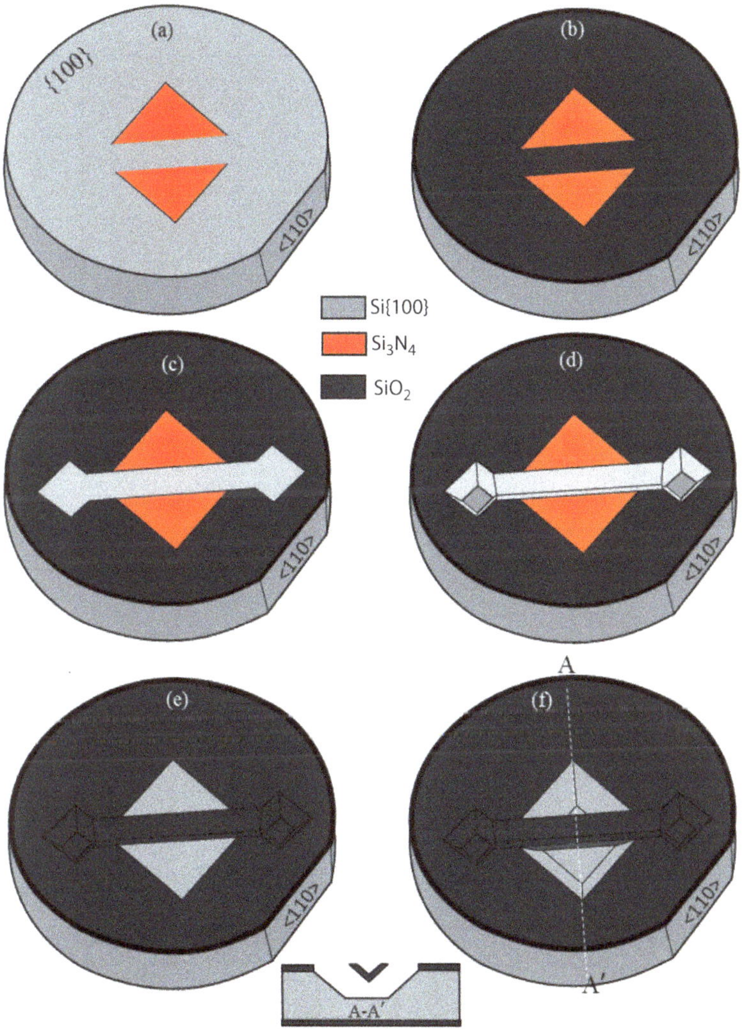

Figure 8.26 Fabrication steps for the realization of overhanging SiO_2 microfluidic channels: (a) deposition and patterning of Si_3N_4, (b) thermal oxidation (LOCOS), (c) patterning of oxide, (d) Si anisotropic etching in 25 wt% TMAH+surfactant, (e) thermal oxidation for the deposition of a thick oxide for microfluidic channel walls, followed by removal of nitride, and (f) silicon etching in pure 25 wt% TMAH to release the channels.

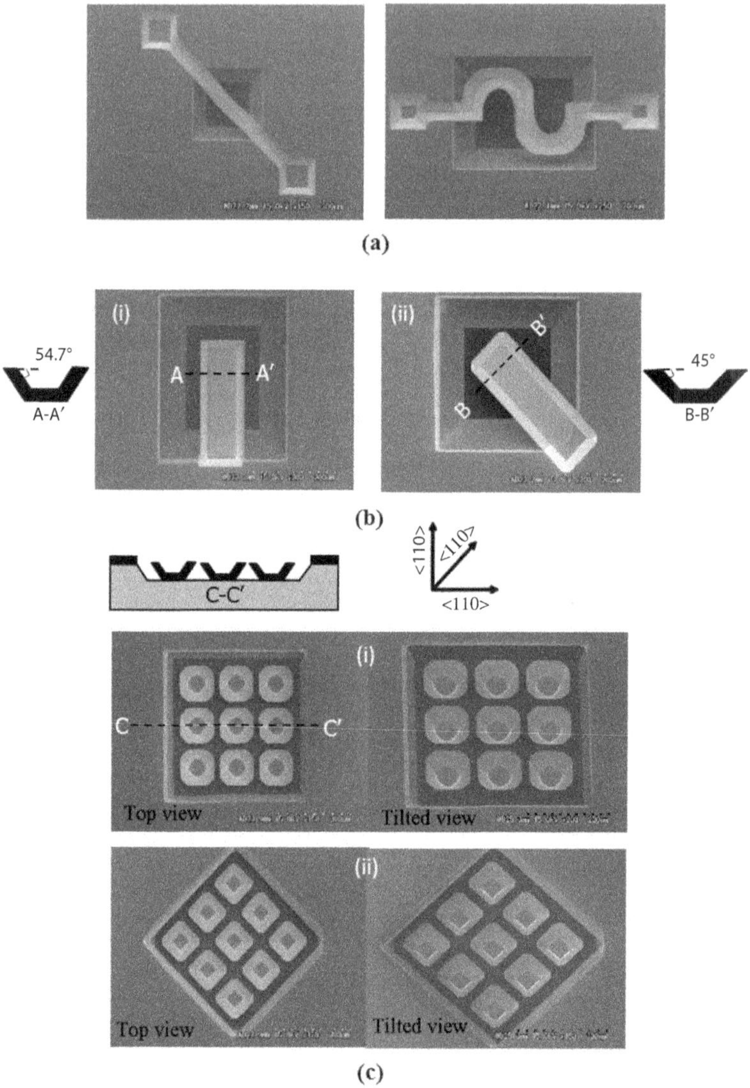

Figure 8.27 SEM pictures of various shapes of SiO_2 freestanding (a) microfluidic channels, (b) tray-shaped cantilevers, and (c) side-released structures (i) circular and (ii) square cups using the process steps presented in Fig. 8.26.

$$W_{\{111\}} = \frac{T}{\sqrt{2}} \tag{8.3}$$

$$W_{\{110\}} = T \tag{8.4}$$

$$P = 3W_{\{111\}} \tag{8.5}$$

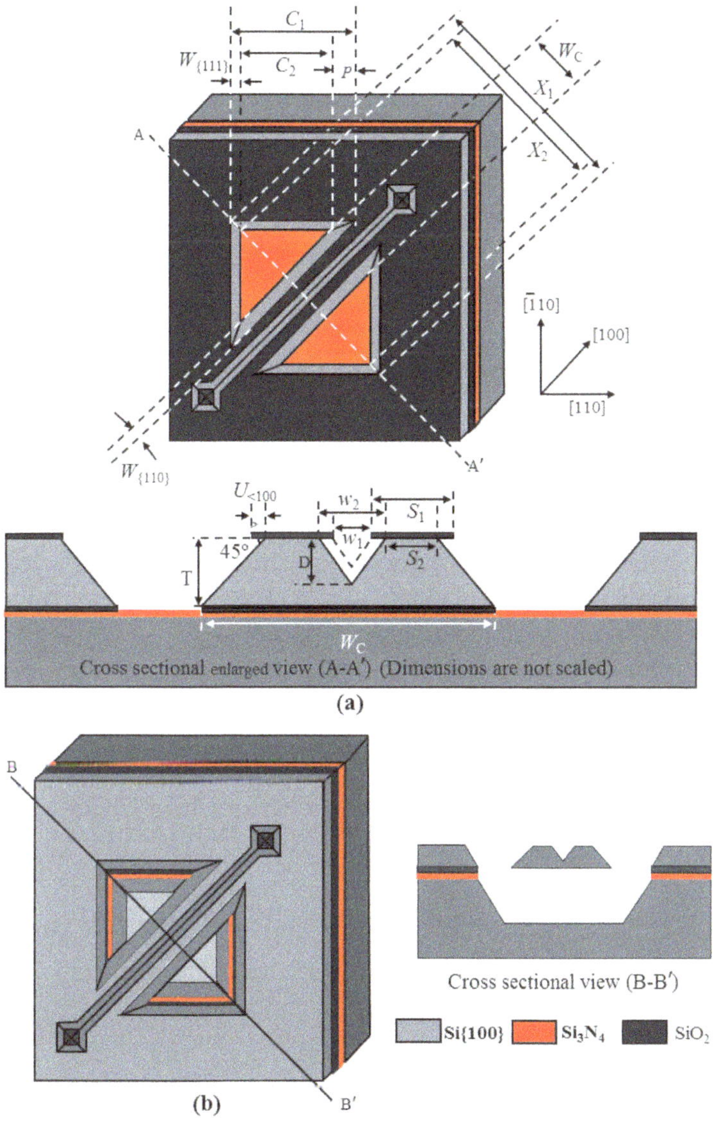

Figure 8.28 Schematic view of a diagonally supported microfluidic channel on a silicon nitride–based {100} SOI wafer: (a) after first anisotropic etching in TMAH+surfactant and (b) released form after second anisotropic etching in pure TMAH, followed by oxide etching.

$$C_2 = C_1 - P - W_{\{111\}} + \frac{U_{<100>}}{\sin 45^\circ} \tag{8.6}$$

where T is the thickness of the structural layer; $W_{\{110\}}$ and $W_{\{111\}}$ are the projected widths of slanted {110} planes at <100> edges and {111} planes at <110> edges on the wafer surface, respectively; and $U_{<100>}$ is the undercutting at <100> edges for structural layer thickness T. The last factor in Eq. 8.6 corresponds to the undercutting at <100> edges.

The channel depth D is given by

$$D = \frac{w_1}{2} + U_{<100>} \tag{8.7}$$

and the width at the bottom (W_C) is given by

$$W_C = 2S_1 - 2U_{<100>} + w_1 + 2T \tag{8.8}$$

The structures after the first anisotropic etching step are released in a pure TMAH solution. Therefore, during the release step no V-grooves (or rectangular cavities) shaped by the slow {111} planes should form. If this happens, the undercutting will stop and the structures will remain unreleased. Thus, the following condition must be satisfied for the successful release of the structure (Fig. 8.28b) after the first anisotropic etching step:

$$\frac{C_2}{\sqrt{2}} > \frac{W_c}{2} \tag{8.9}$$

Figure 8.29 SEM pictures of suspended silicon microfluidic channels fabricated in silicon nitride–based SOI wafers using the similar process steps illustrated in Fig. 8.16 for the fabrication of freestanding silicon cantilever beams.

The total release time (t_2) depends on the width at the bottom (W_C) and the etch rate of the plane appearing at the edges. The mathematical relations derived above must be considered during the mask design to obtain the desired shape of microstructures and microfluidic channels. SEM images of different shapes suspended silicon microfluidic channels are presented in Fig. 8.29.

8.8 AFM Cantilevers

Atomic force microscopy is a type of scanning probe microscopy (SPM) which is widely used in surface science for mapping the forces near the surface with high spatial resolution on the order of fractions of a nanometer [44–51]. Moreover, it has proven its potential as a tool for surface modification and nanolithography. Thus atomic force microscopy is a powerful tool for the investigation of nanoscale science and technology. The atomic force microscope (AFM) probe consists of a flexible force-sensing cantilever with a nanometric sharp tip at its free end, which is used to scan the specimen surface. As the working principle is based on the detection of force between tip and surface, atomic force microscopy is also known as scanning force microscopy (SFM). A force between the tip and the sample surface leads to a deflection of the cantilever beam. Several methods are used for measuring the deflection of the cantilever beam, including optical deflection (e.g., detection of a laser spot reflected from the top surface of the beam by a position-sensitive detector), optical interference, tunneling current, and electrical technique (capacitive, piezoresistive, and piezoelectric sensing). Sharpness of the tip apex is of great importance. AFM cantilevers are mostly fabricated from silicon and silicon nitride. Silicon bulk micromachining is predominantly exploited for the fabrication of cantilevers and sharp tips. Bulk micromachining is carried out using either dry plasma etching or wet chemical etching or a combination of them. Etching can be isotropic or anisotropic. Each of them has its own advantages and disadvantages. However, wet etching is an economical method and requires a very simple system. Numerous reports based on purely wet anisotropic etchants have described many kinds of silicon tips. Silicon is the preferred material for the fabrication of cantilevers with integrated tips because of its excellent mechanical properties (higher Q factor than the amorphous silicon nitride) and the tip can easily be sharpened through thermal oxidation. Figures 8.30 and 8.31 present the process steps for the fabrication of a silicon AFM probe using a SOI wafer and the P$^+$-Si etch-stop method, respectively. Both types of probes can easily be fabricated using wet bulk micromachining. Figure 8.32 shows SEM images of AFM probes with perfectly sharp edges and corners. The fabrication of these probes involves three steps of wet anisotropic etching and two steps of LOCOS [35].

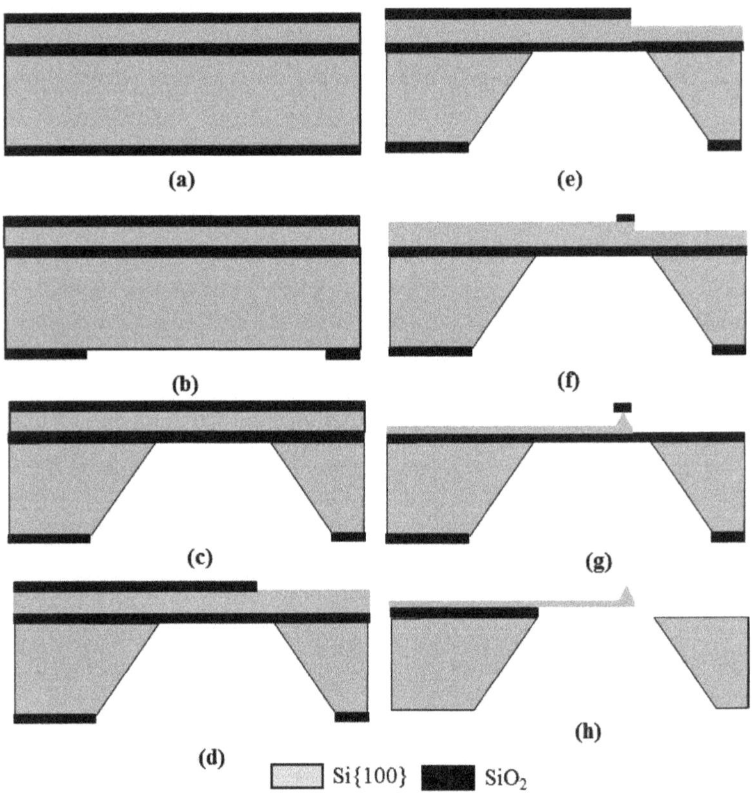

Figure 8.30 Process steps for the formation of a Si AFM probe using a SOI wafer: (a) oxidized SOI wafer, (b) patterning of oxide on the back side, (c) wet anisotropic etching, (d) patterning of a front-side oxide layer, (e) silicon etching to define the thickness of the cantilever, (f) further patterning of the front-side oxide for the tip, (g) silicon etching for the formation of a cantilever beam with a sharp tip, followed by oxide etching.

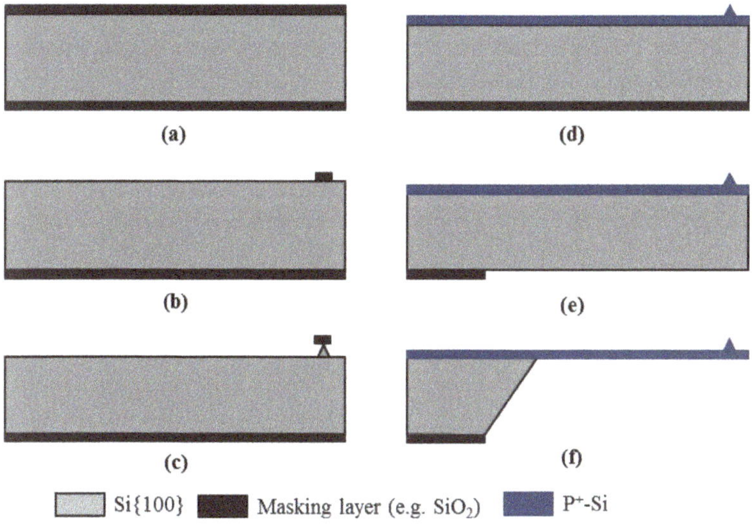

Figure 8.31 Process steps for the formation of a P^+-Si AFM probe: (a) deposition of an etching mask (e.g., SiO_2), (b) patterning of a masking layer, (c) dry (or wet) etching for tip formation, (d) deep boron diffusion for a P^+-Si cantilever, (e) patterning of a mask layer on the back side, and (f) wet anisotropic etching.

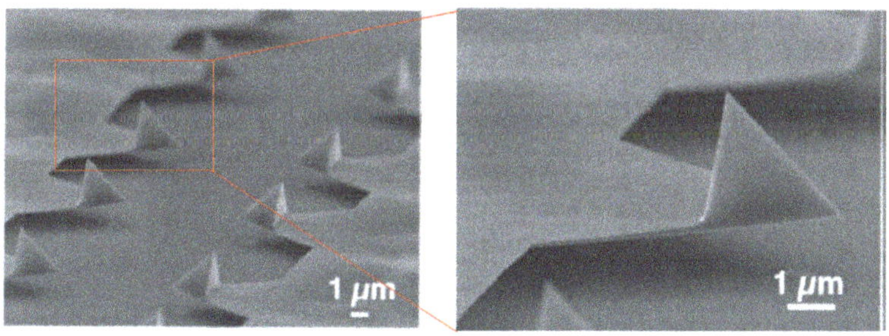

Figure 8.32 SEM micrograph of single-crystal silicon cantilevers with an AFM tip. The tip height is approximately 2 μm. Three steps of wet anisotropic etching and two steps of LOCOS are employed in the fabrication process [35].

8.9 Microvalves and Micropumps

Microfluidics deals with the design and development of miniature devices that manipulate (or process) small amounts of fluids

(10^{-9}–10^{-18} L) using channels with dimensions of tens to hundreds of micrometers [52, 53]. Therefore the main advantages of microfluidic systems are very less consumption of chemical (or biological) reagents and drastic decrease in assay times due to reduced chemical reaction time. Owing to less consumption of sample volume, they are highly suitable for expensive chemical reagents. Microfluidic systems have been increasingly applied to a wide range of fields, including chemical analysis, drug delivery, molecular separation, micromixers, inkjet printers, localized cooling, etc. Microvalves [53–63] and micropumps [64–76] are the key components (or "heart") of a microfluidic system. An automatic microdosing system (or drug delivery system) to dispense drugs (or therapeutic agents) into the human body consists of a drug reservoir, a micropump, microsensors, microchannels, and necessary related circuits. This kind of dosing system is very useful for patients who are suffering from serious chronic diseases such as diabetes, melancholia, malignant lymphoma, etc., or abrupt life threats, such as heart attack, stroke, etc. The integrated microsensors enable real-time measurement to calculate the appropriate and effective amount of drug to be injected by the micropump. Silicon bulk micromachining is a promising method for the fabrication of silicon-based microvalves and micropumps [53–76]. Moreover, silicon micromachining enables the integration of other components such as microfilters and microsensors within the valve. A brief overview of microvalves and micropumps fabricated using silicon wet bulk micromachining is presented in the following sections.

8.9.1 Microvalves

A microvalve is a fundamental part of any microfluidic system. Microvalves, which are studied and fabricated, are divided into two categories, active and passive. An active microvalve consists of a flexible diaphragm actuated by an external force. The passive valves (or check valves) do not include any actuation. The active valves are used to control the flow rate under a certain pressure difference, while in the case of a passive valve, the flow rate is controlled by the pressure difference between the inlet and the outlet of the valve. Moreover, on the basis of their initial mode, microvalves can be termed as "normally open," "normally closed," and "bistable" microvalves. Figures 8.33 and 8.34 present schematic diagrams of passive and active microvalves, respectively. Both types of structures are created using wet chemical-based silicon bulk micromachining. Passive valves exhibit diode-like characteristics as they only open to forward pressure, as presented in Fig. 8.33. As discussed in the next section, passive microvalves are commonly used as part of micropumps due to their simple design and easy fabrication. In active microvalves, various actuation schemes such as piezoelectric, electrostatic, electromagnetic, thermopneumatic, bimetallic (or thermal bimorph), and thin-film shape memory alloy (SMA) are adopted to actuate mechanical moving parts. Each actuation method has its own merits and demerits in terms of response time (i.e., the transition time during open-to-close or close-to-open), power consumption, force delivered by the actuator (i.e., actuator performance), etc. Figure 8.34 shows an active microvalve in which the bimetallic actuation scheme has been used [59].

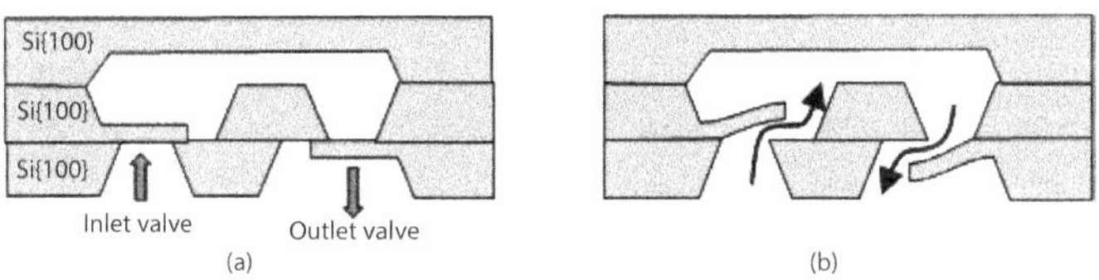

Figure 8.33 Schematic diagram of a check valve (or passive valve) at (a) idle position and (b) under forward pressure.

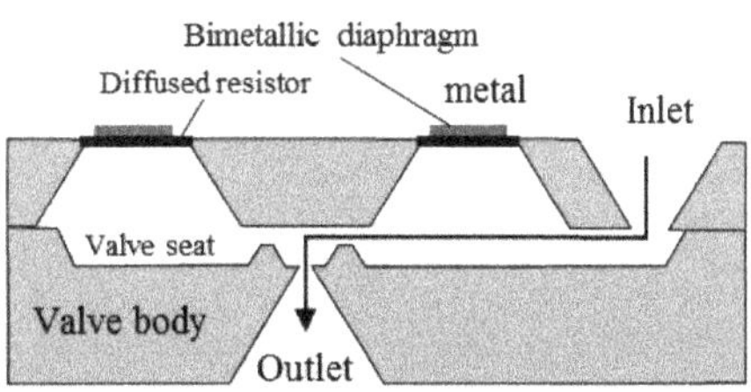

Figure 8.34 Schematic view of an active microvalve with an integrated bimetallic actuator [59].

8.9.2 Micropumps

Micropumps are classified into two groups, mechanical and nonmechanical. The micropumps with moving parts such as a diaphragm/

actuator and a microvalve are referred to as mechanical micropumps, while those without moving parts fall into the category of nonmechanical micropumps. The nonmechanical micropumps transform nonmechanical energy into kinetic energy so that the fluid in microchannels can be driven. On the basis of their working principle, they are named as ultrasonically driven micropumps, electrohydrodynamic micropumps, electro-osmosis micropumps, surface-tension-driving micropumps, etc. In mechanical micropumps, a physical actuator is needed to perform the pumping function. They can be further subdivided into displacement pumps and dynamic pumps according to whether the mechanical energy is added periodically to increase pressure to move the fluid or continuously to increase the velocity of the fluid [66]. In these two categories, displacement pumps, which consist of a pressure chamber having a flexible diaphragm driven by an actuator and passive valves, are commonly used. The check valves and the actuator determine the performance, such as the flow rate and the output pressure. The output pressure of a micropump directly depends on the available force the actuator can deliver and the flow rate is determined by the dynamic characteristics of the actuator diaphragm. Analogous to active microvalves, various kinds of actuation mechanisms are developed for micropumps, such as electrostatic, piezoelectric, thermopneumatic, SMA, bimetallic, etc. Each technique of actuation has its own benefits and drawbacks in terms of output pressure, flow rate, and fabrication. A schematic illustration of an electrostatically actuated membrane micropump consisting of four wet-bulk-micromachined chips is presented in Fig. 8.35 [70]. The actuation unit is made from two silicon chips that embody the flexible pump diaphragm and a rigid counterelectrode in a capacitor-like configuration. When a supply voltage is applied between the pump diaphragm and the counterelectrode, the diaphragm is driven by electrostatic forces (or capacitive force) and bends toward the counterelectrode. Thereby fluid is sucked in through the inlet valve, as shown in Fig. 8.35b. When the supply voltage is turned off, the diaphragm is released (or springs back), forcing fluid out through the outlet valve, as presented in Fig. 8.35c. Two passive check valves direct the flow from inlet to outlet. As the electrostatic pump is actuated by the electrostatic attraction force, the deflection of the diaphragm can be easily controlled by the applied voltage. Similarly, the working principle of other actuation methods can be understood. The flexibility of the diaphragm determines the volume stroke of the micropump. Therefore it is important to design a flexible diaphragm.

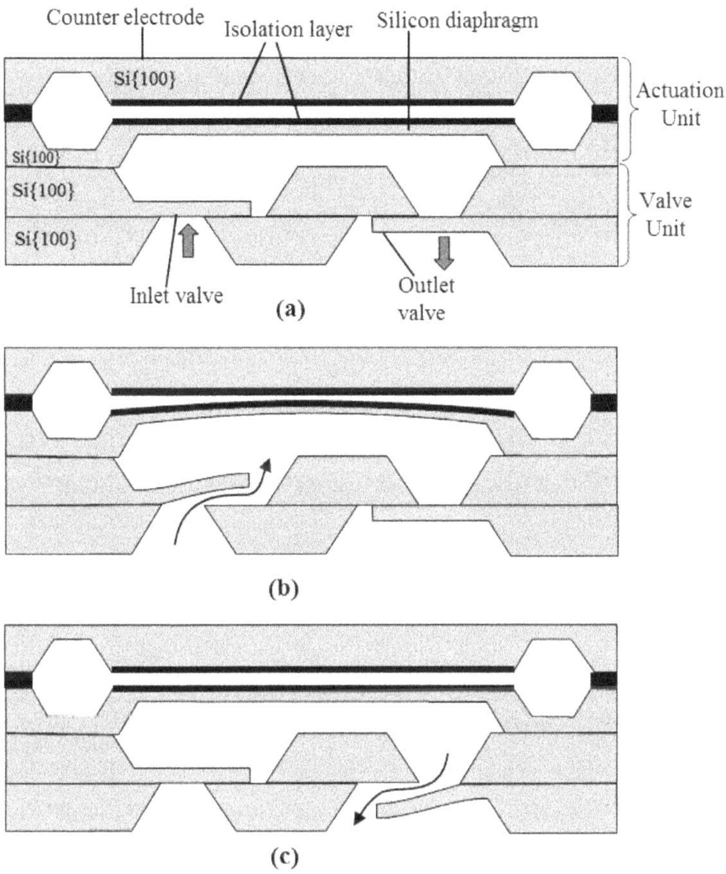

Figure 8.35 Schematic view of an electrostatically driven diaphragm pump: (a) initial (or off) position, (b) suction stroke, and (c) discharge stroke [70, 71].

8.10 Microstructures inside a Silicon Wafer

The moving parts of MEMS, such as the cantilever beam, microbridge, and suspended proof mass, often need to be enclosed in a cavity for a variety of reasons, including pull-down electrode formation for electrostatic actuation and, in some cases, for protection from the environment [77–81]. If the moving parts are recessed in the substrate, these can be secured easily by a flat surface (or planar wafer) of any type of substrate using the planar wafer–bonding process, as shown in Fig. 8.36. This provides a pull-down electrode for electrostatic actuation of the suspended structure or an electrode for capacitive sensing of the moving microstructure, if required.

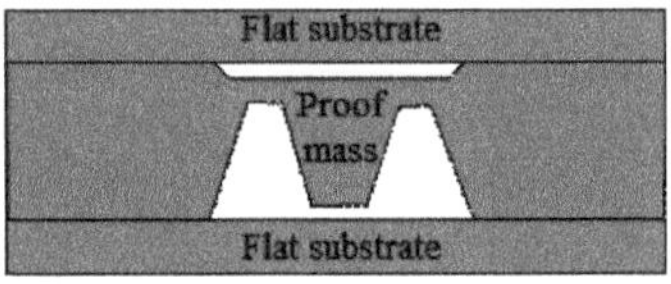

Figure 8.36 Schematic view of a recessed suspended proof mass enclosed by flat substrates.

Owing to uneven surface topography, photolithography on anisotropically etched cavities is a difficult task in micromachining technology [82–88]. There are three major problems: (i) coating of the photoresist with uniform thickness, (ii) undesired ultraviolet (UV) exposure of the photoresist due to reflection from the sloping sidewalls, and (iii) pattern transfer due to the uneven gap between mask and substrate. The use of a standard liquid photoresist, deposited by a spin-coating process, results in nonuniform thickness, especially on sloping sidewalls and convex edges/corners, as illustrated in Fig. 8.37a. At the convex edges, the film may even be discontinuous, as the photoresist reflows from these edges due to surface tension. The very thin or discontinuous photoresist coating at the convex edges fails to protect the masking layer at the sharp edges during the etching process. This, in turn, leads to etching of silicon if the silicon etching is carried out after patterning of the mask layer. Moreover, due to reflow of the spin-coated photoresist, an excessive amount of photoresist is accumulated on the bottom of the cavities/trenches. To coat a uniform-thickness photoresist on an uneven surface, electrodeposition and spray-coating methods can be employed, but they have their own limitations [83–89]. The electrodeposited photoresist needs a conductive base on which the photoresist electrodeposition takes place, and therefore it cannot

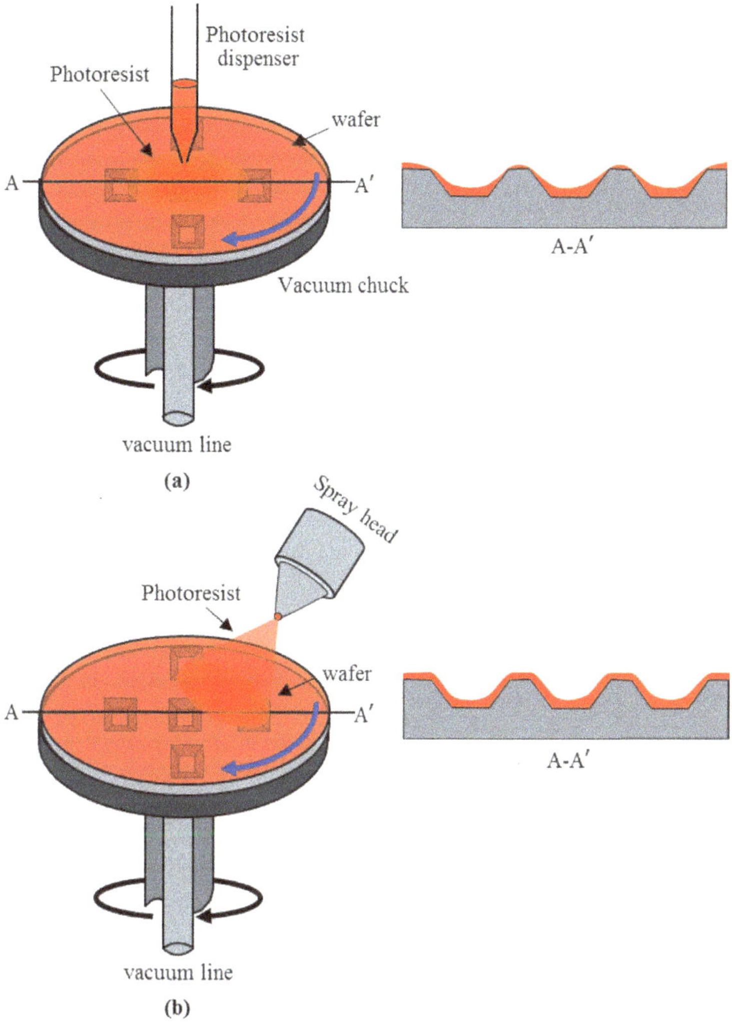

Figure 8.37 Schematic representation of a photoresist coating using (a) spin-coating and (b) spray-coating methods on a silicon wafer comprising anisotropically etched patterns.

be used at all stages of the process. In the case of the spray-coating technique, photoresist thickness, uniformity, and surface roughness depend on several parameters such as viscosity of the photoresist, orifice of the spray nozzle, and the air pressure applied to the nozzle

[88, 89]. Although it provides better results than the spin-coating method, as presented in Fig. 8.37b, it is also not an ideal solution to coat a uniform-thickness resist on an uneven wafer surface, such as cavities with a large difference in size/depth on the same sample. The second problem is the reflection of UV light from tilted sidewalls during UV exposure. This unwanted reflection exposes the photoresist of opposite sidewalls and/or the bottom surface. Although it can be avoided using polarized UV light, it requires extra hardware in the exposure tool [83]. Another way to overcome this problem is to make suitable changes in the mask design. In the mask, the area which comes on top of the tilted sidewall is kept dark [77]. The photoresist (positive or negative) is chosen accordingly. The third issue is the uneven gap between the mask and the wafer surface. It may cause a diffraction problem and can change the feature size transferred from mask to substrate.

Figure 8.38 presents the fabrication steps of a recessed seismic mass with perfect convex corners for accelerometers using silicon wet bulk micromachining [90]. Supporting arms of the seismic mass are realized by the P^+-Si diffusion process. The process starts with the patterning of an oxide layer on a double-side polished Si{100} wafer. The patterns are aligned to each other using the front-to-back alignment technique, as shown in Fig. 8.38a. The size of the square-shaped mask openings depends on the dimensions of the microstructure to be fabricated. In Fig. 8.38, the dimensions of the front- and the back-side cavities are taken to be identical for simplicity. After this step, anisotropic etching is performed to etch the cavity up to the required depth, as shown in Fig. 8.38a. This eventually defines the gap between the seismic mass and the counterelectrode, as illustrated in Fig. 8.36. The wafer is now reoxidized thermally to obtain about 1 micron oxide thickness, which acts as the mask layer for diffusion and the next step of silicon etching. The next lithography step is carried out on the front side of the wafer to define the pattern for supporting arms for the seismic mass. Boron diffusion is carried out to make the supporting arms P^+-Si. The thickness of the supporting arms of the seismic mass is controlled by the boron predeposition and drive-in cycles. After the diffusion process, the oxide on the back side of the wafer is patterned.

This is schematically shown in Fig. 8.38c. Thereafter, the silicon nitride is deposited. The nitride can be deposited by any technique used in semiconductor processing, such as low-pressure chemical vapor deposition (LPCVD), plasma-enhanced chemical vapor deposition (PECVD), or photo-CVD, but it should be of good quality so that it can be employed for LOCOS. The wafer is then patterned on the back side by selective etching of silicon nitride by reactive ion etching (RIE) followed by silicon dioxide etching, as illustrated in Fig. 8.38d. After this step, the anisotropic etching of silicon is carried out. The schematic cross-sectional drawing is shown in Fig. 8.38e. After proper cleaning of the wafer, LOCOS is carried out. In this oxidation process, only sidewalls of anisotropically etched cavities will be oxidized, as shown in Fig. 8.38e, because the rest of the wafer is covered by silicon nitride. Now, the silicon nitride is globally removed in orthophosphoric acid or by dry etching. Figure 8.38f shows the cross-sectional and the back-side view of the wafer after this step. The second anisotropic etching is then performed. Finally the oxide is removed in BHF. Figure 8.38g shows the schematic and SEM images of the fabricated structure. A proof mass with perfect convex corners supported by four symmetric cantilever beams fabricated by the same process steps is presented in Fig. 8.39. In this process convex corners of the proof mass are realized by two-step etching and LOCOS. However, the same can be achieved by the corner compensation method, but edges of the corners cannot be ideally sharp.

Figure 8.40 presents schematic diagrams of particle filters inside a Si{100} wafer [91]. These types of particle filters are fabricated by creating pores in a recessed membrane of silicon or silicon nitride. The fabrication process involves silicon anisotropic etching on both sides of the wafer. In the fabrication process, first deep cavities of the required depth are formed by wet anisotropic etching. Thereafter, the pattern containing arrays of circular pores (diameter 10–50 μm) is transferred onto the bottom of the cavity, followed by dry etching up to the required depth. A second wet anisotropic silicon etching step is performed from the back side of the wafer to form the membrane. Both types of particle filters have their pros and cons. In the case of silicon-based filters a thicker membrane can be fabricated, while in nitride-based filters only a thin diaphragm (<1 μm) can be realized.

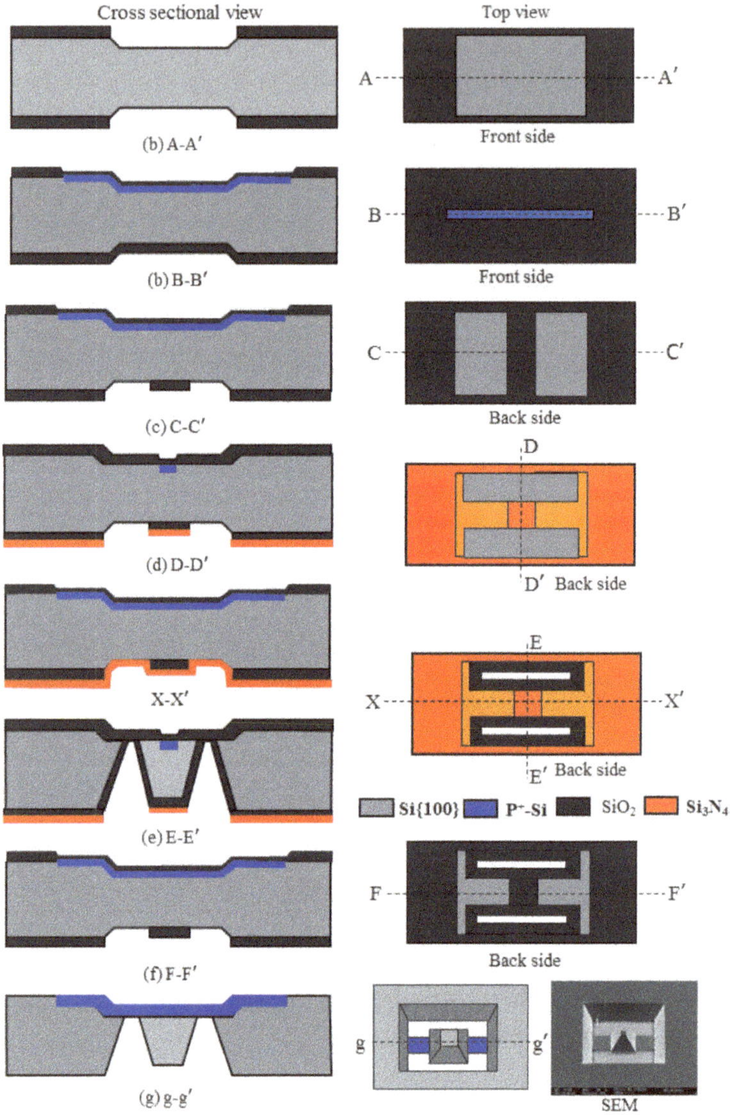

Figure 8.38 Process sequence for the fabrication of a proof mass with perfect convex corners inside a Si{100} wafer: (a) creation of cavities on both sides of the wafer in an oxidized wafer using anisotropic etching, (b) lithography followed by P^+ diffusion for supporting arms, (c) patterning of SiO_2 on the back side, (d) silicon nitride deposition followed by patterning on the back side by Si_3N_4 and SiO_2 etching, (e) anisotropic etching followed by LOCOS, (f) removal of the nitride layer, and (g) anisotropic etching followed by removal of oxide.

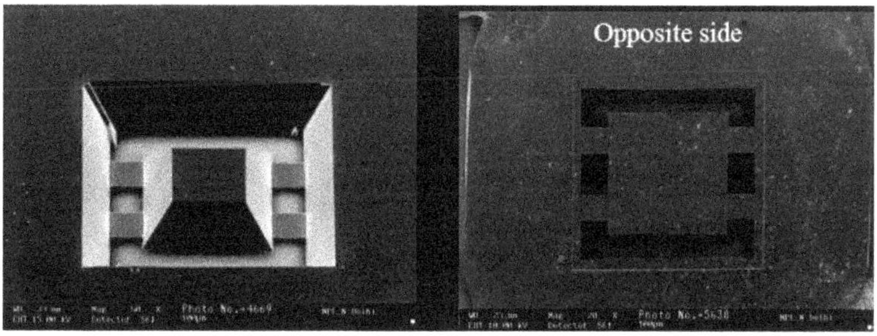

Figure 8.39 SEM photographs of a proof mass with perfect convex corners supported by four symmetric cantilever beams

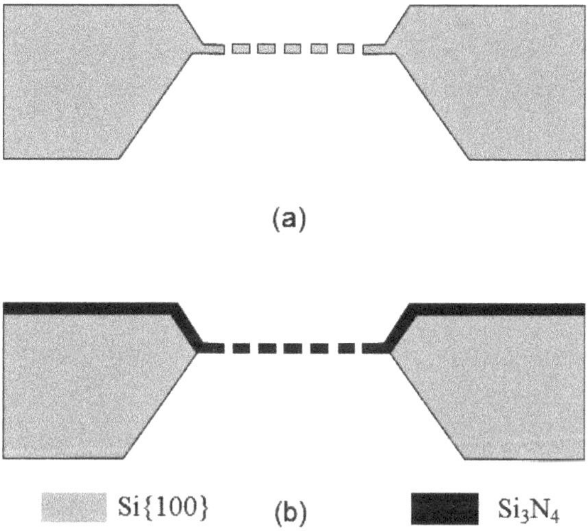

Figure 8.40 Schematic view of particle filters integrated inside a silicon wafer: (a) silicon and (b) silicon nitride membranes [91].

Figure 8.41 illustrates the process steps for the formation of single-crystal silicon membranes oriented in {111} planes [92]. The membranes are realized by patterning of a mask layer on both sides of a Si{100} wafer, followed by simultaneous anisotropic etching on the front and back sides of the wafer through rectangular openings in the masking layers. The thickness of the membranes is determined by the etching time and the relative position of the openings in the masking layers. The resulting membranes make a 54.78° angle with the plane of the wafer.

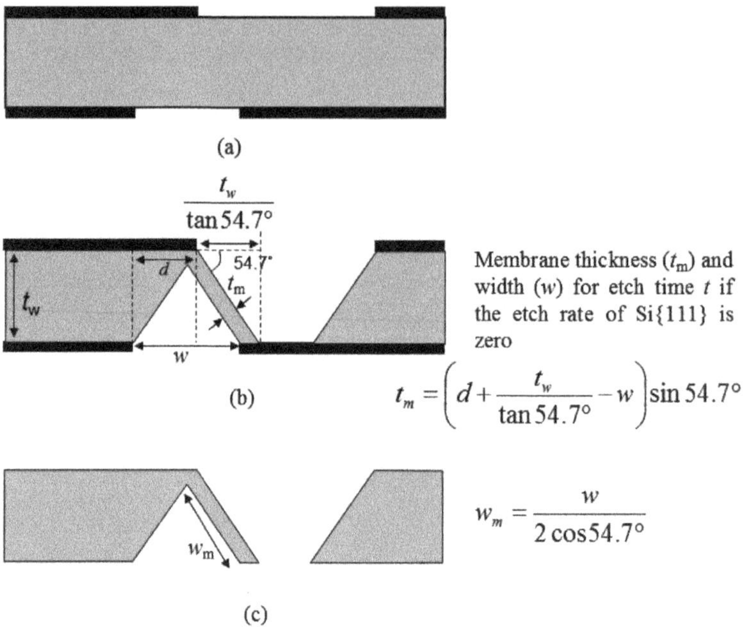

Figure 8.41 Process steps for the formation of a silicon membrane oriented in the {111} plane in a {100} silicon wafer: (a) patterning of a masking layer, (b) anisotropic etching, and (c) removal of the mask layer [92].

8.11 Dry-Assisted Wet Etching

It is well known that wet silicon anisotropic etching cannot be used for forming an arbitrary-shaped mesa/cavity/groove (e.g., circular/rectangular mesa/cavity with vertical sidewalls) in a silicon wafer as the crystallographic properties restrict the size and shape of the silicon microstructures. The deep reactive ion etching (DRIE) process can be used for the fabrication of these types of structures because the etching rate of DRIE is almost crystallographic direction independent. However, DRIE is not suitable for the fabrication of microstructures with slanted sidewalls and the suspended structures if only one side of the silicon wafer (i.e., front side) is used for the lithography process. By combining the LOCOS, DRIE, and wet anisotropic etching processes, one can realize arbitrary shapes of mesa structures with suspended microstructures [93]. The fabrication steps for the formation of a microstructure with suspended parts are illustrated in Fig. 8.42. First, the thermally oxidized Si{100} wafer is patterned. The silicon nitride layer is then deposited. This silicon nitride works as a mask in the subsequent thermal oxidation process, that is, to protect the bare silicon from being oxidized (i.e., LOCOS). After this step, lithography is carried out using a high-viscosity (i.e., thick) photoresist, followed by silicon nitride, silicon dioxide, and silicon etching using the DRIE process for defining the mesas' shape. The thermal oxidation is then carried out for depositing an oxide layer on the sidewalls of the mesas' structures. Thereafter, silicon nitride is removed. Finally silicon wet anisotropic etching is carried out to release the structure. During wet anisotropic etching, the non-{111} silicon planes and convex corners are protected by silicon dioxide and the undercutting initiates only at the places where there is no etch mask layer present. SEM photographs of the microstructures fabricated by this process are shown in Fig. 8.43.

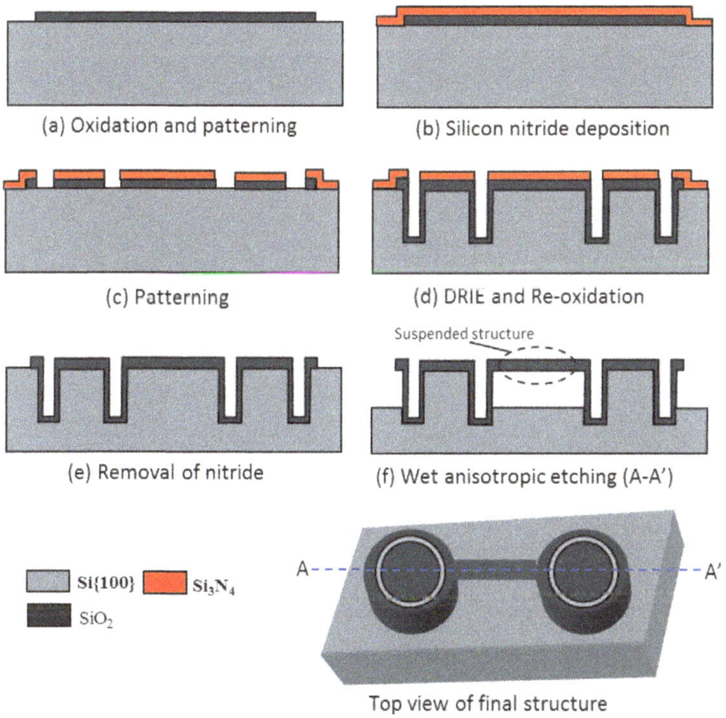

Figure 8.42 Process steps for the fabrication of arbitrary mesas with suspended structures using DRIE-assisted wet anisotropic etching [93].

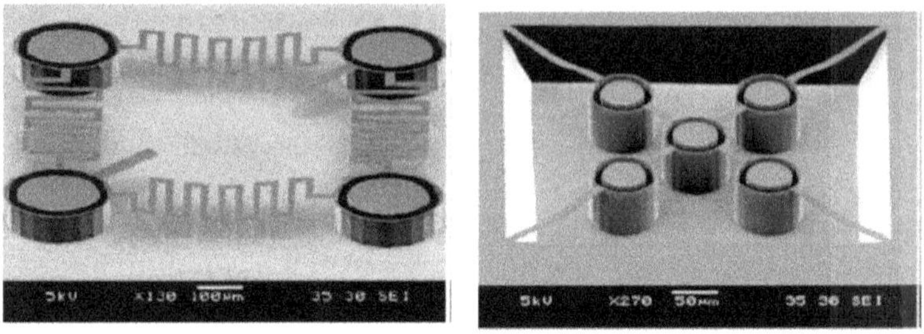

(a) (b)

Figure 8.43 SEM photographs of integrated suspended beams and circular mesas: (a) on the top of a substrate and (b) inside a cavity formed by [111} silicon planes using the process steps illustrated in Fig. 8.42 [93].

The dicing machine is a well-established mechanical cutting tool in semiconductor production to separate chips (or die) from a wafer. This method of dicing is highly automated to ensure precision and accuracy. Figure 8.44 presents the fabrication sequence of microneedles using dicing technology and wet etching [94]. In this method, photolithography is not required for the patterning of the wafer surface. The fabrication process begins with a single-crystal silicon wafer that has a mask layer (e.g., oxide) on one side (or back side). In the case of thermally oxidized silicon wafers, it is achieved by removing the oxide layer on one side in BHF, while protecting the other side by a photoresist or acid-resistant tape. Firstly, step-shaped posts are created using a dicing saw on the unmasked surface (or front surface). This step determines the width and height of the microneedles. Moreover, the spacing between the adjacent structures is defined by the thickness of the saw blade. Thereafter the wafer is properly cleaned and wet anisotropic etching is carried out in KOH. The step-shaped posts are transformed into pen-shaped needle structures during anisotropic etching, as shown in Fig. 8.44c–e. Although wet etching can be carried out in any kind of alkaline solution (e.g., KOH, TMAH), the sharpness of the needle may vary from etchant to etchant.

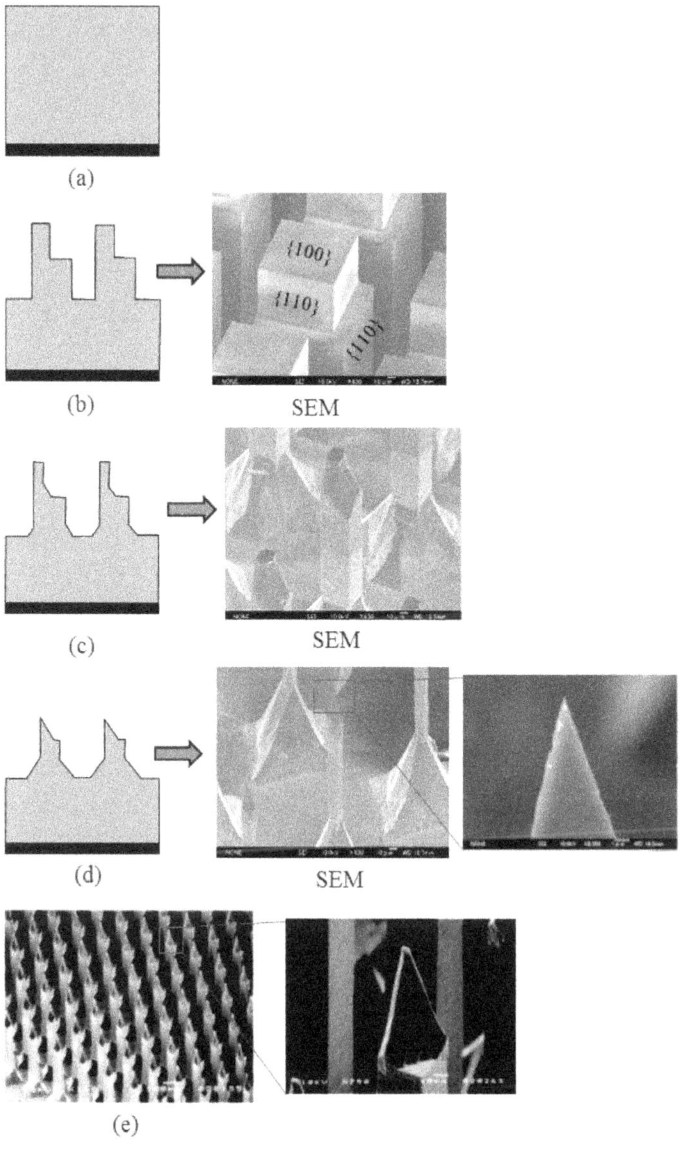

Figure 8.41 Process sequence of pen-shaped microneedle structures: (a) Si{100} wafer with a back-side oxide layer, (b) dicing to form step-shaped posts, (c) anisotropic wet etching, (d) further anisotropic etching, and (e) SEM micrograph of an array of pen-shaped microneedles [94].

8.12 Silicon Molds for the Fabrication of PDMS Structures

Wet bulk micromachining is a promising method for the fabrication of mold patterns for the realization of polydimethylsiloxane (PDMS) stamps for soft lithography. This type of lithography has gained much interest due to its great versatility, high throughput, and potential low cost due to its master and replication system [43, 95–98]. Moreover, silicon is suitable as a master material in a hot-embossing process due to its high mechanical stability and rigidity. The fabrication of pyramidal pits and V-shaped grooves with a sidewall angle of 54.7° can easily be fabricated in a {100} silicon wafer by wet etching in any kind of alkaline solution. These are used for replication in an elastomer (PDMS) and thermoplastic (e.g., polymethyl methacrylate [PMMA] and polystyrene [PS]) materials [95, 96]. However, the fabrication of bent V-grooves, microstructures with 45° sidewalls, sharp-edged corrugations, etc., requires advanced silicon micromachining. By taking advantage of high and low undercutting in pure and surfactant-added TMAH, simple to complex structures can be formed in silicon wafers, as shown in Fig. 8.45b [38, 43, 98]. These can easily be used as molds to realize simple to sophisticated PDMS structures. To fabricate a PDMS structure, a PDMS mixture is prepared by mixing two components (base and curing agent) in the standard ratio of 10:1 and degassed in a vacuum desiccator for 1 h. The degassed PDMS is poured onto silicon mold patterns, followed by degassing again for 1 h to remove bubbles trapped in the sharp patterns. Thereafter PDMS is cured for 24 h at room temperature. The process of curing may vary from lab to lab (or industry to industry). The PDMS is then peeled off very slowly from the silicon mold patterns to prevent any tearing of the PDMS. SEM micrographs of sharp-edged PDMS structures obtained using silicon molds are shown in Fig. 8.45c.

8.13 Complex Patterns for Very-High-Aspect-Ratio Microstructures Using Photoelectrochemical Etching

High-aspect-ratio (HAR) structures in MEMS are usually created by DRIE [99, 100]. The etch rate of the DRIE process depends on several

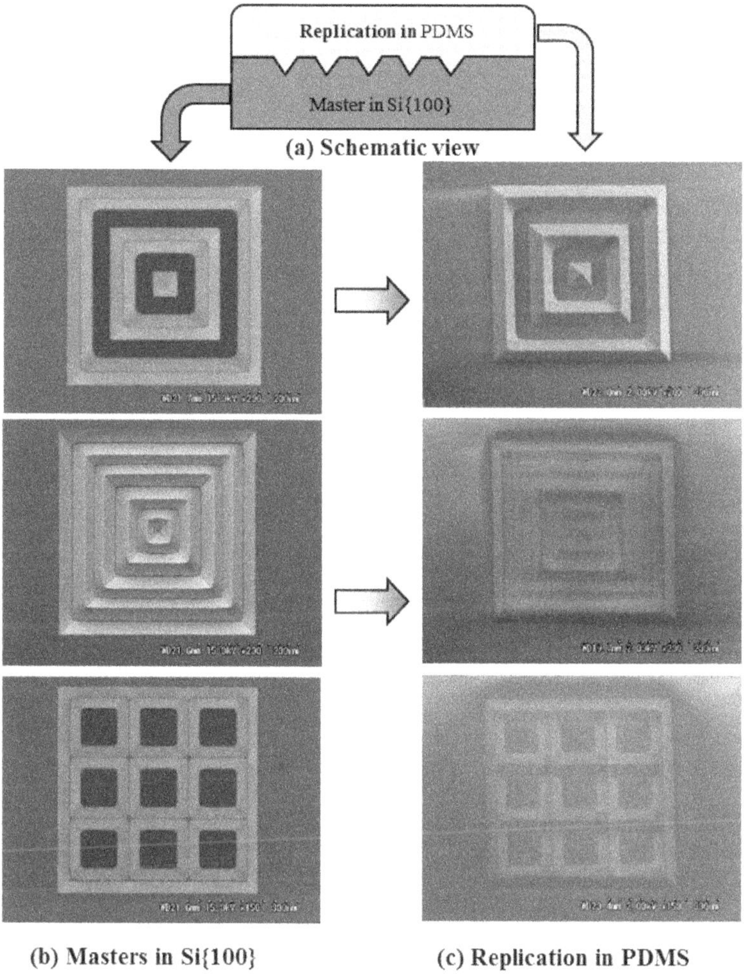

Figure 8.45 PDMS structures fabricated using silicon molds created by wet etching of Si{100} in TMAH+surfactant [43].

parameters such as power, chamber pressure, mask dimensions, etc. Moreover, the formation of periodic corrugations/scalloping on the sidewalls of the structures is a major concern. Photoelectrochemical etching (PEC) of silicon is an alternative technique to fabricate HAR microstructures. This method is very cheap in comparison to DRIE and can produce much-higher-aspect-ratio holes and trenches of smooth sidewalls with a uniform etching depth. However, there is a

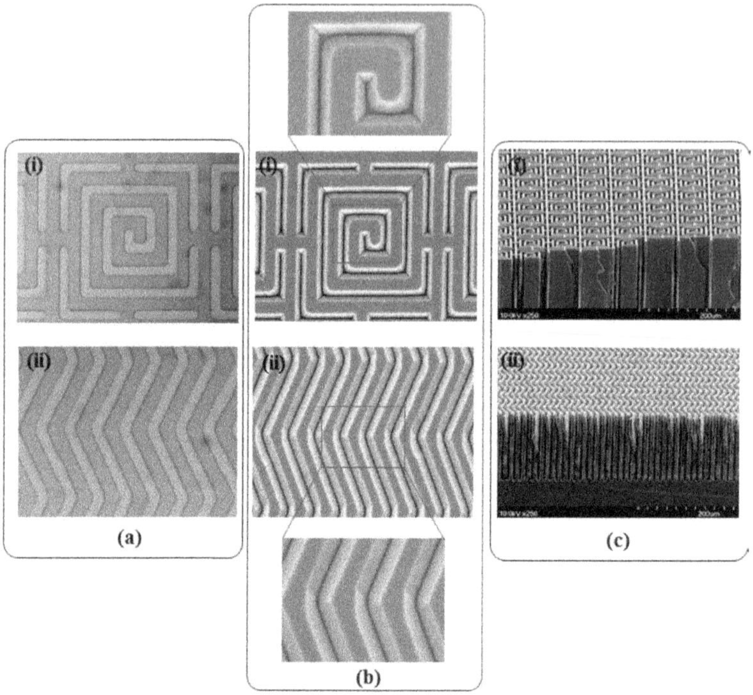

Figure 8.46 Fabrication of very-high-aspect-ratio microstructures in complex patterns by photoelectrochemical etching (PEC) in 2.5 wt% HF: (a) patterned oxide layer (linewidth and spacing = ~4 μm), (b) complex patterns with V-grooves using surfactant-added 25 wt% TMAH, and (c) plates of (i) square-spiral and (ii) zigzag patterns after 5 hours and 3 hours PEC, respectively [100].

limitation in PEC. This technique requires sharp pits (e.g., V-groove) which act as nucleation sites for deep etching. The fabrication of rectangular pyramidal pits (or V-grooves) with a sidewall angle of 54.7° is straightforward. These types of microstructures are created by wet anisotropic etching of mask patterns whose edges are aligned along the <110> directions. To fabricate microstructures of any shape with sharp pits (e.g., microstructures with bent V-grooves or curved edges), wet anisotropic etching must be carried out in modified alkaline solutions (e.g., surfactant-added TMAH or IPA-added KOH). Surfactant-added TMAH is the best choice to suppress undercutting at any kind of mask edge and corner on the Si{100} wafer surface. Therefore it can be employed to create an arbitrary pattern with sharp pits. To create microstructures with sharp pits, a mask layer (e.g., SiO_2) is patterned, as shown in Fig. 8.46a [100]. Thereafter, wet etching is carried out in 0.1% Triton (surfactant)-added 25 wt% TMAH. Wet-anisotropically-etched patterns with sharp pits are shown in Fig. 8.46b. In these structures, the width of the grooves is very small (~4 μm). Therefore, the etching time must be controlled precisely to avoid any undercutting due to the open-circuit potential (OCP) effect, as discussed in Chapter 5 [101]. In other words, the etching must be stopped as soon as the V-grooves are formed. Moreover, any kind of agitation during etching must be avoided. Figures 8.46c-i and 8.46c-ii show ~200 μm tall square-spiral plates and 150 μm tall zigzag plates formed after 5 and 3 h of PEC, respectively [100].

References

1. Yagyu, H., Yamaji, T., Nishimura, M., and Sato, K. (2010). Forty-five degree micromirror fabrication using silicon anisotropic etching with surfactant-added tetramethylammonium hydroxide solution, *Jpn. J. Appl. Phys.*, **49**, p. 096503.

2. Xu, Y. W., Michael, A., and Kwok, C. Y. (2011). Formation of ultra-smooth 45° micromirror on (100) silicon with low concentration TMAH and surfactant: techniques for enlarging the truly 45° portion, *Sens. Actuators A*, **166**, pp. 164–171.

3. Rola, K. P., Ptasinski, K., Zakrzewski, A., and Zubel, I. (2014). Silicon 45° micromirrors fabricated by etching in alkaline solutions with organic additives, *Microsyst. Technol.*, **20**, pp. 221–226.

4. Fruhauf, J., and Kronert, S. (2005). Wet etching of silicon gratings with triangular profiles, *Microsyst. Technol.*, **11**, pp. 1287–1291.

5. David, C., Bruder, J., Rohbeck, T., Grunzweig, C., Kottler, C., Diaz, A., Bunk, O., and Pfeiffer, F. (2007). Fabrication of diffraction gratings for hard X-ray phase contrast imaging, *Microelectron. Eng.*, **84**, pp. 1172–1177.

6. Keller, L. D., Jaffe, D. T., Ershov, O. A., Benedict T., and Graf, U. U. (2000). Fabrication and testing of chemically micromachined silicon echelle gratings, *Appl. Opt.*, **39**, pp. 1094–1105.

7. Graf, U. U., Jaffe, D. T., Kim, E. J., Lacy, J. H., Ling, H., Moore, J. T., and Rebeiz, G. (1994). Fabrication and evaluation of an etched infrared diffraction grating, *Appl. Opt.*, **33**, pp. 96–102.

8. Klumpp, A., Kuhl, K., Schaber, U., Kaufl, H. U., and Lang, W. (1995). Anisotropic etching for optical gratings, *Sens. Actuators A*, **51**, pp. 77–80.

9. Chandra, S., Singh, J., and Chand, A. (1997). Design and development of microstructures for MEMS applications, *Proc. SPIE*, **3226**, pp. 22–30.

10. Ashok, A., and Pal, P. (2015). Synthesis of anodic oxide thin films on Si{100} wafers and their characterization in TMAH for MEMS, *ECS J. Solid State Sci. Technol.*, **4**, pp. Q1–Q7.

11. Pal, P., and Sato, K. (2009). Various shapes of silicon freestanding microfluidic channels and microstructures in one step lithography, *J. Micromech. Microeng.*, **19**, p. 055003.

12. Elwenspoek, M., and Jansen, H. (1998). *Silicon Micromachining* (Cambridge University Press, UK).

13. Lang, W. (1996). Silicon microstructuring technology, *Mater. Sci. Eng.*, R**17**, pp. 1–55.

14. Gad-el-Hak, M. (2002). *The MEMS Handbook* (CRC Press, Boca Raton).

15. Abedinov, N., Grabiec, P., Gotszalk, T., Ivanov, T., Voigt, J., and Rangelow, I. W. (2001). Micromachined piezoresistive cantilever array with integrated resistive microheater for calorimetry and mass detection, *J. Vac. Sci. Technol. A*, **19**, pp. 2884–2888.

16. Wu, Q., Lee, K. M., and Liu, C. C. (1993). Development of chemical sensors using microfabrication and micromachining techniques, *Sens. Actuators B*, **13–14**, pp. 1–6.

17. Pal, P., Kim, Y. J., and Chandra, S. (2006). Front-to-back alignment techniques in microelectronics/MEMS fabrication: a review, *Sens. Lett.*, **4**, pp. 1–10.

18. Pal, P., Tuli, S., and Chandra, S. (2004). A new technique of front-to-back alignment for MEMS, *Sens. Lett.*, **2**, pp. 78–81.

19. Tang, B., Sato, K., Xi S., Xie G., Zhang, D., and Cheng, Y. (2014). Process development of an all-silicon capacitive accelerometer with a highly symmetrical spring-mass structure etched in TMAH + Triton-X-100, *Sens. Actuators A*, **217**, pp. 105–110.

20. Tatic-Lucic, S., Zhang, W. Y., and Navneet, N. (2005). Etch-stop characteristics of heavily B/Ge-doped silicon epilayer in KOH and TMAH, *Sens. Actuators A*, **123–124**, pp. 640–645.

21. Steinsland, E., Nese, M., Hanneborg, A., Bernstein, R. W., Sandmo, H., and Kittilsland, G. (1996). Boron etch-stop in TMAH solutions, *Sens. Actuators A*, **54**, pp. 728–732.

22. Wallman, L., Bengtsson, J., Danielsen, N., and Laurell, T. (2012). Electrochemical etch-stop technique for silicon membranes with p- and n-type regions and its application to neural sieve electrodes, *J. Micromech. Microeng.*, **12**, pp. 265–270.

23. Kloeck, B., Collins, S. D., Rooij, N. F. D., and Smith, R. L. (1989). Study of electrochemical etch-stop for high-precision thickness control of silicon membranes, *IEEE Trans. Electron Devices*, **ED-36**, pp. 663–669.

24. Ashruf, C. M. A., French, P. J., Sarro P. M., Bressers P. M. M. C., and Kelly, J. J. (1998). Electrochemical etch stop engineering for bulk micromachining, *Mechatronics*, **8**, pp. 585–612.

25. Chung, G. S. (2008). Study of electrochemical etch-stop for high-precision thickness control of single-crystal Si in aqueous TMAH:IPA:pyrazine solutions, *Microelectron. Eng.*, **85**, pp. 271–277.

26. Waggener, H. A. (1970). Electrochemically controlled thinning of silicon, *Bell Syst. Tech. J*, **49**, pp. 473–475.

27. Saya, D., Belaubre, P., Mathieu, F., Lagrange, D., Pourciel, J. B., and Bergaud, C. (2005). Si-piezoresistive microcantilevers for highly integrated parallel force detection applications, *Sens. Actuators A*, **123–124**, pp. 23–29.

28. Pruitt, B. L., and Kenny, T. W. (2003). Piezoresistive cantilevers and measurement systems for characterizing low force electrical contacts, *Sens. Actuators A*, **104**, pp. 68–77.

29. Huang, Y., Ergun, A. S., Haeggstrom, E., Badi, M. H., and Khuri-Yakub, B. T. (2003). Fabricating capacitive micromachined ultrasonic transducers with wafer-bonding technology, *J. Microelectromech. Syst.*, **12**, pp. 128–137.

30. Pal, P., and Sato, K. (2009). Suspended Si microstructures over controlled depth micromachined cavities for MEMS based sensing devices, *Sens. Lett.*, **7**, pp. 11–6.

31. Tong, Q. Y., and Gosele, U. (1999). *Semiconductor Wafer Bonding: Science and Technology* (John Wiley & Sons, New York).

32. Huff, M. A., Nikolich, A. D., and Schmidt M. A. (1993). Design of sealed cavity microstructures formed by silicon wafer bonding, *J. Microelectromech. Syst.*, **2**, pp. 74–81.

33. Mack, S., Baumann, H., Gosele, U., Werner, H., and Schlogl, R. (1997). Analysis of bonding-related gas enclosure in micromechanical cavities sealed by silicon wafer bonding, *J. Electrochem. Soc.*, **144**, pp. 1106–1111.

34. Pal, P., and Chandra, S. (2004). A novel process for perfect convex corner realization in bulk micromachining, *J. Micromech. Microeng.*, **14**, pp. 1416–1420.

35. Kawakatsu, H., Saya, D., Kato, A., Fukushima, K., Toshiyoshi, H., Fujita, H. (2002). Millions of cantilevers for atomic force microscopy, *Rev. Sci. Instrum.*, **73**, pp. 1188–1192.

36. Pal, P., Gosalvez, M. A., Sato, K., Hida, H., and Xing, Y. (2014). Anisotropic etching on Si{110}: experiment and simulation for the formation of microstructures with convex corners, *J. Micromech. Microeng.*, **24**, p. 125001.

37. Fruhauf, J., and Kronert, S. (2005). Wet etching of silicon gratings with triangular profiles, *Microsyst. Technol.*, **11**, pp. 1287–1291.

38. Pal, P., and Sato, K. (2009). Complex three dimensional structures in Si{100} using wet bulk micromachining, *J. Micromech. Microeng.*, **19**, p. 105008.

39. Jeong, O. C., and Yang, S. S. (2000). Fabrication of a thermopneumatic microactuator with a corrugated P^+-silicon diaphragm, *Sens. Actuators A*, **80**, pp. 62–67.

40. Kressmann, R., Klaiber, M., and Hess, G. (2002). Silicon condenser microphones with corrugated silicon oxide/nitride electret membranes, *Sens. Actuators A*, **100**, pp. 301–309.

41. Jerman, J. H. (1990). The fabrication and use of micromachined corrugated silicon diaphragms, *Sens. Actuators A*, **21–23**, pp. 988–992.

42. Kwon, J. W., and Kim, E. S. (2002). Multi-level microfluidic channel routing with protected convex corners, *Sens. Actuators A*, **97–98**, pp. 729–733.

43. Pal, P., Gosalvez, M. A., and Sato, K. (2010). Silicon micromachining based on surfactant-added tetramethylammonium hydroxide: etching mechanism and advanced application, *Jpn. J. Appl. Phys.*, **49**, p. 056702.

44. Binnig, G., Quate, C. F., and Gerber, C. (1986). Atomic force microscope, *Phys. Rev. Lett.*, **56**, pp. 930–933.

45. Yu, Q., Qin, G., Darne, C., Cai, C., Wosik, W., and Pei, S. S. (2006). Fabrication of short and thin silicon cantilevers for AFM with SOI wafers, *Sens. Actuators A*, **126**, pp. 369–374.

46. Cho, I. J., Yun, K. S., and Nam, H. J. (2011). A high-speed single crystal silicon AFM probe integrated with PZT actuator for high-speed imaging applications, *J. Electr. Eng. Technol.*, **6**, pp. 119–122.

47. Alessandrini, A., and Facci, P. (2005). AFM: a versatile tool in biophysics, *Meas. Sci. Technol.*, **16**, pp. R65–R92.

48. Tang, B., Sato, K., and Gosalvez, M. A. (2012). Sharp silicon tips with different aspect ratios in wet etching/DRIE and surfactant-modified TMAH etching, *Sens. Actuators A*, **188**, pp. 220–229.

49. Han J., Lu, S., Li, Q., Li, X., and Wang, J. (2009). Anisotropic wet etching silicon tips of small opening angle in KOH solution with the additions of I_2/KI, *Sens. Actuators A*, **152**, pp. 75–79.

50. Zhang, Y., and Zhang, Y. (1996). Formation of single tips of oxidation-sharpened Si, *Appl. Phys. Lett.*, **69**, pp. 4260–4261.

51. Albrecht, T. R., Akamine, S., Carver, T. E., and Quate, C. F. (1990). Microfabrication of cantilever styli for the atomic force microscope, *J. Vac. Sci. Technol. A*, **8**, pp. 3386–3396

52. White, N. M. (2006). The origins and the future of microfluidics, *Nature*, **442**, pp. 368–373.

53. Gravesen, P., Branebjerg, J., and Jensen, O. S. (1993). Microfluidics: a review, *J. Mioromech. Microeng.*, **3**, pp. 168–182.

54. Tiren, J., Tenerz, L., and Hok, B. (1989). A batch-fabricated non-reverse valve with cantilever beam manufactured by micromachining of silicon, *Sens. Actuators*, **18**, pp. 389–396.

55. Bohm, S., Burger, G. J., Korthorst, M. T., and Roseboom, F. (2000). A micromachined silicon valve driven by a miniature bi-stable electro-magnetic actuator, *Sens. Actuators A*, **80**, pp. 77–83.

56. Park, J. M., Evans, A. T., Rasmussen, K., Brosten, T. R., Nellis, G. F., Klein, S. A., and Gianchandani Y. B. (2009). A microvalve with integrated sensors and customizable normal state for low-temperature operation, *J. Microelectromech. Syst.*, **18**, pp. 868–877.

57. Oh, K. W., and Ahn, C. H. (2006). A review of microvalves, *J. Micromech. Microeng.*, **16**, pp. R13–R39.

58. Fazal, I., Louwerse, M. C., Jansen, H. V., and Elwenspoek, M. C. (2006). Design, fabrication and characterization of a novel gas microvalve using micro- and fine-machining, *J. Micromech. Microeng.*, **16**, pp. 1207–1214.

59. Jerman, H. (1994). Electrically activated, normally closed diaphragm valves, *J. Micromech. Microeng.*, **4**, pp. 210–216.

60. Shikida, M., Sato, K., Tanaka, S., Kawamura, Y., and Fujisaki, Y. (1994). Electrostatically driven gas valve with high conductance, *J. Microelectromech. Syst.*, **3**, pp. 76–80.

61. Bien, D. C. S., Mitchell, S. J. N., and Gamble, H. S. (2003). Fabrication and characterization of a micromachined passive valve, *J. Micromech. Microeng.*, **13**, pp. 557–562.

62. Yang, E. H., Han, S. W., and Yang, S. S. (1996). Fabrication and testing of a pair of passive bivalvular microvalves composed of p+ silicon diaphragms, *Sens. Actuators A*, **57**, pp. 75–78.
63. Pramanick, B., Das, S., and Bhattacharyya, T. K. (2014). MEMS based normally closed silicon microregulator for gas and water, *Sens. Actuators A*, **205**, pp. 15–25.
64. Au, A. K., Lai, H., Utela, B. R., and Folch, A. (2011). Microvalves and micropumps for BioMEMS, *Micromachines*, **2**, pp. 179–220.
65. Tsai, N. C., and Sue, C. Y. (2007). Review of MEMS-based drug delivery and dosing systems, *Sens. Actuators A*, **134**, pp. 555–564.
66. Amirouche, F., Zhou, Y., and Johnson, T. (2009). Current micropump technologies and their biomedical applications, *Microsyst. Technol.*, **15**, pp. 647–666.
67. Nisar, A., Afzulpurkar, N., Mahaisavariya, B., and Tuantranont, A. (2008). MEMS-based micropumps in drug delivery and biomedical applications, *Sens. Actuators B*, **130**, pp. 917–942.
68. Laser, D. J., and Santiago J. G. (2004). A review of micropumps, *J. Micromech. Microeng.*, **14**, pp. 35–64.
69. Woias, P. (2005). Micropumps: past, progress and future prospects, *Sens. Actuators B*, **105**, pp. 28–38.
70. Zengerle, R., Ulrich, J., Kluge, S., Richter, M., and Richter, A. (1995). A bidirectional silicon micropump, *Sens. Actuators A*, **50**, pp. 81–86.
71. Zengerle, R., Richter, M., and Sandmaier, H. (1992). A micro membrane pump with electrostatic actuation, *Proc. IEEE Microelectromech. Syst.*, pp. 19–24.
72. Koch, M., Harris, N., Evans, A. G. R., White, N. M., and Brunnschweiler, A. (1998). A novel micromachined pump based on thick-film piezoelectric actuation, *Sens. Actuators A*, **70**, pp. 98–103.
73. Koch, M., Harris, N., Maas, R., Evans, A. G. R., White, N. M., and Brunnschweiler, A. (1997). A novel micropump design with thick-film piezoelectric actuation, *Meas. Sci. Technol.*, **8**, pp. 49–57.
74. Xu, D., Wang, L., Ding, G., Zhou, Y., Yu, A., and Cai, B. (2001). Characteristics and fabrication of NiTi/Si diaphragm micropump, *Sens. Actuators A*, **93**, pp. 87–92.
75. Jeong, O. C., and Yang, S. S. (2000). Fabrication and test of a thermopneumatic micropump with a corrugated p+ diaphragm, *Sens. Actuators A*, **83**, pp. 249–255.
76. Makino, E., Mitsuya, T., and Shibata, T. (2001). Fabrication of TiNi shape memory micropump, *Sens. Actuators A*, **88**, pp. 256–262.

77. Pal, P., and Chandra, S. (2004). Bulk-micromachined structures inside anisotropically etched cavities, *Smart Mater. Struct.*, **13**, pp. 1424–1429.

78. Mandle, J., Lefort, O., and Migeon, A. (1995). A new micromachined silicon high accuracy pressure sensor, *Sens. Actuators A*, **46/47**, pp. 129–132.

79. Yu, J. C., and Lai, F. H. (2000). Design and fabrication of microaccelerometers using piezoelectric thin films, *Ferroelectrics*, **263**, pp. 100–106.

80. Kazinczi, J. R., Mollinger, J., and Bossche, A. (2000). 3-D resonator bridges as sensing elements, *Proc. SAFE-ProRisc-SeSens*, pp. 803–808.

81. Hiltmann, K., Keller, W., and Lang, W. (1999). Micromachined switches for electric load, *Sens. Actuators A*, **74**, pp. 203–206.

82. Craven, D. (1996). Photolithography challenges for micromachining industry, *BACUS Symp.*, pp. 1–10.

83. Linder, S., Baltes, H., Gnaedinger, F., and Doering, E. (1998). Photo-lithography in anisotropically etched grooves, *Proc. IEEE Micromachining Workshop III*, pp. 38–43.

84. Kersten, P., Bouwstra, S., and Petersen, J. W. (1995). Photolithography on micromachined 3-D surfaces using electrodeposited photoresist, *Sens. Actuators A*, **51**, pp. 51–54.

85. Heschel, M., and Bouwstra, S. (1998). Conformal coating by photoresist of sharp corners of anisotropically etched through-holes in silicon, *Sens. Actuators A*, **70**, pp. 75–80.

86. Kutchoukow, V. G., Mollinger, J. R., and Bossche, A. (1999). Novel method for spinning of photoresist on wafers with through-hole, *Proc. 13th European Conf. on Solid-State Transducers (Eurosensors)*, pp. 256–272.

87. Venstra, W. J., Spronck, J. W., Sarro, P. M., and Eijk, J. V. (2009). Photo-lithography on bulk micromachined substrates, *J. Micromech. Microeng.*, **19**, p. 055005.

88. Luxbacher, T., and Mirza, A. (1999). Spray coating for MEMS, interconnects, and advanced packaging applications, *Sensors*, **16**, pp. 61–64.

89. Singh, V. K., Sasaki, M., Hane, K., and Esashi, M. (2004). Flow condition in resist spray coating and patterning performance for three-dimensional photolithography over deep structures, *Jpn. J. Appl. Phys.*, **43**, pp. 2387–2391.

90. Pal, P., and Chandra, S. (2004). Recessed microstructures with perfect convex corners for accelerometers, *Sens. Lett.*, **2**, pp. 226–231.

91. Venstra, W. J., Pham, N. P., Sarro, P., M., and Eijk, J. V. (2005). Particle filters integrated inside a silicon wafer, *Microelectron. Eng.*, **78–79**, pp. 138–141.

92. Venstra, W. J., and Sarro, P. M. (2003). Fabrication of crystalline membranes oriented in the (111) plane in a (100) silicon wafer, *Microelectron. Eng.*, **67–68**, pp. 502–507.

93. Chu, H. Y., and Fang, W. (2004). A vertical convex corner compensation and non {111} crystal planes protection for wet anisotropic bulk micromachining process, *J. Micromech. Microeng.*, **14**, pp. 806–813.

94. Shikida, M., Ando, M., Ishihara, Y., Ando, T., Sato, K., and Asaumi, K. (2004). Non-photolithographic pattern transfer for fabricating pen-shaped microneedle structures, *J. Micromech. Microeng.*, **14**, pp. 1462–1467.

95. Kim, G. M., Kovalgin, A. Y., Holleman, J., and Brugger, J. P. (2002). Replication molds having nanometer-scale shape control fabricated by means of oxidation and etching, *J. Nanosci. Nanotech.*, **2**, pp. 55–59.

96. Michel, B., Bernard, A., Bietsch, A., Delamarche, E., Geissler, M., Juncker, D., Kind, H., Renault, J. P., Rothuizen, H., Schmid, H., Schmidt-Winkel, P., Stutz, R., and Wolf, H. (2001). Printing meets lithography: soft approaches to high-resolution, *IBM J. Res. Dev.*, **45**, pp. 697–719.

97. Bogdanski, N., Schulz, H., Wissen, M., Scheer, H. C., Zajadacz, J., and Zimmer, K. (2004). 3D-Hot embossing of undercut structures: an approach to micro-zippers, *Microelectron. Eng.*, **73–74**, pp. 190–195.

98. Pal, P., Sato, K., Gosalvez, M. A., Tang, B., Hida, H., and Shikida, M. (2011). Fabrication of novel microstructures based on orientation dependent adsorption of surfactant molecules in TMAH solution, *J. Micromech. Microeng.*, **21**, p. 015008.

99. Wu, B., Kumar, A., and Pamarthy, S. (2010). High aspect ratio silicon etch: a review, *J. Appl. Phys.*, **108**, p. 051101.

100. Sun, G., Zhao, X., and Kim, C. J. (2012). Fabrication of very-high-aspect-ratio microstructures in complex patterns by photoelectrochemical etching, *J. Microelectromech. Syst.*, **21**, pp. 1504–15012.

101. Pal, P., Sato, K., and Gosalvez, M. A. (2012). Etched profile control in anisotropic etching of silicon by TMAH+Triton, *J. Micromech. Microeng.*, **22**, p. 065013.

Index

<100> beam, 246–48, 250–52, 254–57, 276, 325
<110> beam, 244–45, 254, 259
<112> beam, 266
<100>-oriented beam, 237, 246, 248, 250, 255–56
<110>-oriented beam, 243, 259
<100>-oriented beam, 268, 276

abrasive grinding machine, 75
absolute indexing, 231
absorption peaks, 184–85
  monohydride, 187
acceleration energy, 61
accelerometers, 180, 194, 219, 336, 364
accuracy, 295–97, 299, 318, 370
acid-resistant tape, 370
acoustic lens, 75
acoustic microscopy system, 75–76
activation energy, 90, 165, 192
activeness, 72
  chemical, 69
actuation, 290, 358–60
actuation unit, 360
actuation voltage, 339
actuator performance, 359
actuators, 290, 359–60
  integrated bimetallic, 359
  physical, 360
additive, 49, 87, 107, 114, 121, 123, 164, 168, 177, 194, 198, 214–16, 219, 229–30, 323–25
  alcohol, 113, 118, 219, 222
  soluble, 114
adhesion promoter, 43
adsorption, 45, 129, 131, 138, 146, 149, 167, 184, 188
  enhanced, 120
  orientation-dependent, 187, 224
  preferential, 189
  selective, 187
adsorption behavior, 189
adsorption density, 127
adsorption layer, 126–27, 138, 142, 144–46, 148
adsorptivity, lower, 224
AFM, *see* atomic force microscope
AFM probes, 355–56
AFM tip, 357
aggregates, 144, 147
  large, 143
  micellar, 143, 147–48
  multimolecular alcohol, 143
aggregation, 147
aggressive acidic etchants, 68
agitation, 34, 48–49, 76, 194–95, 199, 324, 375
  ultrasonic, 194
airborne particulates, 33
airflow patterns, 33
air motion, 33
air pressure, 33, 270, 363
air showers, 33
alcohol additives, 113, 118, 219, 222
alcohol concentration, 109, 114–15, 125–31, 133, 138, 143–48
alcohol concentration range, 147
alcohol film, 109
alcohols, 101, 106–7, 109–15, 117, 119–23, 125–39, 141–44, 146–51, 156, 180, 191
  dissolved, 143
  heavier, 115, 120–21

light, 143, 147
one-hydroxyl, 122
pure, 132
alignment, 43, 287, 289, 291, 295–96, 298, 300, 306, 310–12, 318, 320, 346
accurate, 291, 301
high-precision, 217
alignment accessory pattern, 307
alkaline etchants, 165
common, 191
alkaline solutions, 49, 71–72, 82, 192, 323, 370, 372
modified, 374
pure, 219
alphabet letters, 223
wet-bulk-micromachined, 223
aluminum, 26, 57
aluminum metal patterns, 164
amorphous structure, 1
analytical plane, 105
analyzer, 60
mass, 60
anchor points, 292
angle of inclination, 111, 118, 330
angular deviation, incremental, 104
angular interval, 298, 300, 313
angular pitch, 297, 306, 308
anisotropic etching, 49–50, 74–75, 79, 83, 163, 165, 269–71, 274, 331–34, 348–49, 351, 364, 366, 368, 370–71
back-side, 335
chemical, 71
first-step, 274
perfect, 46
simultaneous, 367
two-step, 271
wet silicon, 368
anisotropic etching system, 72
anisotropic material, 71
anisotropic nature, 69, 79, 82
anisotropy, 46, 51, 69, 82, 89–90, 101–2, 105, 113–15, 124, 129, 147, 163, 213, 323
high, 52
reversal, 113
strong, 82
anisotropy ratio, 108, 119–21, 123, 151, 153–55, 166, 214
annealing, 61–65
APCVD, *see* atmospheric pressure chemical vapor deposition
APF, *see* atomic packing factor
area ratio, 198
relative, 196
argon ambient, 27
arrangement
atomic, 1–4
symmetrical, 15
tetrahedral, 23
two-dimensional, 3
arrays, 290–92, 365, 371
cold emitter, 75
periodic, 2
tall needle, 74–75
Arrhenius equation, 90
atmosphere, 35
inert, 27, 36
special gaseous, 63
atmospheric pressure chemical vapor deposition (APCVD), 39
atomic force microscope (AFM), 97, 118, 154–55, 355–57
atomic force microscopy, 355–56
atomic packing factor (APF), 12
atomic position, 1
atomic steps, 99–104
bunched, 100
atomistic etching behaviors, 101
atomistic scale, 221
atomistic step-etching behaviors, 101
atom removal, 100
atoms, 1–4, 12–14, 57, 79–80, 99–100, 110, 222, 224–25, 295

concave-edge, 226
dopant, 59
evaporated, 40
identical, 13
ionized, 40
low-coordination, 222
neighboring, 1, 23, 79–80, 226
neutral, 60
ATR-IR, *see* attenuated total reflection infrared
attachment sites, 39
attenuated total reflection infrared (ATR-IR), 184
autoignition, 36
axes, 6, 9–11, 18, 101, 228
corresponding, 11
crystallographic, 9–10
primitive, 5
vertical, 90
axis directions, 5

back bond, 80–81, 165
backbone, 51, 331
back-side etching, 290, 338–39
time-controlled, 335, 338
band, 231, 237
barrier, 261
effective, 105
batch process, 338
bath time, 188
BCC, *see* body-centered cubic
beam, 61, 244, 246, 249, 255, 257, 266, 268, 290–92, 355
asymmetric, 244
asymmetric-shaped, 244
incoming, 308
integrated suspended, 370
narrow, 255
optical, 294
parallelogramic, 263, 268, 276, 325
reflected, 308
same-size, 268, 276, 325
suspended microcantilever, 55
beam compensation geometry, 245
beam length, 249
beam neutralization, 60
beam width, 268
beveled direction, 235, 257
beveled shapes, 264
beveling, 184, 240, 244, 259
significant, 242, 244, 259
beveling directions, 259
bevels, 184, 222
BHF, *see* buffered hydrofluoric acid
BHF, *see* buffered hydrogen fluoride
bias, 46
BiCMOS, *see* bipolar complementary metal-oxide semiconductor
BiCMOS devices, 325
BiCMOS processes, 25
bio-MEMS, 288
biosensors, 25, 290
bipolar complementary metal-oxide semiconductor (BiCMOS), 25, 325
bird's beak, 55
blocker, potential, 105
body-centered cubic (BCC), 12–13
body diagonal, 13, 23
boiling point, 91, 109, 115, 121, 138, 149
high, 123, 125, 133, 138
boiling points, low, 114, 121, 129
Boltzmann's constant, 90
bonded step
single-dangling, 103–4
two-dangling, 104
bonding, 62–64, 184
direct, 62
direct wafer, 62–63, 340
fusion, 62
silicon-to-silicon, 62
weak, 63
bonding agent, 62
bonding energy, 65

bonding method, 62
bonding process, 62, 65, 362
bonding strength, 63–64
bonding surfaces, 62
bonds, 79–80, 110, 226
  break, 222
  chemical, 62–63
  hydrogen-terminated, 80
  hydrogen-terminated silicon, 82
  identical covalent, 23
  weaker, 295
boron concentration, 191–92
boron diffusion, 357, 364
boron diffusion depth, 93
boron doping, 93–94
  heavy, 92–94
  high, 191
boron-doping technique, heavy, 95
boron predeposition, 364
boron solubility limit, 191
Bosch process, 53, 170
bottom
  flat-shaped, 90
  wedge-shaped, 89
boundaries, 1
  grain, 1
branched carbon chains, 130
Bravais lattices, 7–8
bubbles, 372
  hydrogen gas, 79, 82, 190
buffered HF, 48
buffered hydrofluoric acid (BHF), 44, 47, 48, 271, 343, 347, 349, 365, 370
buffered hydrogen fluoride (BHF), 70
bulk etchant, 45, 167
bulk gas, 51–52
bulk micromachining, 52, 55–56, 84, 323, 355
  wet anisotropic etching–based, 211, 294
  wet anisotropic etching–based silicon, 288
  wet chemical–based, 29
  wet chemical–based silicon, 359
  wet isotropic etching–based silicon, 287
by-products, 51–52, 79
  volatile, 51

cages, 143
cantilever beams, 56, 94, 97, 180, 268, 288, 290, 292, 331, 335, 339, 341, 348, 355–56, 361
  freestanding silicon, 354
  overhanging, 219
  simple silicon, 340
  suspended, 334, 339
  suspended silicon, 336, 350
  symmetric, 365, 367
cantilevers, 73, 93, 192, 290, 326–27, 333, 336, 338, 355–56
  force-sensing, 355
  polycrystalline silicon, 73
  tray-shaped, 349, 352
cantilever structure, 97
capacitance, 339
carbon atoms, 115, 121–22, 156
carbon chain, 128, 130
catalyzer, 69, 165
cathode, 51
  RF-powered, 52
cavities, 55, 69, 75, 179, 215–16, 269–72, 289, 298, 325–26, 339–40, 361–62, 364–66, 368, 370
  back-side, 364
  long, 288
  micromachined, 339
  neighboring, 299
  polygon-shaped, 217
  rectangular-shaped, 326
  sealed, 339
  simple, 191, 323
  square-shaped, 326

truncated-shape, 326
cells
galvanic, 196
high-efficiency solar, 174
minimum-volume, 4
chamber, 51
evacuated, 40
chamber pressure, 373
channels, 288–91, 295, 326, 346, 348, 351, 358
microfluidic, 353
oversized, 290
patterned, 290
serpentine-shape, 180
silicon microfluidic, 350
suspended, 348
characteristics
anodic passivation, 192
dynamic, 360
orientation-dependent, 184
characterization, 85
thin-film, 187
chemical analysis, 358
chemical bindings, stable, 64
chemical etching, 32, 50, 72
pure, 50
wet, 45, 55, 212, 215, 271, 275, 355
chemical-mechanical polishing (CMP), 33, 340
chemical processes, 45
chemical reaction, 37, 39, 43, 45, 51–52, 71, 79, 88, 167
chemical reaction time, reduced, 358
chemicals, 49, 83, 89
chemical structure, 62
chemical vapor deposition (CVD), 25, 38–40, 54, 70, 92, 269
atmospheric pressure, 39
low-pressure, 39, 365
plasma-enhanced, 39, 92, 365
chemisorption, 194
chip
wet-bulk-micromachined, 360
wood, 26
chip isolation, 254, 259–62
chip isolation grooves, 180, 219, 253
chronic diseases, 358
circuit elements, 57
circuits, 25, 358
electronic, 83
integrated, 23, 289
semiconductor, 94
signal-conditioning, 56
circular openings, 297–299, 306, 309–11, 314, 318–19
circular window features, 297
classifications, common, 33
clean-1, standard, 34
clean-2, standard, 35
cleanliness, 33
cleanrooms, 33–34, 43
cleavage, perfect, 295
CMC, *see* critical molecular concentration
CMOS, *see* complementary metal-oxide semiconductor
CMOS circuitry, 163
CMOS compatibility, 164
CMOS devices, 163, 166
CMOS fabrication process, 164
CMP, *see* chemical-mechanical polishing
coating, 270, 362
antireflection, 141
conformal, 39
spin, 43
cohesive strengths, 65
compatibility, 214
compensating beam, 268
compensating design, 238, 240, 244, 253, 261
triangular-shaped, 241
compensating geometries, 237, 240, 243, 261, 264–65, 276

optimal, 276, 325
square-shaped, 243
triangular-shaped, 240
compensating pattern, 238, 241, 250, 254–55, 264, 266
triangular-shaped, 240
compensating structures, 236, 240, 247, 250, 253, 255, 257, 269
superimposed, 253
compensation, 253, 259
complete, 240
perfect, 259
triangular-shaped, 243
compensation beam, 244
compensation design, 241–42, 244, 247, 259, 276, 325
compensation geometry, 237, 241, 245, 247, 250, 252, 255, 257, 259–61, 263, 325
beam-shaped, 261, 268, 325
rhombus, 267
square-shaped, 244, 259
triangular-shaped, 239, 276
compensation pattern, 236, 246, 259, 264
beam-shaped, 268
square-shaped, 267
compensation strip, 259
compensation structures, 243, 246, 255, 259, 263–64, 267, 325
simple, 254
triangular-shaped, 262
compensation technique, 269
complementary metal-oxide semiconductor (CMOS), 163, 214, 323, 325
components, 38, 56, 69, 101, 104, 358, 372
chemical etching, 52
electrical, 25
key, 358
mechanical, 25, 55
structural, 219
compressive stresses, 34, 37, 39
concentration, 48–49, 88–90, 122, 124–26, 129–31, 133–38, 141, 143, 146–47, 149–50, 167–68, 174–75, 214, 323, 332–33
boron-doping, 93
diol, 134–37, 140
high, 57, 182, 198, 324–25
hydroxyl, 114, 129
limit, 126
low, 57, 127, 132, 138, 149, 214
maximum, 57, 143
molar, 126
saturated, 115
stress, 72
volume, 176
concentration dependence, 88
concentration range, 88, 153
high, 94
low, 89
concentration ratio, 69
conditions
boundary, 59
equilibrium, 57
harsh, 71
initial, 80
optimal, 139
total constant dopant, 59
conductive base, 270, 362
conductivity, opposite, 62
conductor, 24, 94
constant
dielectric, 185
linear rate, 38
parabolic rate, 38
consumer electronics, 25
consumption, power, 359
contact
electric, 96
electrical, 95–96, 377
contact angles, 62, 127–31, 146, 149–51
contact printing, 41–42
contact probe, 85–86, 173

contamination, 28, 33, 35, 62, 83–84, 170, 338
convex corners, 112–13, 115, 194, 211, 214–29, 231–32, 235–45, 249–55, 257, 259, 261–65, 267–73, 276, 280–81, 324–25
  etched, 276
  fabricated, 257
  obtuse, 224, 263, 265–66, 268
  protected, 212, 236, 239, 270, 273–74, 281, 325, 378
  right-angled, 224
  sharp-edge, 246, 249, 251, 255, 269, 276
  undercut, 227, 229, 231, 234–35
  well-defined, 236, 259, 261
  well-protected, 249, 261
  well-shaped, 236, 243, 250
convex structures, 111–12
cooling, localized, 358
corner compensation, 194, 236, 252
corner compensation designs, 238, 253
corner compensation geometries, 237, 262, 265
corner compensation method, 194, 236, 250, 269, 276, 325, 365
corner compensation structures, 258, 260
corrugation
  circular, 343, 345
  periodic, 373
  rectangular, 343, 345
  sharp-edged, 372
counterelectrode, 337, 360, 364
critical molecular concentration (CMC), 149
crystal, 1, 3, 5–6, 15–18, 28–29, 57, 80, 102, 111
  cubic, 11, 16, 18
  damaged, 61
  seed, 27, 29
crystal directions, 287, 298–299, 301, 304, 306–7
  precise, 295, 312
crystal faces, 15
crystal-growing puller, 27
crystal growth process, 27
crystal lattice, 295
crystalline grain boundary, 64
crystalline lattice slippage, 94
crystalline structure, 1
crystallites, 1
crystallization, 28
crystallographic direction, 9, 24, 29, 46, 111, 151, 169, 215, 217–18, 227, 287–89, 293–97, 308, 320, 368
crystallographic orientations, 88, 90, 98, 107–9, 111, 117
crystallographic planes, 15, 167, 169–70, 172–73, 214
crystallography, 15
crystal orientation, 27, 29, 69, 82, 84, 89, 97, 101, 218, 304, 306
crystal orientation dependence, 168, 215
crystal planes, 9, 15, 117
crystal rotation, 27
crystal specimens, 101
crystal structure, 2–3
crystal surface, 80–81, 99
  flat, 100
crystal symmetry, 101
crystal zone, 107, 110
curing agent, 372
CVD, *see* chemical vapor deposition
CZ, *see* Czochralski
Czochralski (CZ), 26–29

damage, 32, 42, 61, 269
dangling bonds, 64, 80–82, 90, 103, 222, 224–25
DC sputtering, 40
decomposition, 39

deep reactive ion etching (DRIE), 52–53, 75, 170–71, 368–69, 372–73
defects, 32, 295
deflection, 355, 360
  optical, 355
deformation, 259
de-ionized (DI), 32, 170, 186
density, 127, 189, 222
  bond, 295
  break-bond, 222
deposition, 40, 52–53, 55, 141, 269, 273–74, 293, 332, 335, 344, 346–47, 349, 351, 357
  thin-film, 39
deposition conditions, 39
deposition rate, 41
deposition techniques, 36
depth
  channel, 354
  desired junction, 59
  etch, 46, 249, 256, 263, 328
  etched, 92, 98
  impurity, 61
depth direction, 102–3
design methodology, 237, 263, 348
desorption, 39, 45, 167
detector, position-sensitive, 355
deviation, 99, 311, 315–16, 318
  electrostatic, 60
  equal, 315
  large, 303
  lateral, 316–18
  negative, 104
  standard, 61
  unequal, 315
deviation angle, positive, 104
device fabrication, 29, 32
device quality, 26
devices, 24, 38, 45, 48, 83, 163, 338–39, 348
  bipolar, 62
  field emission, 74
  miniature, 357
  semiconductor, 111, 325
DI, *see* de-ionized
DI water, 32, –36, 48, 170, 184–88
diameters, 18, 26, 29–30, 58, 103, 185, 297, 302, 308, 313–14, 319, 365
  inner, 29
  pit, 102
diamond lattice structure, 23, 79–80
diaphragm, 83, 93, 191–92, 288, 290, 292–94, 326, 331–33, 348, 359–61
diaphragm-based sensor, 294
diaphragm length, 294
diaphragm material, 292–93
diaphragm pump, 361
diaphragms
  actuator, 360
  corrugated, 343
  deflectable, 94
  flexible, 358, 360
  freestanding, 331
  oversized, 293
  suspended, 292–93, 331–32, 336
dicing, 184, 370–71
dicing machine, 370
dielectric constant, 185
dielectrics, 36
diffraction effect, 42
diffraction gratings, 330
  etched, 331
  high-quality, 330
diffusion, 37–38, 45, 51, 57–60, 71, 167, 189–90, 193, 364, 366
  thermal, 93
  vertical, 59
diffusion depth, 338
diffusion furnace, 59
diffusion-limited case, 38
diffusion-limited process, 71, 76, 90
diffusion mask layer, 58

diffusion mask material, 58
diffusion process, 58, 364
  two-step, 58–59
diffusion system, 58
  solid-source, 58
diffusivity, 38, 59
dimensional change, 85
dimensioning, 256
diode, 193, 289
diols, 121–25, 129, 133–38, 141–42, 149, 156
direct wafer bonding (DWB), 62–64, 193, 340
discs, 58
  solid, 58
dislocation, 32, 100, 102–3
displacement, 3
displacement pumps, 360
dissociation, 52
dissolution, 45, 167, 336
distillation, 26
dopant dose, 61
dopant redistribution, 61
dopants, 24, 27, 57–61
  impurity, 57
dopant sources, 57–58
dopant species, 58
doping, 26, 57, 73, 93–94
  heavy, 93
  impurity, 57, 93
  uniform, 57, 61
doping concentration, 27, 29, 95, 191
doping control, 59
doping level, 27
  low, 95
  normal, 93
doping profile, 59
dose, 61
  uncontrolled, 60
draft bench, 84
DRIE, *see* deep reactive ion etching
drug delivery, 358
  transdermal, 75
drug delivery system, 358
drug reservoir, 358
dry etching, 44–45, 50, 56, 271, 287, 341, 365
  chemical, 45, 51–52
dry etch process, 55
du Nouy method, 126
DWB, *see* direct wafer bonding
dynamic etching process, 127
dynamic pumps, 360

e-beam lithography, 41
edge contouring, 32
edges, 178, 223–25, 227–28, 232–35, 240–41, 249–50, 272, 295, 298–99, 303, 313, 327, 345–46, 349–50, 354–55
  aligned, 299
  beveled, 241
  beveling, 234
  cleaved, 295
  curved, 327, 374
  long, 246, 318
  masked, 228
  misaligned, 289
  pattern, 303
  sharp, 270, 325, 342–44, 346, 356, 362
  short, 184
  undercut, 231
EDP, *see* ethylenediamine pyrocatechol
effects, dielectric, 185
EGS, *see* electronic-grade silicon
EGS polysilicon, 27
electrical conductivity, 24
electrical discharge, 40
electrical isolation, 38
electric field, 38
electric leakage, avoiding, 96
electrochemical etch stop, 92, 94–96, 192–93, 326, 331, 333, 336, 338

electrode
  grounded, 51
  platinum, 193
  powered, 51
  pull-down, 362
electrode configuration, 51
electrodeposition, 270, 362
electrodes, 51, 75, 193, 338, 362
electrolyte, 196
electronegativity, large, 165
electronic-grade silicon (EGS), 26–27
electronics, integrated, 83
electrons, 52, 61, 95, 185
  free, 51–52
  light, 52
electrostatic actuation, 362
electrostatic potential, 195
electrostatic pump, 360
ellipsometric investigation, 188
ellipsometry, 183, 187–88
emitters, 57
  field, 74
environment, local, 185
equilibrium, 95
  thermodynamic, 100
equipment, 24, 300–1, 303, 312
  automatic processing, 29
  diffraction, 295
error, 95, 153, 301, 304, 310
  associated, 320
  theta, 303, 310, 312, 314
error function, 59
etch, 45, 47–48, 83, 93, 100, 170, 301, 364
  self-limiting, 326
etch anisotropy, 85–86, 89
etchant, 44–46, 48–49, 163–64, 167–69, 177, 189–93, 211, 213–15, 224, 226–27, 229–31, 239, 250, 263–64, 323–27
  new, 261
  strong, 70
  wet, 170
etchant composition, 177
etchant concentration, 192, 194, 198, 226
etchant inhomogeneities, 191
etchant species, 51
etchant type, 214, 216, 331, 346
etch depth, 85–86, 181–82, 214, 217, 223, 226, 237, 239–41, 243–45, 249, 252–53, 255, 258, 268–69, 336
etched cavities, 69–70, 327, 362, 365
etched channels, 291
etched dimensions, 92
etched patterns, 50, 181–83, 228, 269, 298–300, 304–5, 310, 314, 319, 363
etched profiles, 49, 67, 87, 89, 178–80, 212–13, 226, 228–30, 234, 246–48, 257, 259, 302–3, 305, 315–17
  final, 300
  progressive, 243, 248
  stable, 178, 211
etched recesses, 67
etched sidewalls, 273
etched silicon lens, 76
etched structures, 111, 113, 120
etched surface morphology, 49, 134, 163, 174–78, 180, 198, 323–24, 326
etched surface quality, 91
etched surface roughness, 91–92, 97, 99, 119, 125, 153, 156, 174–75, 177
etched surfaces, 83, 89, 92, 97–98, 111, 119, 125, 132, 134, 147, 153, 156, 294
  smooth, 325
etched surface textures, 98
etch front, 196, 198–99, 227–28, 235, 250, 324
etch front lines, 250
etch front planes, 231

etching, 44–46, 99–101, 108–10, 125, 138–40, 164–65, 184–87, 193–96, 211–13, 249–50, 269–72, 297–300, 313–14, 334–38, 340
  alkaline, 94, 99
  chemical, 50
  electrochemical, 192, 333
  finite, 215
  front-side, 290, 339
  groove, 105
  ion beam–based, 45, 50
  laser beam–based, 45, 50
  one-step, 328
  photoelectrochemical, 372–74
  plasma-based, 51
  prolonged, 303
  selective, 272, 365
  side, 67
  sputter, 50
  time-controlled, 272, 276, 331–32, 335–36, 339, 341
  time-controlled back-side, 335
  time-delayed, 194, 241, 269
  undercut, 73
  uniform, 48
  vertical, 196
etching agents, 146
etching anisotropy, 82, 111, 129
etching bath, 76
etching behavior, 180, 197, 224
etching characteristics, 164–65, 180, 183, 189, 194, 198, 211, 215, 219, 223–24, 237, 246, 263, 323, 327
  fast, 88
  orientation-dependent, 183
  superior, 164
etching conditions, 48, 86–87, 89, 264
  normal, 83
  single, 84
etching depth, 257
  uniform, 373
etching experiment, 45
etching front, 52, 95–97, 195
etching mask, 83, 95, 357
etching mask openings, 95
etching mechanism, 71, 79, 142, 165
etching MEMS structures, 119
etching parameters, 198, 217, 231, 250, 324
etching plane, 49, 231
etching process, 45, 52–53, 55, 113, 115, 117, 126, 129, 132, 134, 146, 198–99, 224, 226, 324
etching process temperature, 156
etching properties, 69–70
etching rate, 67, 71–73, 84, 88, 104–5, 166, 224, 234, 368
  orientation-dependent, 69
  slowest, 49
  uniform, 67
etching reaction, 69, 71, 79–80, 104, 143, 165
  chemical, 72
  moderate, 69
  total, 90
etching results, 69, 129
etching selectivity, 72, 213
  high, 192
etching solution, 71–72, 76, 80, 82–83, 91, 94–95, 100–101, 105, 111, 121, 128, 133, 139, 147, 156
  alkaline, 69, 72
etching species, 71–72
etching stages, 142
etching step, 45, 51, 58, 172, 263, 269, 271, 326, 343, 350
  first anisotropic, 269, 354
  post-CMOS anisotropic, 163
  second anisotropic, 272, 350
  second wet anisotropic silicon, 365
etching systems, 71

etching techniques, 92
etching temperature, 49, 70, 72, 88, 90, 92, 168, 177, 198, 214, 226, 336
etching time, 165, 177, 214, 226, 300, 304, 307, 310, 318, 326, 331, 336, 367, 375
etching time control, 73
etch mask, 83, 164, 342
etch mask layer, 332, 335, 369
etch parameters, 163, 323
etch patterns, 240, 336
etch pit profiles, 102
etch pits, 102–4
  shallow, 103
etch profile analysis, 87
etch profiles, 87, 89–90, 105–6, 252
  rounded, 67
etch profile simulation systems, 87
etch rate anisotropy, 83–84, 87, 90, 101–2, 106, 180
etch rate characterization, 105
etch rate contour maps, 85–86, 101, 104, 235
etch rate contours, 231, 234
etch rate data, 87, 228
etch rate database, 87
etch rate difference, 241
etch rate distributions, 105–6, 204
  angular, 105
  planar, 105
etch rate maximum, 85
etch rate planes, slow, 191
etch rate ratios, 111, 119, 124, 164, 191, 247, 255
etch rates, 46, 87–93, 100–101, 104–5, 107–11, 117–21, 123–24, 131–34, 136–38, 145–47, 153–56, 165–73, 189–92, 213–15, 223–24
  absolute, 169
  aluminum, 164
  anisotropic, 99
  averaged, 100
  correct, 84
  electrochemical, 198
  equal, 48
  fast, 240
  finite, 178, 289
  finite oxide, 269
  high, 88, 91, 198, 226
  higher, 88, 224
  lateral, 46, 169, 215, 227, 232, 238
  lateral undercutting, 226
  local maximum, 235
  low, 111, 119, 138, 147, 153, 156, 263
  lowest, 106, 138, 147, 215
  macroscopic, 101
  measured, 87
  multiple, 170–71
  negligible, 338
  reduced, 187
  relative, 104, 107, 167, 169, 214, 308
  significant, 164
  step, 100
  symmetry-equivalent, 172
  vertical, 46
etch rate selectivity, 83
etch release step, 341
etch resistance, 43
etch selectivity, 92
  high, 164, 331–32
etch stop, 57, 92–93, 338
  geometrical, 326
  heavy boron-doping, 95
etch stop characteristics, 94, 163, 323
etch stop layer, 97, 191, 272, 326, 331, 333, 336
etch stop method, 332
etch stop surface, 96
etch stop technique, 93, 95, 97
  electrochemical, 96
etch technology, 51

etch time, 139–40, 170, 336
ethylenediamine pyrocatechol (EDP), 79, 83, 236, 272
evaporation, 40, 109
  e-beam, 40
  rapid, 138
  thermal, 40
excimer laser, 42
exposure, 41–42, 193, 247, 362
exposure tool, 41, 271, 364
extrusion, 241, 261

fabrication, 72–75, 198–200, 219, 257–59, 287, 323–26, 331–33, 338–40, 342–44, 346–48, 354–56, 368–69, 372, 374–75, 378–82
  IC, 23–25, 33
  MEMS, 23, 25, 33, 38, 166, 199, 212, 217, 219, 236, 263, 287, 323, 325, 338, 372
fabrication sequence, 273–74, 335, 370
fabrication steps, 269, 334, 339, 346–47, 349, 351, 364, 368
face-centered cubic (FCC), 12–14, 23
faces, close-packed, 172
facets, 98–99, 195, 199, 216, 229–30, 324
  developing, 198
  sloping, 222
fan-shaped groove patterns, 104
FCC, *see* face-centered cubic
FET, *see* field-effect transistor
field-effect transistor (FET), 57
film quality, 41
films, 36, 62, 70–72, 83, 270, 362
  deposited, 36, 39
  passivation, 95
  photographic, 43
  resist, 41
  solid, 38
float zone (FZ), 27, 29
flower pattern, 170
flow rate, 36, 358, 360
fluctuation, 76
force balance, 128
forces, 355, 359–60
  capacitive, 360
  capillary, 63
  dispersion, 128, 146
  electrostatic, 63, 360
  electrostatic attraction, 360
  external, 358
  flexible, 355
  physical, 63
forces of attraction, 62
Fourier transform infrared (FTIR), 183–84, 188, 206
fractions, packing, 13–14
freestanding cantilever beams, 333, 335
freestanding microstructures, 56, 219
frequency shifts, 185
front side, 332, 335–36, 338, 364, 368
FTIR, *see* Fourier transform infrared
FTIR spectroscopy, 187
furnace, 26, 36
  resistance-heated, 36
furnace tube, 58
FZ, *see* float zone
FZ growth method, 27
FZ process, 28–29

galvanic protection, 198–99, 324
  gradual, 196
gas cylinders, 36
gaseous by-products, 39
gaseous form, 60
gaseous sources, 58
gaseous state, 58
gases, 1, 36, 39, 60, 79
  dopant, 29, 57
  inert, 29

inert carrier, 58
reactive etch, 51
gas flow control units, 51
gas mixture, 57
gas-phase etchants, 45, 50
gas source, 58, 60
gas species, 39
Gaussian distribution, 59
geometry, 184, 227, 236, 240, 250, 263, 269, 293, 297, 303–4, 310
etched, 306
overetched, 293
patterned, 292
reservoir, 346
triangular-shaped, 238, 241
glass, soda lime, 41
grain boundary, 1–2
gratings, 313, 346
echelle, 330
linear, 330–31
grinding, 29–30
groove fabrication, 90, 260
groove profiles, 105–6, 331
grooves, 83, 89, 104–5, 191, 198–99, 255, 259–62, 300–1, 324, 326, 331, 343, 345–46, 368, 375
blazed, 330–31
flat-topped, 331
growth, reactive, 36
growth rate, fast, 38
Gutmann's electron–acceptor number, 185
Gutmann's electron–donor number, 185

hardware, extra, 271, 364
heat filament, 60
heating coils, 39
height, needle, 73
hexamethyldisilazane (HMDS), 43
high-index planes, 190
fast-etching, 249
high-performance devices, 290, 292, 294
high-power RF coil, 28
hillocks, 92, 109–10, 117, 119, 132, 137–38, 147, 153
bulky, 117
single, 110
solitary, 151
uncontrolled, 113
HMDS, *see* hexamethyldisilazane
HNA, *see* hydrofluoric, nitric, and acetic acids
HNA isotropic system, 75
holes, 69, 111, 336
hot-embossing process, 372
hydrocarbons, 64
hydrofluoric, nitric, and acetic acids (HNA), 32, 48, 69, 71, 74–75
hydrogen bond network, 147
continuous, 143
hydrogen bonds, 143–44
strong, 143
hydrogen bubbles, 95
hydrogen fluoride, 64
hydrophilic bonding, 62–63
hydrophilicity, 63
higher, 62
hydrophobic bonding, 64
hydrophobic chains, 126
hydrophobic hydration, 143–44, 146–47
hydroxylation, 80

IC, *see* integrated circuit
IC, silicon-based, 25
ID, *see* inner diameter
illumination, 70
light, 94
image analysis, 170
impingement, direct, 40
implantation, 38, 57, 60–61
implantation energies, 60

impurities, 29, 38, 57, 59–60, 93, 168, 180, 214
accepter, 57
carbon, 27
donor, 57
implanted, 61
localized, 177
organic, 34
impurity activation, 61
impurity concentration, 94
large, 338
indices, 9–11, 231
industrial processes, 91
ingot, 29–30, 57, 79, 294
cylindrical, 27
inner diameter (ID), 29–30
insulator, 24, 38, 40
integrated circuit (IC), 23–24, 32, 36, 289
silicon-based, 25
integration, 25, 56, 163, 358
ultra-large-scale, 50
interaction, 138, 189
galvanic, 195–98
interface bubbles, 62
interface, 37, 95, 184
air–liquid, 126
bonded, 65
etched, 189
solid–liquid, 27, 127
inversion, 5–7
ion bombardment, 72
ion collisions, 61
ion concentration, high, 91
ion distributions, 60
ionic contamination, 164
potassium, 164
ion implantation, 60–61, 193
ion implanter, 60
ionization, induced, 52
ion path, 60
ions, 41, 49, 52, 60–61, 79, 81, 83–84, 90, 95, 101, 105, 163, 165, 167, 185
alkali, 34–35, 164
bombarded, 52
charged, 60
complex, 95
dopant, 60
energized, 61
immobile, 163
implanted, 61
metallic, 87
positive, 51
total implant, 61
water-soluble hydroxide, 79
IPA, *see* isopropyl alcohol
IPA-added KOH, 235–36, 250, 267, 327, 329
IPA concentration, 114, 131
IPA solutions, 108, 114, 132, 141, 145
islands, 142, 152
island structures, 142
isopropyl alcohol (IPA), 107, 109, 112–14, 123, 125–26, 129–30, 137–38, 142, 152–53, 164, 214, 230, 325, 327
isotropic etchants, 48
isotropic etching, 48–49, 67–74
wet, 48, 75, 287
isotropic etching system, 71–73
chemical, 71
isotropic plasma etching, 75
isotropic process, 55
isotropy, 46, 70

kinetic energy, 52, 360
kinetics, basic, 37
KOH-based anisotropic etching, 79
KOH-based etching, 229, 325
KOH-based wet anisotropic etching, 49
KOH–water solution, 83, 91, 98,

lapped wafers, 32
laser spot, 355
laser tuning, 330

lateral diffusion, 61
  high, 59
lateral etching, 46, 178, 215, 249, 251, 255, 263
lateral etch rate, 227, 232, 238, 240, 244, 248
lateral motion, 100
lateral undercutting, 228
lateral underetch rate, 169, 228
lateral underetch rates, orientation-dependent, 169
lattice constant, 9, 23
lattice damage, 61
lattice parameter, 189
lattice points, 2–5, 7–8
lattices, 2–4, 6–7
  cubic, 9–11
  infinite, 6
  orthorhombic, 10
  semiconductor, 61
  two-dimensional, 4–6
lattice sites, 61
lattice translation operation, 3
lattice translation vectors, 2–3
lattice vectors, 12
layer, 38, 41, 43, 47, 54, 56, 93, 96, 125, 143, 184, 190, 224, 338–39, 349
  conducting, 38, 40
  dense, 224, 226
  epitaxial, 62
  etch-resistant polymer, 53
  insulating, 40
  intermediate, 340
  molecular, 62
  n-doped epitaxy, 96
  n-type epitaxial silicon, 95
  passivation, 35
  resist, 41–42
  sacrificial, 56, 72–73
  structural, 56, 327, 333, 348–49, 354
  thin-film, 62
light reflectance coefficient, 142
liquid etchants, 45
liquid phase, 164
liquids, 1, 58, 348
liquid sources, 57–58
lithography, 41, 43, 45, 61, 269, 271, 326, 340, 366, 369, 372
  back-side, 339
  e-beam, 41
  ion beam, 41
  optical, 41, 43
  soft, 372
  two-mask, 194, 276
local oxidation of silicon (LOCOS), 54–55, 271–72, 274, 325, 331, 340–44, 346–47, 349, 351, 356–57, 365–66, 368–69
locations
  back-bond, 80
  flat, 29
  spatial, 80
LOCOS, *see* local oxidation of silicon
logarithmic scale, 90
low-pressure chemical vapor deposition (LPCVD), 39, 68, 365
LPCVD, *see* low-pressure chemical vapor deposition

macroscopic differences, 101, 105
magnetic field, 60
magnifications, 312, 319
magnitude, 24, 315, 317–18
map, contour, 85, 102
mask, 38, 41–43, 45, 67, 83–84, 110–11, 269–73, 292–94, 299, 320, 346, 348, 362, 364, 369
  back-side, 335
  channel, 346
  circular, 142
  circular disk-shaped, 73
  external, 255
mask aligner, 295, 297, 300
mask apertures, 68–69

mask damage, 42
mask design, 273, 355, 364
mask dimensions, 373
mask edge direction, 331
mask edges, 163–64, 178–79, 194, 196, 198–99, 215–17, 228, 230–31, 234, 287–94, 308–10, 312–13, 323–24, 328–30, 346
  alignment of, 217, 294, 331
  misaligned, 217
  straight, 84
  subsequent, 300, 319
mask generator, 304
mask geometries, 212–13
mask image, 42
masking, 46
masking layer, 47, 61, 163, 198, 215, 269–70, 272, 298, 208, 326, 335, 357, 362, 367–68
masking materials, 54, 67, 83, 95, 213, 335
mask inversion, 343, 349
mask layer, 46–47, 183, 193, 270, 273, 329, 331–33, 335, 357, 362, 364, 367–68, 370, 375
  front, 170
  front-side, 332, 335
mask material, 71, 92, 170, 323
mask opening, 67, 95, 179, 196, 216–17
  back-side, 332, 336
  large-size, 339
  large-size back-side, 339
  long, 168–69
  square-shaped, 364
mask patterns, 110–12, 214–15, 217–18, 258, 260, 288, 290, 292–94, 296, 300, 303, 327, 329, 343, 346
  different-shaped, 180
  groove, 104
  pre-etching, 304
  square-shaped, 247
  wagon-wheel, 85
mask pattern shape, 84
mask stripes, 331
mask undercut, 67, 73
materials, 14, 23–24, 39, 45–46, 48–49, 51, 54–55, 68, 71, 180, 213, 219, 331–33, 338–39, 343
  building, 36
  bulk, 55, 287
  insulating, 40
  master, 372
  nonsilicon, 331
  polycrystalline, 26, 71
  radiation-sensitive, 41
  raw, 26
  sidewall-protecting, 53
  single-crystal, 71, 84, 295
  structural, 73
  target, 40
  work, 71
"Material Safety Data Sheet" (MSDS), 84
material transfer, 71–72
maximum adsorption, 138
maximum density, 138, 146
measurement-free technique, 310, 319
measurements, 85, 126, 130, 134, 146, 149, 184, 303, 306, 310, 312, 318–19
  dynamic, 150
  ex situ ellipsometric spectroscopy, 187
  in situ, 185
  real-time, 358
  side-etch, 84
  spectral, 140
  spectroscopic, 184
mechanical agitation, 177, 180, 194–95, 199, 324
mechanical energy, 360
mechanical internal stresses, 94
mechanical machining, 72
mechanical profiler, 98
mechanical support, 29

megasonic agitation, 34
melting temperature, 27, 40
membranes
  recessed, 365
  thin, 97
MEMS, *see* microelectromechanical systems
  integrated, 84
  optical, 174, 308
  polymer, 25
MEMS applications, 39, 155, 200, 202, 284–85, 376
  optical, 297
MEMS-based devices, 24–25, 343
MEMS-based sensing devices, 331
MEMS components, 52, 342
MEMS devices, 290
MEMS fabrication, 25, 33, 219, 236
MEMS functions, 95
MEMS microfluidic channel, 349
MEMS structures, 25, 114, 138, 194, 263, 292, 325, 327
  freestanding, 348
  silicon-based, 217
  suspended silicon, 335
mesas, 111, 255, 325, 369
  circular, 84, 370
mesa structures, 180, 194, 219, 224, 231, 235, 240–45, 252–54, 256, 261, 326, 368
metal contacts, 333
metallic contaminations, 35
metallization, 34
metallurgical-grade silicon (MGS), 26
metal-oxide semiconductor (MOS), 38
metal-oxide-semiconductor field-effect transistor (MOSFET), 60
metals, heavy, 34
method
  cleaved-edge, 295–96
  deposition, 36
  etched-back, 339
  evaporation, 40
  lithographic, 41
  pre-etched, 320
  spin-coating, 270, 364
  spray-coating, 270–71, 362–63
  sputtering, 40
  wagon-wheel, 84
MGS, *see* metallurgical-grade silicon
microbiology, 261
microbridge, 361
microcantilever, 220
microchannels, 288–89, 358, 360
  high-aspect-ratio, 313
microelectromechanical systems (MEMS), 23–26, 32, 36, 38, 55, 57, 62, 163, 166, 212, 287–89, 323, 325, 327, 331
microelectronics, 39
microfabrication, 37, 48, 51, 55, 221, 270
microfilters, 358
microfluidic channels, 275, 345–48, 350, 352, 355
  freestanding, 349
  overhanging, 350
  overhanging SiO_2, 351
  suspended silicon, 354–55
microfluidics, 261, 348, 357
microfluidic systems, 358
micromachining, 25, 55, 72, 223
micromasks, developing, 147
micromechanical structures, 55–57, 62, 93
micromirrors, 111, 113, 118, 297, 308
micromixers, 358
microneedles, 370
  pen-shaped, 371
micro-opto-electromechanical systems (MOEMS), 119, 325
microphones, 343
micropumps, 357–61
microscope, 41, 299–1, 303

high-magnification, 318
optical, 70, 303, 311, 319
microscopy, optical, 170
microsensors, 358
microstereolithography, 25
microstructures, 95–97, 193–94, 196, 219–21, 249–51, 253, 258, 262, 269–75, 324, 326–29, 338–39, 342, 368–69, 372–75
microsystem technology (MST), 25
microvalves, 357–60
active, 358–60
passive, 359
silicon-based, 358
Miller indices, 9–10, 87
minimization, 312
minimum, local, 101
MIR, *see* multiple internal reflection
misalignment, 215, 217, 288–93, 298, 301, 303, 305–6, 311–14, 319
crystallographic, 294
degree of, 217, 288, 303
misalignment angle, 294
misorientation, 217
modes, 51
asymmetric stretch, 187
vibrational, 187
MOEMS, *see* micro-opto-electromechanical systems
molar mass, 115–16, 122, 149
monochromators, 330
MOSFET, *see* metal-oxide-semiconductor field-effect transistor
MOS, *see* metal-oxide semiconductor
MOS transistors, 60
neighboring, 54
MSDS, *see* "Material Safety Data Sheet"
MST, *see* microsystem technology
multienergy implants, 61
multiple internal reflection (MIR), 184

nanoelectromechanical systems (NEMS), 25
nanoelectronics, 40
nanolayer, 147
nanolithography, 355
nanomachining, 55
nanoscale science, 355
nanostructures, 287
native oxide, 63–64, 170
nearest neighbor, 12
needle fabrication, 75
needle tip, 73, 75
wet-etched, 75
negative photoresist, 44, 71, 154
negative resist, 43
NEMS, *see* nanoelectromechanical systems
nitride-based filters, 365
nitride etching, 274, 341
nitride etch rate, 272
nonionic surfactants, 82, 148, 164
nonmechanical energy, 360
nonvolatile, 38
notches, 31, 310–12, 319–20
n-type epitaxial layer, 193, 338
n-type silicon material, 27
nucleation sites, 374

obtuse corners, 231, 264, 266–68, 276, 325
OCP, *see* open-circuit potential
one-mask process, 326
open-circuit potential (OCP), 94–95, 192–93, 195, 324, 375
openings
back-side, 339
front-side square, 336
oversized, 300
optical exposure, 41

optical images, 98, 170, 310–12, 318–19
optical interference, 355
optical microscopy images, 97, 102
optimization, 125
organic compounds, 126
organic matter, 34
organics, 34–35
orientation, 29, 84–91, 98–101, 105–6, 108, 110, 172, 175, 177, 198, 224, 229–31, 236, 317, 324–25
orientation dependence, 85, 173
orientation groups, 88
orientations
  crystalline, 72
  fast-etching, 85
  flat, 99, 101
  high-etch-rate, 85
  principal, 166, 174–75, 215
  rotational, 102
  seed, 27
  slow-etching, 85
  triangle, 102
outlet valve, 359–60
out-of-plane reflection, 308
overetching, 46–47, 270, 272, 310, 331
oversaturation, 109, 138
oversizing, 288–89, 292
oxidation, 37–38, 48, 165, 273–74
  dry, 36
  local, 54, 271, 325
  pyrogenic, 36
  thermal, 36–37, 54, 272, 370
  wet, 36
oxidation temperature, 36
oxide, 32, 37–38, 59, 63, 184, 188, 193, 269, 273–74, 340–44, 346–49, 351, 356, 364–66, 370
  buried, 97
  front-side, 356
  pad, 54
  thin, 184
oxide etch, 271
oxide etching, 340, 342, 344, 348, 353, 356
oxide film, 73, 83
  native, 72
oxide layer, 36–38, 43, 45, 54–55, 58–59, 63–64, 271–72, 274, 339, 342–43, 346–49, 364, 369–70
  back-side, 371
  blocking, 193
  buffer, 54, 347
  buried silicon, 96
  front-side, 356
  high-quality, 55
  patterned, 184, 334, 374
  thermal, 37
  thin, 35, 184, 348
oxide thickness, 37–38, 364
oxidization, 69
  thermal, 70
oxidizer, strong, 34
oxidizing agent, 202
oxygen diffusion, 54–55
oxygen plasma, 44, 206

packing density, 189–90
parabolic stage, 38
parasitic effects, 163
particle filters, 365, 367, 382
particles, 33–35, 40, 57, 62
  charged, 60
passivation, 192
passive valves, 358–60
patterning, 43, 45, 47, 270–71, 273–74, 292–93, 302–3, 332–33, 335, 340, 342–44, 346–49, 351, 356–57, 366–68
PDMS, *see* polydimethylsiloxane
  degassed, 372
PDMS mixture, 372
PDMS structures, 372–73
PEC, *see* photoelectrochemical etching

PECVD, *see* plasma-enhanced chemical vapor deposition
PEG, *see* poly(ethylene glycol) 164
pen-shaped microneedle structures, 371, 382
pen-shaped needle structures, 370
percolation, 143
periodicity, 1, 6
periphery, 75, 84, 296
phase, 57
  acceleration, 60
phenomena, corrosion, 195
phosphorous disc, 57
phosphosilicate glass, 73
  boron-doped, 73
photoelectrochemical etching (PEC), 372–75
photolithography, 41–42, 54, 57–58, 340, 343, 348, 362, 370
photomask, 41, 271, 341
photoresist (PR), 34, 41, 43–44, 170, 270, 362–64, 369–70
  electrodeposited, 270, 362
  spray-coated, 270
  standard liquid, 270, 362
  uniform-thickness, 270, 362
photoresist coating, 270, 363
photoresist electrodeposition, 270, 362
photoresist thickness, 363
photosensitive polymer, 43
physical etching, 50
physical vapor deposition (PVD), 40
physisorption, 194
piezoelectric, 355, 359–60
piezoresistive, 355
pinhole mask aperture, 68
Piranha bath, 34
pit, 103–4
pit profile, 104
pit sidewall velocities, 104
planes
  beveled, 230, 236
  beveling, 231, 238
  etched, 132
  fast-etching, 222, 241, 250
  highest-etch-rate, 235
  high-etch-rate, 221, 226, 235
  high-indexed, 232
  minimum-etch-rate, 228
  recessive, 233
  sidewall, 170, 172
  slowest-etching, 215, 235
  slowest-etch-rate, 178, 211
  slow-etch-rate, 191
  standing, 231, 235
  tangent, 224–25
  undercut, 231
  vertical, 228, 257
plasma, 34, 40–41, 45, 50–52
plasma-enhanced chemical vapor deposition (PECVD), 39, 365
plasma etching, 45, 50, 52
  dry, 355
plasma mode, 51
plasma mode wafers, 51
plasma treatment, 64
plates, electrostatic deflection, 61
PMMA, *see* polymethyl methacrylate
p-n diode, 62, 325
p-n junction, 95–96, 192–93, 338
polarity
  highest, 122
  significant, 165
polarization, 63
polishing, 32–33, 62
  chemical, 184
  chemical-mechanical, 33, 340
polycrystalline, 1–2, 36, 39, 48, 55, 68–69
polycrystalline structures, 1
polydimethylsiloxane (PDMS), 372
poly(ethylene glycol) (PEG), 164
polymethyl methacrylate (PMMA), 372
polymerization, 53

poly-Si, 39, 72–73
polysilicon, 192
polystyrene, 372
pores, 365
  circular, 365
power supply, 51, 338
PR, *see* photoresist
prebonding, 62–63
precision, 293, 304, 370
  high, 296
predeposition, 59
prediffusion, 332
prefabricating, 312
pressure, 41, 339, 360
  forward, 359
  output, 360
  positive, 34
  reduced, 51
  room, 34
pressure chamber, 360
pressure difference, 358
primitive cell, 4–5, 12–14
printing, shadow, 42
process compatibility, 163, 323
  post-CMOS, 164
processing, 84, 269, 272, 333, 336–37, 340
  back-side, 339
  batch, 59, 338–39
  high-temperature, 94
  semiconductor, 365
processing time, 91, 314
proof masses, 180, 194, 219, 365–67
  recessed suspended, 362
  suspended, 336–37, 361
proximity printing, 42
p-type impurities, 338
p-type silicon, 192–93
  doped, 57, 191, 338
p-type surface, 193
p-type wafers, 193, 338
pull-down electrode formation, 361
pump diaphragm, 360
  flexible, 360
purity, high, 26
PVD, *see* physical vapor deposition
pyramidal aperture, 259, 261, 263
pyramidal pits, 372
pyramids, 138, 142, 298–99

quartz, 1, 27, 41, 69, 84
quartz boat, slotted, 36
quartzite, 26
quartz tube
  cylindrical fused, 36
  high-temperature, 57
quartz water bubbler, 36–37

radial diagonal, 309–10
Radio Corporation of America (RCA), 33–35, 63
radio-frequency (RF), 25, 41, 51–52
radius, atomic, 13
random motion, 57
ratio
  etch-depth-to-width, 46
  standard, 372
RCA, *see* Radio Corporation of America
reactants, 39, 51, 71, 79, 88, 163, 189–91, 323
reactive ion etching (RIE), 51–53, 75, 170–71, 271, 365, 368
recombination, 52
refractive index, 187
reoxidation, 340, 349
repeatability, 76
replication, 372
reproducibility, 295, 310
reservoirs, 346, 348
  multiple, 275, 345, 347, 350
residual mass, 259
resist, 41, 43–44
  uniform-thickness, 271, 364
resistivity, 29, 38

RF, *see* radio frequency
RF field, 28
RF generator, 51
RF sputtering, 25, 40
RF voltage, 52
ridges, 300–1, 304–6
  etched, 307
  tapered, 305
RIE, *see* reactive ion etching
RIE mode, 51
RIE system, 51, 53
room temperature (RT), 1, 12, 35, 62, 64, 70–72, 75, 126, 129, 164, 188, 372
rotation, 5–7, 300
  crucible, 27
rotational angle, 106
rotational symmetry, 6
rotation axes, 6
rotation rate, 27
rotation symmetry, 6
roughness, 99, 106–7, 109–11, 113, 117–21, 125, 131, 136, 138, 147, 151–52, 154, 156, 177
roughness map, 98
roughness profile, 99
RT, *see* room temperature

sample agitation, 195
sample holder, 192
samples, 43, 170, 184, 186, 188, 199, 271, 324, 364
saturation, 122–23, 125–26, 129, 131, 138, 147, 176
saturation level, 114, 147, 156
sawing, 29, 32
SC, *see* simple cubic
SC crystals, 12
scanning electron microscopy (SEM), 170, 316, 326
  cross-sectional, 89
scanning force microscopy (SFM), 355
scanning probe microscopy (SPM), 355
scanning tunneling microscopy (STM), 99
SD, *see* step dihydride
seismic mass, 364
  recessed, 364
selectivity, 46–47, 51, 72, 92, 163–64, 192, 323
  crystal plane, 163, 323
  high, 46
self-aligning feature, 312
SEM, *see* scanning electron microscopy
semiconductor material, 23–24
semiconductor production, 370
semiconductors, 36, 40, 57, 60
sensing devices, 292
sensors, 62, 290
  capacitive pressure, 343, 345
  chemical, 25, 376
separation, 42
  molecular, 358
SFM, *see* scanning force microscopy
shape memory alloy (SMA), 359–60
sharp-edged corrugation, 372
sharpness, 272, 355, 370
side-etched planes, 84
sidewalls, 52–53, 110–11, 113, 119–20, 151, 153, 228, 232–35, 308, 310, 313, 329–30, 365, 369, 372–73
  slanted, 328–29, 368
  smooth, 294, 373
silicon, 23–24, 36–37, 48–50, 53–54, 62–65, 68–69, 71–72, 84, 93–96, 166–67, 172–74, 191–92, 270–72, 330–31, 364–65
  bare, 35, 369
  bulk, 65
  doped, 192, 337–38

electronic-grade, 26
exposed, 55, 269, 271, 274, 343, 347
extruded, 243
metallurgical-grade, 26
n-type, 192–93, 338
oxidized, 63
single-crystalline, 295
soluble, 49
silicon acoustic lens, 75
silicon AFM probe, 356
silicon anisotropic etching, 131, 273, 332, 335, 342, 347, 365
silicon atoms, 24, 37, 79, 81, 90, 102, 167, 222, 224–26
hydrogen-terminated, 81
silicon-based filters, 365
silicon bulk micromachining, 212, 287, 355, 358
silicon cantilever beams, 342
silicon chips, 25, 360
silicon crystal, 79
silicon crystal lattice structure, 79
silicon crystal orientations, 102
silicon doping, 49
silicon etching, 51, 53, 142–43, 145, 147–48, 184–85, 272, 335, 342–44, 346, 349, 351, 356, 362, 364
silicon etch rate, 173, 191
silicon etch stop, 191
silicon gratings, 330–31, 346–47
silicon ingot, 27
silicon membranes, 368
single-crystal, 367
silicon MEMS components, 191
silicon micromachining, 25, 75, 358
silicon microstructures, 192, 343, 368
freestanding, 334
suspended, 339
silicon molds, 372–73
silicon needle arrays, 74
silicon-on-insulator (SOI), 62, 92, 272, 326, 338
silicon prisms, 184–85
silicon structures, spatial, 110
silicon substrate, 49, 54, 57, 75, 96, 108, 142, 308
silicon surface, 32, 35, 37, 45, 71, 82, 95, 98, 148–50, 167, 184, 186–87, 189–90, 194, 262
bare, 64
hydrophilic, 63
wet-etched, 97
silicon surface atom, 81
silicon tips, 355
silicon wafer manufacturing, 25
silicon wafers, 34, 36, 41, 54, 57–58, 62, 83–84, 92–94, 169–71, 212, 214–16, 295–96, 345–46, 367–68, 372
bare, 63
oxidized, 370
p-type, 95
single-crystal, 26, 79, 370
simple cubic (SC), 12
single-crystal silicon, 69, 71, 79
single-crystal silicon cantilevers, 357
Si-Si bonds, 64
Si-Si covalent bonds, 64
sloped sidewalls, 271
sloping sidewalls, 270, 362
SM, *see* step monohydride
SMA, *see* shape memory alloy
SOI, *see* silicon-on-insulator
solar cell applications, 138
solubility, 109, 113–14, 125, 137, 147, 149
solubility limit, 114–15, 122, 127
solvents, organic, 34
spectra, 184, 186–87
absorbance, 184
background, 184
spectrographs, 330
spectroscopy, 183–84, 207

spectrum, 184
spin coater, 25, 43
spin-coating process, 270, 362–63
SPM, *see* scanning probe microscopy
spokes, 170–71
spray-coated resist thickness, 271
sputtering, 40
  DC, 40
  RF, 25, 40
square compensation geometry, 242, 244
square compensation patterns, 243
square V-groove, 310
stability, 100, 104
  high mechanical, 372
statistical analysis, 304
steam, pure, 36
step atoms, 104
step density, 100
step dihydride (SD), 110
step edges, 100
  atomic, 100
step length, 100
step monohydride (SM), 110, 113, 118
stereographic projection, 14–16, 18–21, 110–11, 172
stirring, 72, 76, 139–40, 156
  magnetic bead, 180
  mechanical, 131, 134, 139
stirring motion, 71
STM, *see* scanning tunneling microscopy
stress, 39, 54, 94
  internal, 338
  residual, 338
substitutional sites, 61
substrate assembly, 43
substrate crystallographic orientation, 111
substrate orientation, 129
substrate surface, 39–40, 45, 52, 59, 73
substrate texturing, 142
surface activation, 62
surface atoms, 45, 51, 80, 101
surface bonds, 110
surface contaminants, 33
surface diffusion, 65
surface layer, 32
  work-damaged, 72
surface micromachining, 55–56, 72, 287
surface mitigation, 39
surface modification, 355
surface morphology, 101, 174, 176, 189–90, 326
surface passivation, 38
surface profiles, 72, 85
surface roughening, 89
surface roughness, 92, 98, 113, 137, 147, 151, 154, 174–76, 190–91, 270, 363
surfaces
  cleaved, 295
  hydrophilic, 62, 64
  hydrophobic, 64, 189
  mirror-finished, 75
  unmasked, 370
surface tension, 126–27, 129–35, 138, 143–45, 149, 270, 362
  liquid–gas, 128
  solid–gas, 128
  solid–liquid, 128
surface tension of water, 149–50
surface topography, 270
  uneven, 270, 362
surfactant, physisorbed, 187
surfactant-added TMAH, 180–81, 184–85, 194, 198–99, 224, 226, 235–36, 250, 252–53, 261, 263, 327, 350, 372, 374
surfactant adsorption, 194, 226
surfactant bath time, 188
surfactant concentration, 150–51

surfactant layer, 126, 187–90
  adsorbed, 194
  packed, 189
surfactant molecules, 126, 150, 165, 172, 187–90, 194, 224, 226
surfactant treatment, 187–88
symmetry, 5–6, 101–3, 272
  fourfold, 101, 103, 107, 110, 172, 180
  mirror, 102
  twofold, 262
symmetry operations, 5–7
system
  automatic microdosing, 358
  optical, 42
  pump, 51
  replication, 372
  three-component, 69
  three-electrode, 192
system-on-a-chip, complete, 25

tall needles, 74
taper, 29, 32
tapered groove effect, 300
tapered ridge effect, 300
tapered shape, 75
techniques, 25, 38–41, 54, 95–96, 194, 269–73, 294–95, 297, 301–2, 304–5, 309–10, 312, 318–20, 338–39, 373–74
  electrical, 355
  electrochemical, 96
  nondestructive, 187
  spin-coating, 270
  spray-coating, 363
technologies, 25, 36, 55, 355
  dicing, 370
  microsystem, 25
  semiconductor device, 106, 286
  silicon semiconductor, 271
  surface-micromachining, 56
Teflon, 192
Teflon cell, 184
  custom-made, 184
temperature, 33, 48, 87, 90–91, 114, 126, 131, 139, 141, 156, 166, 168, 191–92, 227, 230
  absolute, 90
  elevated, 57, 71
  high, 36, 54, 59, 62–64, 129, 138, 158, 200
  melt, 27
  surfactant bath, 188
temperature dependence, 91
temperature sensors, 290
tensile stress, 94
  induced, 94
  strong, 54
tensioactive compounds, 125, 128–29, 146, 148, 150, 156
tensioactive substance, 126
terrace monohydride (TM), 110
terraces, 99–100
terrace surface, 99–100
  flat, 100
tetramethylammonium hydroxide (TMAH), 83–84, 102–5, 147–48, 163–68, 171–78, 180–84, 190–93, 229–32, 234–36, 240–45, 250–53, 261–62, 327–29, 340–44, 348–51
textures, 98–99
  step-and-terrace, 99
texturing, 138, 141
  optimized, 139
texturization, 139
therapeutic agents, 358
thermal conductivity, 24
thermal energy, 63, 90
thermal evaporator, 25
thermal oxidation, 36–37, 54–55, 83, 274, 343–44, 348, 351, 356, 369
  dry, 184
thermal oxidation furnace, 25

thermal oxide, 58, 164, 170, 269, 273
thermoplastic, 372
thermopneumatic, 359–60
thickness, 29–30, 37–38, 95–96, 187–88, 192, 194, 331–32, 334–36, 338–39, 343, 354, 356, 364, 367, 370
  desired, 95, 335, 339–40
  film, 46
  membrane, 92–93
  nonuniform, 270, 362
  required, 56, 339–40
  saturated, 188
  standard, 29–30
  total, 272
  uniform, 270, 338–39, 341, 362
  wafer-to-wafer, 339
thickness change, 84
thickness variation, 29, 272
thin films, 36, 47, 284, 293, 376
thin-film shape memory alloy, 359
through-hole, 75, 272
throughput, 41–42, 166, 214
  high, 41, 372
tilted sidewalls, 270–71, 364
tilting angle, 99
time delay, 191
tip, 75, 355–56
tip apex, 355
TM, *see* terrace monohydride
TMAH, *see* tetramethylammonium hydroxide
  aqueous, 185, 377
  low-concentration, 166, 176, 180, 198, 324–25
  pure, 166, 180, 199, 223, 229, 324, 327, 340, 343–44, 353
TMAH-based etching, 101, 343
TMAH-based wet anisotropic etching, 49
tolerance, 303
total thickness variation (TTV), 29
toxicity, 83
trace amounts, 188
trace contaminants, 34
transformation, 252
transistors, 289, 325
  bipolar, 57
  field-effect, 57
translation operation, 3
translations, 5–6
  primitive, 3
translation symmetry, 6
translation vectors, 3–4
  nonprimitive, 4
  primitive, 4
transport
  limited diffusion, 191
  slow diffusion, 191
trenches, 52, 197, 270, 362, 373
triangular compensation, 240, 261, 265–66
triangular compensation geometry, 240
Triton, 146–47, 149–51, 153, 156, 164, 172, 175–76, 187, 195–96, 375
Triton adsorption, 149
T-shaped cantilever beams, 341
T shaped silicon cantilevers, 343
TTV, *see* total thickness variation
tube, acceleration, 60
two-step etching, 194, 269, 271, 273–76, 325, 331, 336, 342, 346, 348, 365

ULSI, *see* ultra-large-scale integration
ultra-large-scale integration (ULSI), 50
ultraviolet (UV), 41, 362
undercut, symmetric, 301
undercut direction, 255
undercut profiles, 255
undercut structures, 228

undercutting, 163–64, 180, 190–91, 194–95, 197–99, 211–24, 226, 235–36, 248–49, 257, 261–65, 288–89, 323–25, 343, 354
  convex-corner, 220, 232, 290
  corner, 180, 219, 227, 234, 250, 254
  largest, 196
  lateral, 236, 246, 249–50, 255, 268, 273
  low, 180–81, 372
  maximum, 289
  minimal, 198
  smallest-measured, 196
  symmetric, 300–1
undercutting directions, 231–32, 234–35, 248, 265
  lateral, 235
undercutting length, 238–39, 264, 268
undercutting lines, 264
undercutting planes, 230
undercutting rate, 199, 211, 235, 300, 324
  high, 219
undercutting ratio, 182, 217, 236–37, 239, 242
undercutting results, 195
underetching, 178, 213, 215, 227, 292, 309
  smallest, 115
unit cell, 4, 9, 12–14, 23, 189
  cubic, 263–64
  nonprimitive, 4
UV, *see* ultraviolet
UV exposure, 43, 271, 364
UV light, 42–43
  polarized, 271, 364
  reflection of, 270–71, 364

vacuum, 33, 63–64
vacuum chamber, 51
vacuum desiccator, 372
vacuum pump, 52
vacuum system, 51
valence band, 69
valves, 358–60
  active, 358
van der Waals forces, 62–64
vapor, 60
vapor bubble, 34
vapor transport, 58
vectors
  crystal translation, 3
  translational, 2
velocity, 40, 259, 360
versatility, great, 372
vertical grooves, etched, 89
vertical sidewalls, 52, 172, 268, 368
V-grooves, 113, 180, 198–99, 219, 253–63, 276, 324, 326, 354, 372, 374–75
  111-delimited, 196
  crossed, 253, 259
  sequential, 256
vibration, 33, 186
  ultrasonic, 75
vicinal, 104, 172, 190, 226
vicinal orientations, 99, 101, 105, 183, 199, 324
vicinal planes, 109, 180–81, 214, 224, 235, 323, 327
vicinal surfaces, 99, 106, 183, 189
viscosity, 270, 363
visual inspection, 311–12, 319
visual investigation, 301
visualization, 298
voids, 62, 65
  hydrogen-induced, 64
voltage, 41, 51, 192
  applied, 360
  supply, 360
  threshold, 60, 163
voltage bias, 60
volume fraction, 71
  typical, 68

volume ratio, 72
volume removal, 100
volume removal rate, 100
volume stroke, 361
V-shaped grooves, 326, 372
V-shaped profile, 49

wafer bonding, 339
wafer breakage, 295
wafer curvature, 29
wafer diameter, 29
wafer disintegration, 295
wafer flat (WF), 288, 294, 296–97, 305–6, 310, 313, 320
wafer holder, 193, 338
wafer lapping, 32
wafer orientation, 172, 330–31, 346
wafer pair, 63
wafer parameters, 29
wafer patterns, 41
wafer periphery, 96
wafers
  bonded, 64, 329
  large-size, 58
  nitride-deposited, 343
  nitride-patterned, 348
  normal, 97
  oxidized, 366
  patterned, 300
  planar, 362
  quartz, 84
  supporting, 339
wafer shaping, 33
wafer slicing, 30
wafer surface, 32, 34–35, 37, 62, 64, 67, 72–73, 94–96, 169–70, 172, 212–13, 217, 230, 295–96, 329–31
wafer thickness, 272, 293
  uniform, 269
wagon wheel (WW), 168, 171–72
wagon-wheel method, 171
walls
  grounded chamber, 52
  microfluidic channel, 351
  mirror, 308
  shallow pit, 103
water network structure, 144
water rinse, 64
water structure, 146–47
  internal, 143
water surface, 103
water vapor, 36
wavelengths, 42–43
wave propagation, 103
wax bond, 33
wedges, 89, 105, 170–72
  elongated, 170
  retracted, 171
wedge tip, 170
weight, molecular, 115, 119–20
wet anisotropic etching, 49, 178, 180, 211, 213, 236, 287–88, 293–94, 296–97, 327–28, 330, 340, 356–57, 368–70, 374
wet bulk micromachining, 55–56, 180, 199, 288, 323, 342, 356, 358, 364, 372
wet-etched recesses, 68
wet-etched shape, 169
wet etching, 45, 50–52, 55, 72, 75, 90, 190, 195, 216–17, 338–39, 341, 346, 355, 370–73, 375
wet etching process, alkaline-based, 57
wet etching systems, 68
wet etch process, 55
wet process, 96
wettability, higher, 62
wetting, 128, 150
WF, *see* wafer flat
Wigner–Seitz cell, 5
wire sawing, 29–30
Wulff–Jaccodine method, 105–6, 228–31

WW, *see* wagon wheel
WW pattern, 169–70
WW structure, 169–72
  micromachined, 170

X-ray diffraction, 295
X-ray lithography, 41

Young's modulus, 24, 65

zigzag chain, 189
zone axis, 18, 228–29, 235